Green Manufacturing Technologies in Industry 5.0

Mani Venkatesh • Srimantoorao S. Appadoo
Mustafa Rehman Khan
Editors

Green Manufacturing Technologies in Industry 5.0

Springer

Editors
Mani Venkatesh
Asper School of Business
University of Manitoba
Winnipeg, MB, Canada

Srimantoorao S. Appadoo
Supply Chain Management Department
University of Manitoba
Winnipeg, MB, Canada

Mustafa Rehman Khan
UCSI Graduate School of Business
UCSI University
Kuala Lumpur, Malaysia

ISBN 978-981-95-0362-9 ISBN 978-981-95-0363-6 (eBook)
https://doi.org/10.1007/978-981-95-0363-6

This Springer imprint is published by the registered company Springer Nature Singapore Pte Ltd.
The registered company address is: 152 Beach Road, #21-01/04 Gateway East, Singapore 189721, Singapore

To the visionaries who dare to reimagine industry,
To the researchers and innovators committed to a sustainable future,
And to the next generation of leaders who will inherit this planet.
May your work be guided by wisdom, resilience, and responsibility.

And most personally,
To my beloved parents, Fazal ur Rehman and Shahnaz Sardar.
For your unwavering love, sacrifices, and endless encouragement.
This book is dedicated to you with deepest gratitude and affection.

Foreword

In an era marked by pressing environmental challenges, rapidly evolving technologies, and shifting industrial paradigms, the convergence of sustainability and innovation has never been more vital. *Green Manufacturing Technologies in Industry 5.0: A Multi-dimensional Perspective* arrives at a crucial juncture, one where industry, academia, and policy must work in harmony to foster responsible, human-centric, and ecologically sound production systems.

The transition from Industry 4.0 to Industry 5.0 signifies more than a technological upgrade. It represents a philosophical shift, one that reintroduces the human element into highly automated environments, champions resilience over mere efficiency, and places sustainability at the core of industrial design. In this context, green manufacturing is not just an option, it is a necessity. The pursuit of carbon neutrality, resource efficiency, ethical labor practices, and circular economic models is rapidly redefining how products are conceived, created, consumed, and cycled back into the system.

This book brings together leading scholars, researchers, and practitioners from around the globe to examine the complex intersections of green technologies, digital transformation, supply chain innovation, policy frameworks, and ethical governance. Through well-researched chapters, it offers readers a comprehensive understanding of the enablers and barriers to green manufacturing in the context of Industry 5.0.

What distinguishes this work is its multi-dimensional approach. Rather than viewing green manufacturing through a single lens, it incorporates insights from environmental science, engineering, operations management, economics, ethics, and information technology. The result is a cohesive, interdisciplinary volume that not only informs but also inspires action.

For academics, this book provides a wealth of theoretical frameworks and research pathways. For industry leaders, it offers practical strategies and case insights. For policymakers and sustainability advocates, it outlines the institutional and technological ecosystems required to drive large-scale, sustainable industrial transformation.

As we strive toward a future where technological advancement aligns with planetary boundaries and human well-being, the ideas and perspectives shared in this volume serve as a roadmap and a call to action. The editors and contributors have made a significant contribution to the global discourse on sustainable manufacturing, and I am confident that this book will stimulate further innovation, collaboration, and impact.

Edinburgh Business School,
Heriot-Watt University Edinburgh,
Edinburgh, UK

Muhammad Shujaat Mubarik

Preface

The rise of Industry 5.0 represents a fundamental shift in the philosophy of industrial development—one that places human-centric innovation, environmental sustainability, and technological integration at the core of manufacturing systems. As global industries confront the triple challenge of ecological degradation, resource scarcity, and the urgency to decarbonize, green manufacturing technologies emerge as a transformative pathway to a resilient and responsible industrial future.

This book, *Green Manufacturing Technologies in Industry 5.0: A Multi-dimensional Perspective*, offers an interdisciplinary exploration of how the convergence of advanced technologies, ethical frameworks, and sustainable practices is reshaping the manufacturing landscape. It brings together contributions from global scholars and practitioners who collectively address the complexity, urgency, and promise of green transitions in the context of next-generation industrial systems.

Structured across twelve comprehensive chapters, the book examines critical themes such as eco-design and life cycle assessment, digital transformation, circular economy models, blockchain integration, renewable energy integration, and smart logistics. It also delves into ethical considerations, labor practices, environmental justice, and policy implications—providing a truly multi-dimensional analysis of green manufacturing under the Industry 5.0 paradigm.

A unique feature of this volume is its holistic treatment of both technological enablers and societal impacts. Chapters on Amazon's ESG-driven digital reinvention, green supply chains, smart warehousing, and decarbonization strategies offer applied insights, while systematic reviews and case studies synthesize global best practices and future research agendas.

This book aims to serve as a foundational resource for academics, engineers, policy architects, business leaders, and graduate students who are engaged in shaping the future of manufacturing. Whether read as a strategic reference or an academic guide, the insights provided herein will support informed decisions, foster collaboration, and catalyze innovation in sustainable manufacturing systems.

We extend our deepest appreciation to all contributing authors for their intellectual rigor, and to our institutions and collaborators for supporting this important scholarly endeavor. Most importantly, we thank the readers, visionaries, researchers, and change-makers, for your commitment to building a greener, smarter, and more inclusive industrial future.

Winnipeg, MB, Canada — Mani Venkatesh
Winnipeg, MB, Canada — Srimantoorao S. Appadoo
Kuala Lumpur, Malaysia — Mustafa Rehman Khan

Acknowledgments

This book is the result of the collaborative efforts, intellectual commitment, and shared vision of numerous individuals and institutions. We are deeply grateful to all those who contributed their time, expertise, and support throughout the development of *Green Manufacturing Technologies in Industry 5.0: A Multi-dimensional Perspective*.

First and foremost, we express our sincere appreciation to all contributing authors and researchers whose scholarly rigor and innovative thinking have shaped the depth and quality of this volume. Your diverse perspectives and expertise across disciplines have enriched the multi-dimensional narrative that this book aspires to deliver.

We extend heartfelt thanks to our respective institutions, Universiti Putra Malaysia, Leeds Beckett University, University Canada West, Quaid Azam University, Guru Ghasidas Vishwavidyalaya Central University, and other participating universities, for fostering an environment of academic excellence and interdisciplinary collaboration. Your support in encouraging sustainability-focused research has been instrumental in realizing this work.

Special thanks go to the editorial and review teams for their meticulous feedback and assistance in maintaining the academic integrity and coherence of each chapter. Your dedication ensured the consistency and clarity required of a scholarly publication of this nature.

We are also grateful to the publishing team for their professionalism, responsiveness, and support throughout the publication process.

Finally, we acknowledge the growing global community of researchers, practitioners, and students working at the intersection of green technologies, ethics, and advanced manufacturing. Your commitment to sustainable development and technological responsibility continues to inspire the work reflected in this book.

This book is a collective step toward shaping a future where environmental sustainability and industrial innovation coexist, driven not only by technology but by ethical leadership and global collaboration.

Editors
Mani Venkatesh
Srimantoorao S. Appadoo
Mustafa Rehman Khan

Contents

About the Editors and Contributors

About the Editors

Mani Venkatesh is an Associate Professor of Supply Chain Management at the Asper School of Business, University of Manitoba, Canada. He possesses over 24 years of academic and industrial experience, of which over a decade he had served in fortune 500 companies in various senior management roles. Prior to Asper, he worked as an Associate Professor, Montpellier Business School, France. He is teaching/taught in ISCTE Business School Lisbon, IIM Bodhgaya, TAPMI India, and University of Porto as a visiting professor. He holds his PhD from the Indian Institute of Technology (IIT).

His research examines digital transformation challenges, failures and learnings from global corporations, especially digital transformation of global supply chains, sustainability, and modern slavery from emerging economies. He has contributed many research articles in refereed journal that include, but not limited to: *Harvard Business Review*, *International Journal of Operations and Production Management*, *International Journal of Production Economics*, *International Journal of Production Research*, *Transportation Research Part A*, *Supply Chain Management: An International Journal*, *Annals of Operations Research*, and *Journal of Business Research*.

He edited special issues in *Annals of Operations Research* and *International Journal of Physical Distribution and Logistics Management*. He serves as an Associate Editor of *Journal of Global Information Management*.

Srimantoorao S. Appadoo is a professor and the institutional head of the Department of Supply Chain Management at the University of Manitoba, Canada. In addition to teaching undergraduate and graduate courses, he is a prolific researcher, with his work continually being published in esteemed journals (both national and international) and refereed proceedings. His tenacious and innovative research has garnered global acclaim—frequently cited by academics both domestically and internationally. Professor Appadoo has published in journals such as the *European*

Journal of Operational Research (EJOR), *Annals of Operational Research* (AOR), *Fuzzy Optimization and Decision Making* (FODM), *International Journal of Production Economics* (IJPE), and *Information Science*, among others.

Professor Appadoo's work spans interdisciplinary horizons, exemplified by his partaking in collaborative projects—including working with over fifty researchers from around the world in various academic disciplines. Having published over one hundred and fifty articles in international journals and proceedings, his publications have appeared in internationally revered journals—specializing in management science and supply chain management. His articles were included in the "Top 25 Hottest Articles" list for ScienceDirect, published by Elsevier, and were among the most cited articles on Elsevier. He co-authored a book titled, *More-for-Less Solutions in Fuzzy Transportation Problems* (Studies in Fuzziness and Soft Computing). His research is funded by the National Sciences and Engineering Research Council of Canada (NSERC) Discovery Grant as a Principal Investigator. The Discovery Grants Program promotes and maintains a diverse base of high-quality research capabilities in Natural Science and Engineering in Canadian universities, fostering research excellence and providing a stimulating environment for dynamic research training. Professor Appadoo is a recipient of the Associates' Achievement Award, the Joint UM/UMFA Committees on Merit Award, and several best paper awards at some national and international conferences.

In addition to his academic prowess, professor Appadoo is known for his capabilities as an educator. Students commend Professor Appadoo's teaching methodology and consistently rank his classes and instruction as exceptional. He serves on departmental, faculty, and university committees and contributes extensively to the professional community through review processes—operating as a conference chair and academic reviewer. Owing to Professor Appadoo's significant contribution and efforts in creating the Ph.D. program in the Department of Supply Chain Management, the Asper School of Business is one of the limited institutions offering a Ph.D. program in SCM.

Mustafa Rehman Khan is a seasoned academician and prolific researcher specializing in sustainable business practices, supply chain management, green innovation, and digital transformation. He is currently affiliated with UCSI University, Kuala Lumpur, Malaysia. Dr. Khan holds a Ph.D. in Business Management from Sultan Idris Education University (UPSI), Malaysia. He has published extensively in prestigious international journals, including *IEEE Transactions on Engineering Management, Management of Environmental Quality: An International Journal, International Journal of Productivity and Performance Management*, and the *Journal of Small Business and Enterprise Development*. Dr. Khan has co-edited academic volumes with renowned publishers such as Springer and IGI Global and serves on the editorial boards of *PLOS ONE, Journal of Small Business and*

Enterprise Development, and *Journal of Asia Business Studies*. His research contributions reflect a strong commitment to addressing real-world challenges through evidence-based inquiry. Dr. Khan's work advances understanding at the intersection of sustainability, technology, and organizational performance in both emerging and global contexts.

Contributors

Tehreem Ansari Jinnah University for Women, Karachi, Pakistan

Ahsan Anwar Centre for Business Informatics and Industrial Management (CBIIM), UCSI Graduate Business School, UCSI University, Kuala Lumpur, Malaysia
Advanced Research Centre, European University of Lefke, Lefke, Turkey
Korea University, Seoul, South Korea
Department of Business Administration, ILMA University, Karachi, Pakistan

R. Banumathi Acharya Bangalore B-School, Bangalore, India

Gregory Barnes Leeds Beckett University, Leeds, UK

Syed Danish Bukhari Quaid e Azam University, Islamabad, Pakistan

Tat-Huei Cham Swinburne University of Technology Sarawak Campus, Kuching, Malaysia
Faculty of Business, Sohar University, Sohar, Oman
Tashkent State University of Economics, Tashkent, Uzbekistan
International College, Krirk University, Bangkok, Thailand
Universitas Multimedia Nusantara, Tangerang, Indonesia

Blessy Doe Sri Ramakrishna Engineering College, Coimbatore, India

Hajar Fatorachian Leeds Business School, Leeds Beckett University, Leeds, UK

Nikhil Gangadhar Amity Global Business School, Bangalore, India

Arsalan Mujahid Ghouri London South Bank University, London, UK

Revathi Gopal Universiti Pendidikan Sultan Idris, Tanjong Malim, Malaysia

Hadi Kazemi School of Built Environment and Engineering, Leeds Beckett University, Leeds, UK

Ismail Khan Sunway Business School, Sunway University, Kuala Lumpur, Malaysia

Muhammad Rahies Khan Department of Management, Bahria University, Karachi, Pakistan

Naveed R. Khan UCSI University, Kuala Lumpur, Malaysia

Sri Yogi Kottala Symbiosis Institute of Business Management Hyderabad Campus, Hyderabad, India
Symbiosis International University (Deemed University), Pune, India

Risidaxshinni Kumarusamy Putra Business School, Selangor, Malaysia

Agatha Lavinia Tarumanagara University, Jakarta, Indonesia

Shukhrat Mamatkulov Tashkent State University of Economics, Tashkent, Uzbekistan

B. R. Manoj Kumar Xavier Institute of Management and Entrepreneurship, Bangalore, India

Umathevi Munikrishnan Faculty of Business and Management, UCSI University, Kuala Lumpur, Malaysia

Kesavan Nallaluthan Universiti Pendidikan Sultan Idris, Tanjong Malim, Malaysia

Zohaib Razzaq Department of Management Sciences, Lahore Garrison University, Lahore, Pakistan

Atul Kumar Sahu Department of Industrial & Production Engineering, School of Studies in Engineering & Technology, Guru Ghasidas Vishwavidyalaya (A Central University), Bilaspur, Chhattisgarh, India

Norhisham Seyajah Universiti Kuala Lumpur, Kuala Lumpur, Malaysia

Sibtain Sohair Computer Science Engineering and Maths, Middlesex University, London, UK

J. Soonu Aravindan Kumaraguru College of Liberal Arts and Science, Coimbatore, India

Inayat Ul Haq Subhani Department of Management Sciences, Lahore Garrison University, Lahore, Pakistan

K. B. Theja Acharya Bangalore B-School, Bangalore, India

N. Thilaka Acharya Bangalore B-School, Bangalore, India

Muhammad Umar Department of Management Sciences, Lahore Garrison University, Lahore, Pakistan

Muhammad Babar Usman Department of Management Sciences, Lahore Garrison University, Lahore, Pakistan

Prio Utomo Department of Technology Management, Universitas Multimedia Nusantara, Tangerang, Indonesia

Mohammad Yavari Department of Quantitative Studies, University Canada West, Vancouver, BC, Canada

Syed Imran Zaman Jinnah University for Women, Karachi, Pakistan

List of Abbreviations

AI	Artificial Intelligence
ASEAN	Association of Southeast Asian Nations
AWS	Amazon Web Services
BC	Blockchain
BWM	Best-Worst Method
CAD	Computer-Aided Design
CCPA	California Consumer Privacy Act
CE	Circular Economy
DT	Digital Twin
DWR	Durable Water Repellent
EEL	European Economic Letters
EOL	End-of-Life Management
ERM	Enterprise Risk Management
ESG	Environmental, Social, and Governance
ESMA	European Securities and Markets Authority
ETS	Emissions Trading Systems
EU	European Union
FPS	Factors Performance Scores
GDPR	General Data Protection Regulation
GM	Green Manufacturing
GOTS	Global Organic Textile Standard
GRI	Global Reporting Initiative
IAM	Identity and Access Management
IDS	Intrusion Detection Systems
IEA	International Energy Agency
IM	Intelligent Manufacturing
IoT	Internet of Things
IPP	Integrated Product Policy
IUU	Illegal, Unreported and Unregulated
JIT	Just in Time
KMS	Key Management Service

LCA	Life Cycle Assessment
LDVs	Diesel Light Duty Vehicles
LEED	Leadership in Energy and Environmental Design
MADM	Multi-Attribute Decision-Making
MDVs	Medium Duty Vehicles
NFTs	Non-Fungible Tokens
RFID	Radio-Frequency Identification
RoHS	Restriction of Hazardous Substances
ROI	Return on Investment
SMMM	Smart Manufacturing Maturity Model
TBL	Triple Bottom Line
UNEP	United Nations Environment Programme
UNSDGs	United Nations Sustainable Development Goals
US	United States
VSM	Value Stream Mapping
WEEE	Waste Electrical and Electronic Equipment

Green Manufacturing in Industry 5.0: Towards Sustainable Industrial Transformation

Syed Danish Bukhari

Abstract As awareness about environmental sustainability has grown, countries from all over the world have begun to adopt green manufacturing technology in their production techniques. Industry 5.0 is assisting the reengineering of industrial processes with the aid of the integration of robotics, IoT, and AI. This chapter explores the diverse ways through which green manufacturing technologies can augment Industry 5.0 and ensure the sustenance of the environment. One advantage that may be claimed of Industry 5.0 is the ability of overseeing and managing industrial activities using a number of technologies that may help in conserving energy and waste. These industrial technologies are causing more transparency as far as the production and consumption processes are concerned. Industry 5.0 has made it easier to optimize the use of resources along with recycling. New technology consumed in using fewer resources gives an opportunity to understand the trash in a more intelligent way through this digital revolution. Technological breakthroughs and green production practices can help companies achieve a more sustainable future. However, the success of transforming lies in the effort of companies to overcome the challenges and include environmental concerns within their business planning.

1 Introduction

Developing an inexperienced economy has changed into a worldwide consensus. The advancement of a green financial system requires the help of green innovation to ensure the organized improvement of climate and economic systems. The innovation can considerably ease the unnatural and mechanical shade of commonplace innovation and efficiently oppose the unfavourable effect of innovation on the everyday biological system. Subsequently, without green innovation advancement, there may be no achievable improvement from a genuine angle. Green development has become an unavoidable choice for future mechanical development. One applicable

S. D. Bukhari (✉)
Quaid e Azam University, Islamabad, Pakistan
e-mail: s.danish.bukhari@hotmail.com

M. Venkatesh et al. (eds.), *Green Manufacturing Technologies in Industry 5.0*,
https://doi.org/10.1007/978-981-95-0363-6_1

differentiation of improvements for sensible development is between sluggish advancement and extremist development. Green production is a massive manner for ventures to perform a forward jump and plausible turn of occasions. For the assembly enterprise, the accessibility and stage of green assembly capacity will become the key for ventures to achieve intensity. Moreover, the section spotlight verifies the green exhibition of endeavours. Green assembling is a complicated version of modernization. Green manufacturing is a specific procedure, in which environmental constraints form an intrinsic part and may not be separated off the economic benefits, social benefits, and capacity development that are the outcomes of its implementation. Therefore, it is very important to evaluate green manufacturing capacities of businesses both in theory and in practice. In the hypothetical feel, the examination of green assembling has not shaped an efficient and complete hypothetical machine. A global warming alteration is rising, and people involved are focusing more on a spotless and green climate, even though Lean enhancements execution in the SMEs yet would not think about any natural factors during its execution. Green manufacturing is a cost-efficient assembly, an enterprise technique that returns gains through proactive, natural, well-disposed running strategies (Rame et al. 2023). This chapter is structured as follows; Industrial Technology, Industry 5.0 Technology, Green Manufacturing and Industry 5.0 Technology, Challenges, Opportunities, and Future Direction.

1.1 Industrial Technology

1.1.1 Industry 1.0

The first industrial revolution, which involved the mechanization of industries, occurred from the 1760s to the 1780s. It mainly involved a change of direction from producing manual things. Its prime purpose was the promotion of commerce and venture. Besides water and steam energy, Industry 1.0 brought the start of several machines for other industries, such as the textile industry (Verma 2024).

1.1.2 Industry 2.0

The second industrial revolution dealt with assembly lines, mass production, and division of labour in 1870. The primary characteristic of this revolution was industrialization and standardization. In this revolution, the Manufacturing Industry used the concepts and creations of Industry 1.0 to form new methods of different manufacturing types. Besides, this revolution acknowledged electrification, which brought about the substitution of steam and water engines for electricity, growth in the utilization of electric power, and the invention of many new devices, including the typewriter, automatic signalling, the automobile industry, and electricity. In comparing the two forms of the Industrial Revolution, this one seemed to be more economical and efficient (Verma 2024).

1.1.3 Industry 3.0

The third industrial revolution, generically known as the 'Digital revolution', has generated interesting and helpful gizmos such as computers, smartphones, integrated circuits, and the Internet. Another important point concerning this revolution is that it paved the way for the advent of automation. Besides initiating the producer revolution in industrial methods, it also facilitated hastened communication conditioning more than before (Verma 2024).

1.1.4 Industry 4.0

The fourth industrial revolution started to manifest in developed nations in early 2011. This revolution has three attributes: Automation, data interchange, and manufacturing technology—many new technologies developed during the fourth industrial revolution were found in workshops, homes, and industries. The technologies during this revolution were robotics and artificial intelligence, among others, that had greatly enhanced the joy of work. Today, there is a machine for every job, and given the degree of sophistication and functionality, they can find applications in any segment of the economy. This revolution aims to create phenomena called intelligent and innovative machines, such as 3D printing, robotics, wireless technologies, automobile cars, and the industrial Internet of Things (Verma 2024). Industry 4.0, which is the digitalization of the manufacturing industry, may encourage some aspects of sustainability. Industry 4.0 also helps manufacturers communicate the sustainable impacts of their industrial contexts at the microeconomic level and environment. Some of Industry 4.0's consequences can be decreased waste, energy, or emission, possibly due to the enhanced puritanism of processes with the help of additive manufacturing. Industry 4.0 could enhance sustainability at the microscopic (firm) analysis level, but the specifics of socio-environmental sustainability at macroscopic (industry) analysis levels have remained inconclusive (Ghobakhloo et al. 2023).

1.1.5 Industry 5.0

The credit for creating the term Industry 5.0 goes to Michael Rada. The key difference between Industry 4.0 and Industry 5.0 is the integration of human and machine capabilities. This interaction lets people make relevant, self-expressive buys from customized articles and services. Industry 5.0 makes it possible to offer differentiated products and services driven by the user's active involvement. Industry 4.0 is gradually giving way to Industry 5.0 because of the growing use of human–

machine integration and a higher focus on user- and human-oriented products. Industry 5.0 is based on three core values: The study's hypotheses include human orientation, sustainability, and resilience. The first change process is the transition from technology-centred development to post-industrial development that is human- and society-centred, aiming to achieve people's ontological security needs and aspirations. Industry workers must acquire new knowledge and skills for better job opportunities and optimal work–life balance. The second one is preserving the planet, which will, inter alia, need a circular economy to enhance resources' utility and productivity through recycling, restyling, and reducing damaging elements and waste, together with envisaging effects on ecological systems. The third value of resilience is designed to enhance industrial production capacity to resist disruption that may deny necessary product support and infrastructure. There are numerous scientific publications investigating and comparing the opportunities and threats of Industry 5.0, the enabler of technology and industry (Youssef and Mejri 2023).

1.1.6 Industry 5.0 Technology

Industry 5.0 is relatively new, and its main focus is on collaboration between humans and machines, with the establishment of a partnership where people work together with robots based on what they are programmed to do and what they are capable of doing. Industry 5.0 is a revolution that is both resistant, sustainable and human-oriented. The fifth industrial revolution involves controlling robots, machines, and other technologies and products that have been developed over time to ensure people can easily manipulate them and integrate them into industrial processes. This revolution is going to be a values revolution as well as a technological revolution. However, from the above revolutions, it becomes clear that human creativity and skill levels are not raised. Hence, the retention of human resources on the manufacturing grounds became even more critical during this revolution. Now, in many countries of the world, a vast number of industries are turning to the fifth industrial revolution rather than the fourth. In the current world, technology is applied to make human existence more wonderful, to simplify the problems of civilized communities, and to bring more people together. Industry innovation mainly aims to minimize industrial loss by decreasing production costs while promoting intelligent manufacturing. The advancement made possible by modern technology means acquiring superb tools that contribute to industrial growth and boost supply chain and industrial management. Technology undoubtedly catalyses the advancement, but it would be great if this advancement did not bring about a disaster. While accepting that technology impacts people, society, and the environment in many ways, it is wrong to use state-of-the-art technology for the environment. It is used to refer to a loss for both the human race and all animal species worldwide. The ultimate form of automation is where humans and robots can work side by side. It entails the young generation to change and be alert. For there to be any possibilities for a society or a nation to adopt new technology, there must be enough inputs put in place in order to meet the demands of the so-called have-it-all in society. The chances of nations with sound reserves moving towards technical development are higher since the significant economies already have financial backing.

Thus, the society has a tremendous social significance for the country's development. The regular public constitutes society; if any country's public is educated, they will adopt any technology with a more open heart. Otherwise, if the people in society are illiterate, they will never choose/accept a new model where they have to guide a machine and pass commands to a machine. Though Industry 5.0 has contributed much to human well-being, at the same time, society and the environment have suffered a lot due to the Industrial Revolution. The interaction of man, machine, matter, and environment defines the new Industrial Revolution, an innovation apart from the Industrial Revolution. However, the new industry has several concerns, such as nuclear war, destructive AI, cyber security, and bio-tech. Furthermore, as Industry 5.0 societies use robots for labour, the following aspects must be considered: ethical issues and environments for these devices (Verma 2024).

2 Applications of Industry 5.0

On the verge of the new industrial revolution, combining advanced technologies and human-oriented approaches will introduce changes to several industries. Industry 5.0 is, therefore, shifting from traditional operations to the development of industries. Instead, it is a technological evolution. The analysis of Industry 5.0 applicability is supported by evaluating the impact on agriculture, healthcare, and manufacturing sectors, focusing on the potential progress in various technologies and the corresponding shifts in the demand for employees and specific expertise. An understanding of what is on offer and what is at stake in this upcoming phase of the industrial revolution called Industry 5.0. Industry 5.0 transforms industries through relatively effective and sustainable solutions and depersonalization. The organization will support human and machine processes through AI, robotics, and big data analytic-based enhancement (Kovari 2024).

2.1 Healthcare

In healthcare, the potential of Industry 5.0 has primarily transformed. Adopting cobots and artificial intelligence in surgeries and patient care is an outstanding achievement. Most industrial robots are developed to work autonomously. However, cobots, which are developed to work alongside human professionals, can enhance surgeries by providing a surgeon's precision while maintaining a consistent level of performance not easily achievable by surgeons. AI can also analyse large volumes of patient information to help diagnose and develop individualized treatment plans. They not only improve the quality of care, but they can also help healthcare become more available since they can help address human resource issues in areas with limited health human resource capacity (Kovari 2024).

2.2 Agriculture

It realizes the importance of Industry 5.0 in leading to a politically correct concept of precision agriculture. This approach relies on Internet of Things, AI, and big data management to enhance the decision-making process for farming activities. Applications, such as moisture management in soil aiding in irrigation or intelligent pest control systems, can thus make better use of inputs, increase yields, and be environmentally sustainable. Precision farming is a prime example of Industry 5.0, which follows the efficient and sustainable paradigm (Kovari 2024).

2.3 AI and Robotics

As the field grows, future enhancements in AI and robotics will likely bring more engagement between man and machine. New-generation AI can align more closely with human cognition, bringing more extensive and flexible learning. Robotics could become even more adaptable and able to function at a very high level in a variety of different cognitive and creative tasks, which are at the moment considered to be strictly within the machine or human domain (Kovari 2024).

2.4 Internet of Things and Data Analytics

As with the trend in the Internet of Things and big data, it already opens the possibility for industrial processes to be more innovative and environmentally friendly. Improvements in IoT connection will result in intelligent factories that involve data feedback systems for energy, machine downtime, and foreseen servicing. Such a level of operational intelligence provides industries with a foundation for improving productivity and reducing their negative impact on the environment (Kovari 2024).

2.5 Manufacturing

Replacing the ideology of the fourth industrial revolution in the manufacturing industry, Industry 5.0 provides far superior, environmentally positive, and personalized manufacturing processes to the manufacturing sector. This envelopment is marked by the conjunction of AI, robotics, and IoT, where machines can work together with human beings by assisting them in creativity and decision-making rather than working as tools that humans operate. The primary concern is to improve the biomimicry of the human–machine interface, which becomes a flexible production line capable of fulfilling orders explicitly created for every client in a way that minimizes the resources used and maximizes the value added across the supply chain. Therefore, this positive contributes to enhanced responsiveness and responsibility in the manufacturing industry, which has received much emphasis regarding sustainability (Kovari 2024). It is a critical source of support for firms within the manufacturing industry by providing organizational networks that enable material inputs to be converted into finished goods and connect customers to suppliers. Industry 4.0 of the decade of the 2010s became the technological shift of focus. Its mission was to enhance organizational performance and effectiveness by deploying emergent technologies. Such a transition required the establishment of smart manufacturing and the digitization of the whole value chain. However, Industry 5.0 is the consequent development of Industry 4.0 to signify the shift from discrete technological constructs to systematic planning. This approach supports the sustainability of the industrial process by promoting the cooperation of human skills with intelligent, accurate, and optimum machines. Industry 4.0 for supply chain management was earlier called supply chain 4.0 or digital supply chain. These terms capture the use of superior technology in performing supply chain activities and processes to enhance stakeholders in the supply chain processes. At the same time, Industry 5.0 brings a different value approach and new requirements that affect the current industrial relationships of supply chain members and cause a rearrangement of the supply chain. Consequently, Industry 5.0 supply chains may build on the technological legacy of Industry 4.0 and provide an internal/external balanced human–robot collaboration to support mass customization. Furthermore, Industry 5.0 supply chains can shape energy and emissions reduction while advocating for employee needs and enhancing product performance (Dacre et al. 2024).

3 Green Manufacturing and Industry 5.0 Technology

3.1 Green Manufacturing Innovations

Green Manufacturing Innovations alter manufacturing processes, eliminating negativity and improving resource utilization and sustainability of products during production, usage, and disposal. Green manufacturing, more commonly known as the manufacturing technique or approach that is environmentally friendly, may contain methods and actions that aim to reduce industrial emissions. It is different from the conventional manufacturing paradigm, which focuses on environmental considerations only, but green manufacturing is a concept that encompasses the three factors /Manufacturing with a conscience. Green manufacturing's primary goals are Cost Effective: Efficient green production makes optimal utilization of unwrought goods and electric energy besides water. What needs to be done is the enhancement of production processes, reduction of materials' waste, and proper utilization of some materials; Pollution Prevention: Clean technology and pollution preventive measures such as minimizing pollution and deterioration in green manufacturing. Sulfur oxides, nitrogen oxides, particulates, and volatile organic compounds, which currently spew into the air at levels that even the most antiquated estimates classify as hazardous, must be diminished or cleaned up, and waste must be disposed of in ways that do not harm the biosphere; and Product Lifecycle Management: Green manufacturing entails the generation of manufacturing designs for products right from acquiring raw materials to disposal. We minimize the environmental footprint through principles of sustainable design, promoting the reuse and remanufacturing of products and recycling of materials (Yerram 2021).

3.2 Green Manufacturing

Green manufacturing takes place with ideas that translate into decision-making and operations. Green manufacturing follows several principles: Design for Environment (DfE): Green manufacturing entails designing for sustainability in the production process of goods that will be produced and marketed. DfE minimizes the environmental burden of energy, materials, and resources used in product design, manufacturing process, usage, and end-of-life disposal. According to the sustainable design paradigm, new products should be designed and developed to use up the least resources and be easy to recycle while using energy efficiently. Energy Efficiency: Energy is one of the leading environmental concerns. Therefore, as part of green manufacturing, energy efficiency needs to be enhancements. Energy minimizing in green manufacturing refers to the efficient technologies used, developments in manufacturing techniques, and energy from natural sources such as solar and wind (Ali et al. 2023). The following is a list of these standards: These standards cater to many areas, including sustainability management, resource productivity, and social responsibility. The other principle followed is waste, which involves the reuse, recycling, and minimizing the flow of materials within the cycle of the production line. It also continues the manufacturer's authority over waste and resources so they can limit the natural impact and continue their affordability. Several axes of change are described here, including energy efficiency, an elementary concept that refers to the maximization of energy from manufactured greenhouse gases from processes employed in production. In addition, it recommends that implementing renewable power & energy and other energy-saving technologies helpful in process-related visual effects can be minimized. The system's energy management will lower natural impressions and increase operating smoothness. Further sustainable manufacturing fundamentals comprise systematic and disciplined processes of manufacturing essential products by employing environmentally sound material, appropriate packaging, and very keen programmes for disposing of these products in a way that does not harm the global environment. Such perspectives, mainly those with life span characteristics, recyclability qualities, or disassemblability traits in the items they offer, can help avoid their impact on the environment significantly and provide for a circular economy. There is a method of organizing production, supply chain, and design of products. It is impossible to cause as much harm to the natural environment as possible while developing and discussing ideas for selected/a few/chased products and their usage and disposal. A practical application is a material choice concentrating on eco-design, which focuses on energy productivity and recyclability; this makes products more environmentally friendly or sustainable and has positive impacts. Green manufacturing aims to embrace technical procedures that assist firms in operating with minor emissions while not contracting high resource utilization. For example, lean manufacturing may reduce waste and improve production through continuous improvement. Specifically, for sustainability purposes, industries should involve customers and other stakeholders to consider the prospects of reducing environmental degradation around the construction supply chain. The supply chain is a good way to see where attention is required and what action will help minimize the additional use of resources, emissions, and waste production. There are many companies with different segments nowadays, and all of them have faced the implementation of green manufacturing measures and shown that such solutions could be used (Kumar et al. 2024).

3.3 Sustainable Manufacturing

Sustainable manufacturing evolves from the green manufacturing point of view that emanates from the lean manufacturing practice. The fact that lean manufacturing only considered waste as anything that does not add value to the economic value stream and, therefore, got rid of it without hesitation, while green manufacturing acknowledged the expenses of the earth by reducing productivity. Sustainable manufacturing goes ahead of green manufacturing by invoking the time perspective and valuing the future sustainability consequences of present manufacturing regimes. Sustainable manufacturing today is widely seen as a dynamic process state since many of its constitutive components, values, assets, and practices are themselves change constants (Ching et al. 2021). Sustainable manufacturing is a broad concept, and it has changed with the improvements in sustainable concerns and socio-environmental factors. Sustainable manufacturing is defined as reducing the adverse effects of manufacturing processes on the environment. However, it is now considered an appropriate initiative necessary for achieving sustainable development of its economic, environmental, and social aspects. The technological innovations and their industrial applications are congruent with the key goals of sustainable manufacturing (Ghobakhloo et al. 2023).

Sustainable manufacturing is understood as creating a set of technologies, policies, and practices for manufacturing tangible goods and services for which the economic and ecological costs of creating value for stakeholders (employees, customers, and communities) and safeguarding the worth of their consumption are optimized and the consequences on the natural environment are benign. Sustainable manufacturing entails extending manufacturing sustainability and creating new and better sustainable products that deliver the intended economic, environmental, and social value. Similarly, sustainable manufacturing encompasses a systemic understanding of all product life-cycle stages together with the whole value of delivering and creating channels for expressing the sustainability. In its broader sense, the economic-manufacturing sustainability pillar suggests that manufacturers must be financially sustainable to achieve other sustainability objectives. However, this pillar does not need to go the profit at any cost route; it also cares for consumers and value partners. The manufacturing-environmental sustainability mainly relates to decreasing the total adverse impact of manufacturing processes on the natural environment, primarily in cutting actual or potential CO2 emissions, hazardous emissions, energy consumption, and waste output. The manufacturing-social sustainability pillar can be described as the poorest understood and most frequently ignored kind of manufacturing sustainability, which addresses the welfare of stakeholders and the community in which a manufacturing value chain resides. For the manufacturing industry, the most recognized elements of manufacturing-social sustainability are customer experience enhancement, workplace environment enhancement, and designing fair employment conditions. However, this may encompass many micro-social considerations like child labour or fair wage questions in any corner of a supply chain network on a global scale. Nonetheless, companies suffer incremental costs when implementing sustainability strategies in the short run. In the same respect, manufacturers worldwide focus more on sustainability because of competitiveness and regulatory pressures. Sustainable manufacturing activities can offer following benefits to manufacturers once adequately planned and well executed. Financial savings depending on material, energy, and resource conservation are amongst the apparent advantages of sustainability and risky business opportunities that sustainability might bring to manufacturers, including brand image, public credibility, and competitiveness. The Social and Socio-economic Life Cycle Assessment (S-LCA) guideline issued under the United Nations Environment Programme is useful in dealing with the vagueness of the definition of the social sustainability concept. S-LCA provides a step-by-step procedure checklist both to the internal and external stakeholders for evaluating the social and socio-economic life cycle of a product (Ching et al. 2021).

3.4 Green Manufacturing and Industry 5.0

Industry 5.0 adheres to the policy of sustainable development. Digital green innovation management activities have been identified as the key components of low-carbon intelligent development of prefabricated construction enterprises for sustainable urban development. Industry 5.0 as part of sustainable development with a focus on the impact on the promotion of technological advances in green technology. Exploring digital green innovation in photovoltaic building material enterprises is an invaluable means of addressing the challenges of high-quality environmental and economic development. This should mean that the production system must satisfy the requirement for economic growth and, at the same time, be capable of sustainable development without compromising the environment or society. Industry 5.0 is a green innovation economy in the digital innovation paradigm that aims to minimize carbon footprinting by embracing green technologies and innovations by enterprises and technology developers. However, this sustainable, development-oriented innovation aids the achievement of the green economy objective. Concerning the strategy of peaking carbon emissions and attaining carbon neutrality, digital green innovation development plays a more significant role in enterprises. Therefore, for enterprise photovoltaic building materials, in addition to enhancing green competitiveness and improving profit levels, they must select digital green innovation projects for investment (Lv et al. 2024).

Sustainability and productivity in industrial operations require progress and innovation in environmental manufacturing. Green manufacturing, in any case, uses various inventions to ensure minimal interference, minimal resource consumption, and a high overall efficiency. One of them is additive manufacturing, also referred to as 3D printing. This particular technology assists manufacturers in creating complex parts with virtually no scrap. Another is smart process controls and systems involving superior robotic operations that enhance production and reduce energy consumption. In expansion, solar panels or wind turbines are installed in industrial facilities and renewable energy sources in a bid to minimize the use of fossil fuels and, in this way, avoid GHG emissions. Besides that, using IoT devices, real-time checking of production processes through sensor innovation has been made possible. This results in growth in productivity levels and resource utilization rates. These latest-generation concepts accumulate the foundation for green manufacturing, enabling firms to convert to environmentally friendly production mechanisms (Kumar et al. 2024).

The comprehensive range of modern and contemporary green or sustainable manufacturing practices is applied, it leads to decreased environmental evidence and increased resource utilization. Some closed-loop manufacturing systems demonstrate how waste material in one process is utilized in the other to eliminate waste and promote circularity. Bio-manufacturing is another method that uses organic systems, such as bacteria or enzymes, to achieve almost zero natural biomaterials and chemicals. Other than that, green chemistry strategies involving solvent-free processes coupled with catalytic reactions ensure that the products produced are less dangerous and result in less waste production. Moreover, such combined manufacturing structures as, for example, shared resources between different companies' output also significantly influence the resource efficiency and collaboration in an offered area to minimize the natural impacts. Here are some occurrences demonstrating various unique approaches to promote sustainability goals in the manufacturing industry. As the world advances, it is foreseen that the nature of green manufacturing will change, reflecting the following patterns and progressions. Artificial intelligence and machine learning to enhance output methods, prognoses maintenance needs, and identify productivity improvement chances. Besides, nanotechnology enhancements can develop superior, efficient materials and coats with way better qualities that barely infringe on the environment. These new and distinctive trends in green manufacturing suggest a shift in a positive, practical, and more assertive approach (Kumar et al. 2024).

3.5 *Greentelligent Manufacturing*

The term '(in)telligent' has been used to describe intelligent structures, software, and networked solutions for sensor, measurement, and control technologies as well as process integration to the manufacturing systems and information and communication technologies (ICT) that can support online tracking and controlling of wastes/ consumptions and GHG emissions in facilities, and online management of energy/ profitability and efficiency across facilities. Intelligent technologies' flexibility allows for improvement in achieving sustainable goals more effectively while continuing competitiveness and delivering additional value to the sophisticated manufacturing industries. Greentelligent manufacturing relates to the term 'green' in three ways: to offer greener products by altering the manufacturing processes and product design. They introduce greener manufacturing processes by reducing emissions and waste, using natural resources, and improving energy and manufacturing efficiency. The green intelligent techniques of the aforementioned to evaluate (Li et al. 2021). Greentelligent manufacturing (GIM) is called an innovative, human-oriented, and environment-friendly manufacturing paradigm to come to the forefront at the synergetic stage.

On the other hand, Greentelligent manufacturing is the strategy of using intelligent systems and green methods to exercise higher control, efficiency, and effectiveness in manufacturing systems. An intelligent manufacturing framework is structured to embed an expanding set of practices and substrates for clean energy generation and green product design. Greentelligent manufacturing divided into three phases: the emerging and positional. In the emerging stage, Green Manufacturing (GM) and Intelligent Manufacturing (IM) have been introduced independently to a certain extent. However, they share similar concerns, such as resource concurrency and environmental impact, and they focus on sustainability and smart manufacturing. At the transitional stage, the player with the GM and IM titles begin to take advantage of one another to achieve their objective. At a synergetic stage, GM and IM interconnect with each other and share objectives, and the tool used is aimed towards the objectives developed for the two paradigms (Li et al. 2021).

3.5.1 Procedure for Greentelligent Manufacturing

Life cycle assessment (LCA), also called life cycle inventory assessment, is a technique within life cycle assessment that provides a quantified model of the environmental loads associated with a product, process, or activity throughout its life cycle. At every stage of the manufacturing process, describing the inputs and wastes received and setting the corresponding costs is necessary. The actual efficiency varies significantly with the level of automation of this process. This type of assessment can be augmented by using artificial intelligence to identify and precisely target costs or specifically track costs back to manufacturing processes to estimate costs in the future. Sophisticated ICTs facilitate monitoring CO2 from fuel combustion and airflow measurement in several compartments of the furnaces. Based on IoT, it is possible to design energy and CO2 emission networks focusing on improving the manufacturing process services and satisfying the given and somewhat volatile consumptions. Greentelligent manufacturing can use carbon emission as one of the ingredients to map the supply chain. From a consumer perspective, they are not incentivized to pollute less or take steps to decrease the amount of emissions. In contrast, consumers dictate the best action by supporting environmentally friendly producers. IoT techniques and a digital supply chain involve all the tiers of the total supply chain, such as raw material sellers and customers, and consequently price the products using environmental costs to transform the market in this dynamic world. Other pricing mechanisms that influence development towards environmentally friendly practices include Carbon taxes and Emissions Trading Systems (ETS). A carbon tax more directly puts a price on carbon in a sense that defines the taxation rate for each CO2 emission. ETS sets a maximum on the overall level of greenhouse gases and gives those industries that emit less to sell their excess credits to high-emitting sectors. Unlike the carbon tax, the ETS of a carbon tax is not usually predetermined but results from the carbon price some enterprises bargain for. It contributes towards achieving the required emission reduction goal by constantly reminding the emitters of the investment cost of low-carbon technologies and carbon purchases. Carbon pricing mechanisms are currently in place in over 40 countries and applied in 20 cities, states, or provinces, and more countries are in the pipeline for the same (Li et al. 2021).

4 Challenges, Opportunities, Research, and Future Direction of Green Manufacturing and Industry 5.0 Technology

4.1 Challenges

In green manufacturing, various hurdles and opportunities must be faced to succeed when going green. Green manufacturing techniques are expensive since they require new technologies, infrastructural changes, and workers' training. Transitioning to more sustainable production may disrupt current workflow patterns and supply chains, leading to opposition from partners who have become comfortable with traditional manufacturing methods. Others arise from the natural rules, regulations, and measures as the manufacturers struggle desperately to retain their international competitiveness within the systems comprising the quick change of regulations. Besides this, measurements used while evaluating the company's environmental efficiency are challenging, requiring elaborate measurement and data analysis tools. All these steps can only be overcome by active authority, involving all the stakeholders and a harmonized procedure for the sustainability measures. Green manufacturing, although vigorous, shows great development prospects and cuts down on costs, making it unique from others. Through innovative methods that are biologically mindful, the manufacturer has an opportunity to reduce the cost of operations by improving energy efficiency, waste and resources. Green manufacturing also fosters innovation since it calls for the innovation of today's techniques, materials, and technologies with negligible environmental influences. Evening activities such as sustainable manufacturing can help enhance the image and competitiveness of brands in the market as consumers focus more on products that will help promote the environment. Aside from this, green manufacturing provides opportunities for affiliation and partnership concerning every sector that seeks to exchange the best practices, achieve economies of scope, or address similar sustainability concerns jointly. It may be that the ability to tolerate these challenges can help corporations to continue in a world with fast-shifting financial and natural conditions. There are a few other reasons why green manufacturing is not only a method for environmental conservation but also bears financial, social, and operational advantages that may be obtained from implementing green manufacturing. Some of them are that natural resources are conserved to a certain degree, such as the prevention of damage to the environment in terms of emission of carbon dioxide, and so on, which should be sufficient against traditional sources (Kumar et al. 2024).

4.2 Opportunities

Economically considered, green manufacturing means a reduction of costs by improving the efficiency of manufacturing, reducing the costs for disposal of wastes, and enhancing the quality of products. The social benefits include job opportunities, better employee relations, and extended synergism. The producers seek to show their dedication to social responsibility and care towards the partners in the community. Furthermore, green manufacturing effectively advances strength in operations through the privation of risks like resource scarcity, failure to conform to rules and regulations, and mauled image. Admittedly, firms can establish green manufacturing processes that point towards additional value beyond a firm's standard economics when comprehending the green ecosystem, which incorporates natural and social elements. Moreover, inexperienced production actions forward electricity in operations via diminishing publicity to risks which includes scarcity of resources, non-compliance with rules and suggestions, and bruised reputations. Firms can utilize inexperienced manufacturing strategies that make it profitable beyond financial measures at the same time as considering herbal and social dimensions that make sustainability necessary (Kumar et al. 2024).

4.3 Research Areas

The five key areas of focus as applied to Industry 5.0 research initiatives are: (i) evaluation and optimization of supply chains; (ii) business evolution and digital transformation; (iii) creation of innovative and environmentally sustainable production processes; (iv) changes through IoT; AI and Big Data; and (v) an integration of man and machine. These themes suggest more topics for future studies and elaborate on the higher focus on Industry 5.0 as the transition to the human–machine relationship—a literature review of the current literature and recommendations for an absolute innovation management framework. Finally, emerging with the change, the vision for firms to be ready for the IoT and Industry 5.0 will call for integrated innovation, design thinking, and strategy. What is presented in a literature review is the characteristics of Industry 5.0 and related issues and enabling technologies for Industry 5.0, as well as problems such as how to achieve the best value to optimize both human and machine productivity. Industry 5.0's scientific-technological field defines when the sector enters the growth stage to map the leading countries. The primary technology enablers of Industry 5.0 include identification, automation of processes and data, and data collection and analysis, which speaks of collaboration, synchronization, and communication in management, technology, evaluation of performance, and organization. He also emphasizes that Industry 5.0 is at the societal, network, and plant level, and he suggests a triple bottom line of resilient value, well-being, and sustainable society—the emergence of Industry 4.0 and 5.0 conceptualization, perception, and genesis. Industry 4.0 is technically based; Industry 5.0 is values-based and fundamental to modern society's needs, values, and responsibilities and relies on Industry 4.0 for technological support and technical solutions. It then defines the differences between Society 5.0 and Industry 5.0 and calls for increased recognition of both concepts to foster progress in the related theories, strategies, and prospects. Goals, values, and organizational and technological characteristics can differentiate Industry 5.0 and Society 5.0. A critical analysis of the impact of New Manufacturing Industries and Skills is that Industry 5.0 fosters collaboration between humans and robots and encourages the use of both humans and robots in production. Industry 5.0 and its impact on the manufacturing sector and the economy, in general, should be viewed through the prism of economics and productivity. The Industry 5.0 will create more jobs than it will dispense. The theoretical research stances: identifies socio-technical systems theory, skill-biased technological change, and variations, which include task-biased technological change, routine-biased technological change, and the political economy of digital transformation. These three perspectives enrich a new and comprehensive theoretical perspective on the future of complex issues and the diverse and complex opportunities presented by Employment 5.0 (Youssef and Mejri 2023).

4.4 Future Direction

Industry 5.0 and the green innovation economy represent the future direction of industrial and economic development. Industry 5.0 aims to improve the efficiency of resource utilization, reduce waste, and reduce environmental impact, which is in line with the core goal of the green innovation economy. Digital technology supports green innovation, and Industry 5.0 relies on advanced digital technologies, such as the Internet of Things, big data analytics, and artificial intelligence. These technologies can be used to monitor and optimize energy use, production processes, and supply chains to achieve more environmentally friendly production methods. In terms of economic growth, competitiveness, policy support, and new business opportunities, Industry 5.0 and the green innovation economy support each other to jointly promote technological innovation, sustainable development, and economic growth. This synergy will create more opportunities for future industrial and economic development, but it also poses a series of new challenges (Lv et al. 2024).

5 Conclusion

It is concluded that Industry 5.0 and green manufacturing system integrate the production process to promoting higher production efficiency and quality. As a human-centred manufacturing paradigm, promoting sustainable manufacturing emphasizes the role of Industry 5.0 in the digital green innovation economy, with special attention to human–machine collaboration, resource optimization, and the reduction of environmental impact. In the digital green innovation economy, this system enables intelligent resource management and optimization and reduces energy and resource waste through real-time monitoring and feedback.

References

Ali A, Khan MR, Ghouri AM (2023) Carbon neutrality concept and progress. In: Recent developments in green finance, green growth and carbon neutrality. Elsevier, pp 85–108. https://doi.org/10.1016/B978-0-443-15936-7.00003-7

Ching NT, Ghobakhloo M, Iranmanesh M, Maroufkhani P, Asadi S (2021) Industry 4.0 applications for sustainable manufacturing: a systematic literature review and a roadmap to sustainable development. J Clean Prod 334:130133. https://doi.org/10.1016/j.jclepro.2021.130133

Dacre N, Yan J, Frei R, Al-Mhdawi MKS, Dong H (2024) Advancing sustainable manufacturing: a systematic exploration of Industry 5.0 supply chains for sustainability, human-centricity, and resilience. Prod Plan Control:1–30. https://doi.org/10.1080/09537287.2024.2380361

Ghobakhloo M, Iranmanesh M, Foroughi B, Tirkolaee EB, Asadi S, Amran A (2023) Industry 5.0 implications for inclusive sustainable manufacturing: an evidence-knowledge-based strategic roadmap. J Clean Prod 417:138023. https://doi.org/10.1016/j.jclepro.2023.138023

Kovari A (2024) Industry 5.0: generalized definition, key applications, opportunities and threats. Acta Polytech Hung 21(3):267–284. https://doi.org/10.12700/aph.21.3.2024.3.17

Kumar P, Yadav D, Salaam A, Devi PS, Nijhawan G, Nagaraju S (2024) Innovative and sustainable transformation of industrial sector as a green manufacturing approach. In: E3S Web of Conferences, vol 552. EDP Sciences, p 01093. https://doi.org/10.1051/e3sconf/202455201093

Li X, Wang B, Peng T, Xu X (2021) Greentelligence: smart manufacturing for a greener future. Chin J Mech Eng 34(1):116. https://doi.org/10.1186/s10033-021-00656-2

Lv X, Wang Y, Liu L, Yin S (2024) Digital green innovation economy for Industry 5.0. Del J 2(1): 8. https://doi.org/10.62617/se.v2i1.8

Rame R, Purwanto NP, Sudarno NS (2023) Global context of Industry 5.0: current trends and challenges in Indonesia. Jurnal Riset Teknologi Pencegahan Pencemaran Industri 14(3):21–32. https://doi.org/10.21771/jrtppi.2023.v14.no3.p21-32

Verma RC (2024) Industry 5.0: moving towards inclusive and sustainable industrialization. Nepal J Manag Sci Res 7(1):88–102. https://doi.org/10.53056/njmsr-2024.7.1.006

Yerram SR (2021) Driving the shift to sustainable Industry 5.0 with green manufacturing innovations. Asia Pac J Energy Environ 8(2):55–66. https://doi.org/10.18034/apjee.v8i2.733

Youssef AB, Mejri I (2023) Linking digital technologies to sustainability through Industry 5.0: a bibliometric analysis. Sustainability 15(9):7465. https://doi.org/10.3390/su15097465

Transforming Malaysia's Textile Industry: Sustainable Manufacturing in the Industry 5.0 Era

Kesavan Nallaluthan, Arsalan Mujahid Ghouri, Norhisham Seyajah, and Revathi Gopal

Abstract The textile sector is currently transforming. This shift has resulted from Industry 5.0 with the fourth Industrial Revolution. The Malaysian textile business has notably strengthened. This research explores how Industry 5.0 technologies, for instance, artificial intelligence (AI), robotics, and the Internet of Things (IoT) can benefit the textile sector. It studies how Industry 5.0 innovations and sustainable strategies are interwoven. The practices feature life cycle analysis along with circular design. This industry has received criticism for high resource consumption together with waste creation. This has negative environmental implications. Life cycle evaluation (LCA) has been useful. Life cycle evaluation assesses environmental effects across raw material acquisition toward product reprocessing. This assists in the reduction of ecological hazards and further aligns directly with Industry 5.0's focus on data gathering plus data analysis. This information helps in asset management and overall planning processes. Eco-friendly design has been similarly critical. Environment design involves making goods produced with much less waste. This also shows reducing resource use plus making items that can be reused. Further, it focuses on correct waste disposal methods. Information services, finance, utilities, education, and healthcare are the five industries of the Industrial Society II based on the use of AI design tools. With the use of this tool, products are made to be more durable and recyclable, which increases the likelihood that they will benefit the human population by extending the lifespan of labor and enabling workers to take part in production processes in a sustainable environment. Because it promotes personal growth and job satisfaction, this also reduces waste and resource consumption. The circular economy, which was first proposed in the 1990s, is another

K. Nallaluthan · R. Gopal
Universiti Pendidikan Sultan Idris, Tanjong Malim, Malaysia
e-mail: kesavan@fpe.upsi.edu.my; revathi@fbk.upsi.edu.my

A. M. Ghouri (✉)
London South Bank University, London, UK
e-mail: arsalan.ghouri@ymail.com

N. Seyajah
Universiti Kuala Lumpur, Kuala Lumpur, Malaysia
e-mail: hishams@unikl.edu.my

M. Venkatesh et al. (eds.), *Green Manufacturing Technologies in Industry 5.0*,
https://doi.org/10.1007/978-981-95-0363-6_2

important idea in guaranteeing sustainability. By including reader suggestions on how Malaysian textile companies can implement strategies like re-manufacturing, biodegradable materials, and recycled materials, the research's sustainability can be improved. The system's adoption of Industrial 5.0 technologies, which include improved goods tracking, sorting, and the advancement of recyclable materials, makes it possible. It offers industry examples and empirical data to illustrate how to incorporate sustainability into Industry 5. According to the findings, the government and industry must acknowledge some issues and that they are environmentally responsible, but they still want to minimize their environmental impact, which entails reducing the effects of the industry sectors in which consumers are also engaged. If this is true, Malaysia's textile sector has the potential to emerge as a global leader in environmentally friendly production.

1 Introduction

Malaysia's textile industry is at a turning point and needs traditional manufacturing methods to develop to satisfy present worldwide needs for sustainability and efficiency. Using automation, AI, and the Internet of Things (IoT), the Fourth Industrial Revolution Industry 4.0 has changed manufacturing (e.g., Akhtar et al. 2024; Bhatti et al. 2023; Khan et al. 2024). But as businesses migrate toward Industry 5.0, production—centered on people and sustainability—combining modern technologies with environmental and social issues takes the front stage. This chapter looks at how the ideas of Industry 5.0 might be included in the Malaysian textile sector to solve current issues and increase efficiency and sustainability simultaneously (Farhana et al. 2022).

1.1 The Evolution from Industry 4.0 to Industry 5.0

Using automation, intelligent systems, and digitalization, Industry 4.0 marked a radical shift in manufacturing that produced sped up, more efficient, data-driven production processes. According to Mizuno et al. (2014) mostly, the focus was on improving efficiency and production using integrated systems, robotics, and AI-driven automation. These developments resulted in job displacement, ethical questions regarding human involvement in production, and increased environmental damage resulting from too much reliance on resource-intensive technologies even while they improved industrial output and reduced human interaction (Deshpande et al. 2022; Ghouri et al. 2022).

Industry 5.0 is an answer to these limitations by associating technological progress with human creativity and sustainability. It is assumed that technology builds on human labor and does not replace it because it re-integrates the human brain and skill into computerized systems. This "significant change" considers that because of the need for more ethically consistent, environmentally responsible, and socially responsible industrial practices.

1.2 Key Characteristics of Industry 5.0 Include

Human-Centric Manufacturing It highlights the harmony between humans and machines. Industry 5.0 enables employees to use IoT devices, robotics, and AI-powered tools to raise productivity and innovation without threatening job's job security and welfare at work. Unlike Industry 4.0, which focuses on automation and minimizing human participation, in Industry 5.0 it is believed that in various manufacturing processes, human intuition, creativity, and decision are crucial.

Sustainable Production Practices Industry 5.0 brings sustainability into the core of industrial processes. This includes using sustainable manufacturing techniques, improving the use of natural resources, and so reducing waste. Sustainable manufacturing ensures products are built for durability, reuse, and recyclability, lowering carbon footprints, encouraging the use of renewable energy, and embracing circular economy ideas.

Customized and Adaptive Manufacturing Consumers are increasingly looking for premium products that are a line to meet their needs with AI-driven customization, on-demand manufacturing, and flexible supply chain management. Industry 5.0 is helping manufacturers meet these needs. This process ensures proper waste produces with consumers' preferences in mind goods and avoiding overproduction.

Therefore, integrating advantageous technologies supported human decision-making and did not replace it. Robotics, IoT, and AI enhance data-driven decision-making, predictive analysis, and real-time monitoring, supporting a more sustainable and efficient manufacturing process.

Hence, digital transformation and sustainability need to be embraced by the Malaysian textile sector with a comprehensive plan that aligns with ethical labor standards in Industry 5.0. With the proper utilization of advanced technology with human creativity, the Malaysian textile sector can become a world leader in sustainable manufacturing.

1.3 Sustainability in the Malaysian Textile Industry

The textile industry adds to world pollution through large raw material consumption, waste generation, and maximum usage of water. In a line a call for changes to sustainable manufacturing practices in the Malaysian textile industry. Some practices could be preferred to achieve this goal:

1.3.1 Life Cycle Assessment (LCA) for Sustainable Manufacturing

Life Cycle Assessment (LCA) is a comprehensive scientific process used to evaluate the environmental impact of a product across its whole life cycle from raw material procurement to disposal or recycling. In the textile industry, this process is essential since it supports companies' understanding and minimizing the environmental effects from their production process. Using LCA, companies can identify the problems that need improvement by producing more environmentally friendly textile manufacturing products (Braun et al. 2021).

Environmental problems arise significantly because of the usage of raw materials, electricity, and water in the textile industry. For example, traditional textiles which used for large-scale agricultural processes, including cotton farming, which require lots of water and chemical insecticides. Can minimize the environmental impact of textile products by sustainable raw materials like biodegradable textiles, bamboo fibers, and organic cotton. These require less chemical inputs, while materials encourage responsible land use and preservation of biodiversity.

With the LCA, it improves production process optimization by reducing energy consumption and carbon emissions. Among the most energy-intensive industries is textile manufacture; spinning, weaving, dyeing, and finishing operations all call for significant energy consumption. Manufacturing can cut its reliance on fossil fuels and lower greenhouse gas emissions by using energy-efficient machinery and renewable energy sources, including solar or wind power. Also, the last steps in the dyeing process, like waterless dyeing and digital printing, offer the possibility of being more environmentally friendly by decreasing chemical discharge in the water and water consumption.

LCA mainly falls under waste management. The problem of textile waste is becoming increasingly higher, with millions of tons of fabric going to landfills every year. Incorporating the principles of effectiveness in life cycle analysis enables companies to map out usage-reuse-recycling policies for textile products. New or recycled fibers or fabrics stemming from textile waste reduce the reliance on virgin resources and diminish environmental degradation. Also, eco-design ideas can be cross-listed to develop products that become ever more durable, repairable, and recyclable.

LCA supports companies and their certifications based on environmental guidelines and includes the OEKO-TEX Standard 100 and the Global Organic Textile Standard (GOTS), amongst others. These certifications assure that the textiles have followed a rigorous round of sustainability criteria so they will embark with people who have invested in the conservation movement. Companies applying LCA-driven approaches will get a market advantage by employing products reflecting ethical and environmentally today conscious values to enhance consumer awareness of sustainable fashion.

Ultimately, the use of LCA in the Malaysian textile industry promotes ethical manufacturing and consumption in line with the world's environmental goals. Businesses can reduce their footprint and business profitability and competitiveness in the field by highlighting sustainable materials, increasing energy efficiency, and applying circular economy principles.

1.3.2 Circular Economy and Waste Management

Through reusing production processes, the cycle economy is an economy that keeps a focus on a reduction of waste and the optimization of resources. Circular economy principles are being put to use in Malaysia's textile industry and they help in waste management and responsible environmental practices.

One simple way to deal with this change is to transform textile residue into manufactured fibers and fabrics. Instead of dumping used cloth in the... This operation decreases contamination and minimizes reliance on good deals.

The manufacturing of biodegradable clothes is also another action to take as, at the end of their cycle, no poisonous things are released from the clothes. Using natural and eco-based materials, Malaysian textile companies could ensure the cloth does not worsen the long-term environmental pollution.

There can be no growth of a circular economy without consumer involvement. Promotion of the rebirth of decoration fabrics destined for upcycling or appropriation properly fills up a policy of sustainability. Textile companies operate take-back systems where consumers can return unwanted garments, and such textiles can then either be recycled properly or transformed into new materials.

In addition, a longer product life cycle through repair and reuse campaigns can ultimately save a lot of textiles from waste. While designers can create strong, timeless clothing that resists quick fashion trends, manufacturers may offer repair services.

Improving these systems calls for Industry 5.0 technologies. Advanced tracking systems driven by blockchain and IoT help to better sort recyclable materials, increase supply chain openness, and ensure adherence to sustainable standards all around the manufacturing process. Through real-time monitoring, manufacturers can manage waste streams, improve performance, and implement data-driven sustainability solutions (Farshadfar et al. 2025).

Using circular economy ideas will help Malaysia's textile industry lower environmental impact and support long-term economic sustainability and innovation. For Malaysia, it is not only a strategic advantage but also a moral to be a leader in ethical textile manufacture and sustainable waste management.

1.3.3 Eco-Design and Sustainable Product Development

Eco-design and sustainable product development strain the creation through the manufacturing process of textiles with low environmental impact through sustainability. This environmentally affected approach starts with choosing recycled fibers, hemp, and organic cotton as raw materials. These materials cut toxins connected to conventional textile manufacture and support slowing down resource depletion, but also cut toxins connected to sustainable textile manufacturing.

The next crucial component of eco-design is the development of durable and easily recyclable fabrics. Hence, fabric waste reduction and the extent of a lifetime give durability top importance to manufacturers. Therefore, sustainability in fashion improved, and incorporating modular design concepts lets clothing be disassembled and used again.

Textile dyeing operations are much influenced by the use of chemicals, and water affects the environment. However, chemical waste and water consumption can be reduced by plant-derived dyes and waterless dyeing technologies and other sustainable dyeing techniques, thus improving the environmental friendliness of the manufacturing process.

AI-powered design tools also enhance eco-design by improving manufacturing techniques and material choices. By evaluating large amounts of data, machine learning techniques can propose the most environmentally friendly material mixture, ensuring low waste generation and energy use. Besides supply chain management, AI can also improve predictive analysis, minimizing overproduction and excessive resource consumption (Chen et al. 2024).

Malaysia's textile industry remains globally competitive while moving toward a more responsible and sustainable future through these eco-design ideas.

1.4 The Role of Industry 5.0 Technologies in Sustainable Textile Manufacturing

Industry 5.0 technologies play a crucial role in achieving sustainability in the textile industry. These technologies include:

1.4.1 Artificial Intelligence (AI) and Machine Learning

Through Industry 5.0, AI and ML are boosting production, cutting waste, and improving efficiency in manufacturing operations. It is improving resource allocation and decision-making process with real-time database monitoring in textile process by AI-powered predictive analytics (Ahmed et al. 2023; Bangwal et al. 2023).

1.4.2 Optimizing Production Processes

Under AI-driven manufacturing processes, by the predictive analysis helps companies explore and incompetence. In addition to minimizing the textile production downtime, ML could analyze data from manufacturing technological equipment, detecting abnormalities which make sure an efficient manufacturing process.

In addition, material usage which minimizing wasted fabric and textile manufacturing costs by implementing AI-driven automation technologies. Further improvement in textile technological process such as dynamic stitch pattern alterations leading to greater efficiency and less wasting time with the real-time data updates from robotic arms and intelligent sewing machine technology.

1.4.3 Enhancing Sustainable Resource Management

To improve sustainable resources management such as water and electricity and raw materials in textile manufacturing, they should AI driven technology to minimize water waste in textile dyeing. Similarly, energy consumption trends and proposes changes to decrease carbon emissions, which can be evaluated by smart energy technological systems.

Ethical procurement of raw materials through supply chain optimization by AI-powered with blockchain-integrated AI, by tracking fabric origins, ensures adherence to sustainability requirements, hence minimizing the environmental impact of the textile production process.

1.4.4 Reducing Textile Waste and Promoting Circular Economy

ML may foster circular economy principles by identifying ways to recycle and reuse textile waste materials. AI-powered sorters recognize waste textiles by material, progressing textile recycling and the upcycling process.

AI-powered demand forecasting enables manufacturers to produce only what's needed, minimizing overproduction and waste from excess inventory. By syncing production with actual market demand, the textile industry could reduce its environmental footprint.

AI and ML provide transformative solutions for Malaysia's textile industry, boosting production efficiency, minimizing waste, and promoting sustainability. By adopting AI, manufacturers can ensure the industry's sustainability while maintaining profitability and environmental responsibility (Vijay Kumar and Shahin 2025).

1.5 The Role of Collaborative Robotics in Enhancing Efficiency and Sustainability

In recent years, the creation of advanced robots has brought many positive developments to sectors all over. These "smart" machines have improved procedures and made plenty of things more sustainable, but above all, they have upped our productivity. One way to do all that is through the cooperative robots, or "cobots," that have appeared in various workplaces. These robots work with humans and, to quote a GE Worker bee, "make them superhuman." Unlike the robots of old, which were often designed to do the things that humans don't want to do (or can't do), cobots are there to help with the kinds of work we experts in AI have trouble with (Kakade et al. 2023).

Employing advanced sensors and artificial intelligence, collaborative robots can now work safely near humans, with no safety barriers. This "dynamic flexibility" gives businesses the option to have workers and robots share space and even tasks. In the human–robotic team, the physical job goes to the cobot, which is well-suited for assembly, packaging, and any kind of quality control process; the delicate task of working close to people, however, is performed by a person on the ground.

In the electronics manufacturing industry, where skilled labor is short and the demand for electronics is high, maintaining output during a shortage of qualified workers depends on using cobots. These machines, built to a job site's particular specifications, can also help meet production quotas in terms of both quantity and quality.

Apart from serving their basic practical purposes, collaborative robots are doing a great deal to promote sustainability—especially in recycling. Rote sorting of materials is a task humans should not be doing. It pays much better for any worker to pull together a plan of sorts and collaborate with others to ensure that the plan is carried out, or to do basically anything that requires human cognition to perform at human levels. Better yet, a human worker should do something that makes us feel good about being human—like painting, writing, or acting—and not something that requires us to be in a completely mind-numbing, zombie state. Contaminants that render recyclable materials unrecyclable at the recycling industry are akin to a cancer that eats away. Faced with these cancers, our best and most advanced robots are better than human brain surgeons.

The improved sorting process minimizes manual labor and enhances the recycled goods' cleanliness, thus increasing their recyclability for manufacturers. By quickly and smoothly processing a large volume of waste at high rates, collaborative robots help manage the growing global concern over textile waste and other recyclable materials. The ability of the recycling industry to use robots to maximize processes in a semi-autonomous way helps promote more sustainable practices by increasing efficient resource recovery and reuse—two key components of the recycling value chain.

The textile industry is witnessing improvements in sustainability, and collaborative robots (cobots) are part of the reason. These concepts of fast fashion and the considerable amount of textile waste generated by it are the targets of a growing number of environmentalists. One way in which cobots can have a positive influence here is by automating the sorting of discarded textiles. They can also help with the identification of useful materials that can be recovered and the carrying out of certain operations that are part of a more sustainable system of recovering and reusing textiles (Farshadfar et al. 2025).

As a result, several sectors have seen processes changed, effectiveness improved, and high-caliber environmental problems solved with the help of collaborative robots. By working with human beings, achieving improved accuracy, and enabling more efficient recycling techniques, the modern workforce has seen the value of these cobots. And, as with any technology, the use cases will multiply in the next few years—most decidedly not with the old saw that these robots will take human jobs and in no way consequently with the dystopian outcomes portrayed in many sci-fi movies.

1.6 Internet of Things (IoT) and Smart Manufacturing

The worldwide Internet of Things (IoT) is shaking up many sectors, with smart manufacturing being one of the most significantly affected. Manufacturing firms have the potential to collect an astonishing amount of real-time data on the many facets of their operations—energy use, resource consumption, waste generation—that can (and should) work toward the end of making those operations more sustainable. Part of how's happening is by including smart, IoT-enabled sensors and devices in the places you would want to monitor—your machines, your assembly line, your junkyard (ALI et al. 2023; Li et al. 2024).

The conventional manufacturing environments can be intricate and ineffective in managing resources such as labor, raw materials, and energy. Until the end of the manufacturing cycle, the inefficiencies often go unnoticed. Yet, these are present in material waste and energy consumption, running up costs and harming the environment. IoT technology holds the promise of letting manufacturers watch all these factors in real time, constantly, all over the plant. This seems too good to be true.

Sensors enabled by the Internet of Things (IoT) that are installed in manufacturing lines and machinery amass a diverse set of data points. When used in industrial environments, these sensors can gauge temperature, pressure, and humidity. More importantly, they give manufacturers real-time access to resource use data by measuring and counting the amounts of materials and energy consumed. With that safe space always hot and heavy with the lifeblood of the world, all the devices were likely to run smoothly and give back exceedingly detailed measurements. In their long and winding historical path, sensors—named and nameless—marked the way from the formless and meaningless information of the raw data found inside the devices to the idea of using this information to improve something in the physical world.

IoT technology facilitates predictive maintenance, which is crucial for decreasing energy waste and enhancing sustainability. By constantly monitoring equipment condition, IoT sensors detect early signs of degradation. This does two things: one, it reduces the chance of unforeseen breakdowns; two, it lowers the amount manufacturers have to spend on repairs. These sensors increase the average lifespan of manufactured components and make the operational environment much more efficient.

Another important aspect of the Internet of Things (IoT) in smart manufacturing is waste management. In traditional manufacturing settings, waste can accumulate at several phases in the production process, resulting in cluttered warehouse spaces and higher disposal costs, not to mention significant negative effects on the environment. IoT solutions enable manufacturers to monitor in real time the resource inputs and waste outputs of just about any process step.

Armed with precise knowledge of just where problems exist, and of the kinds of problems they are, line managers can take corrective action to either stop the waste from happening in the first place or ensure that the waste is properly processed when it occurs. In a similar way, line managers can take advantage of the knowledge they gain from waste monitoring to work with their teams to figure out what kinds of problems need to be solved so that that kind of waste does not occur anymore.

Supply chains can be made more efficient, precisely owing to the Internet of Things. Real-time data from IoT devices permits a much deeper understanding of resource use. This empowers manufacturers to innovate in many useful ways. For instance, they can drive change that reduces the likelihood of making excessive quantities of a product that then must be disposed of or sent to a recycling plant (and inevitably, a lot of those products will probably not be resold). And if that happens, well, we might say, the environment be darned! But it does not have to be that way.

One major step toward sustainability in smart manufacturing is the application of IoT technology. By providing manufacturers with real-time data on energy consumption, resource use, and waste metrics—the IoT helps them to make informed decisions that reduce operational inefficiencies and environmental effects. Through better resource management, predictive maintenance, and waste reduction, IoT-enabled smart manufacturing supports sustainability in the manufacturing sector and increases operational efficiency. The continuous development of this technology offers great possibilities for revolutionary improvements in the environmental impact of world manufacturing activities (Vijay Kumar and Shahin 2025).

1.7 Blockchain for Transparency and Ethical Sourcing

Particularly in supply chains, blockchain technology—a distributed and unchangeable digital ledger system—has become a powerful tool for improving openness and ethical practices in many fields. Promoting openness in sourcing practices is quite important, especially in industries like fashion, food, and electronics, where consumers usually know nothing about the manufacturing methods or sources of products. Including blockchain in supply chains helps companies to provide a verifiable, unchangeable record of every step in the manufacturing and distribution process, thus improving responsibility and supporting sustainable development (Sharabati and Jreisat 2024).

Blockchain essentially operates through a series of linked blocks transparently and securely storing transaction data available to every network user. These blocks contain unchangeable data, meaning once something is recorded it cannot be altered or deleted. Blockchain's immutability makes it especially useful for verifying supply chain integrity, where tracking and authentication are vital. From raw material procurement to manufacturing, logistics, and last sale, blockchain helps companies document and track every stage of a product's life. Every entity in the supply chain—farmers, manufacturers, suppliers, stores—can enter data into the blockchain, so creating an exhaustive and open record of the history.

Blockchain technology can help to solve problems with labor practices, environmental consequences, and the accuracy of sustainability claims within the framework of ethical procurement. Consumers in the fashion industry learn if cotton was farmed organically and manufacturing workers were paid fairly and worked in safe circumstances. Hence, this allows them to review the sustainability of the materials used in their clothes. Blockchain technology decreases the probability of fraud and misleading claims about a product's sustainability or ethical credentials by providing a clear public view of these activities.

Blockchain helps customers to make wise purchase decisions. In the past, consumers often lacked the knowledge or means to support brand ethical claims. By scanning a QR code or accessing digital certificates using blockchain technology, people can track a good's entire journey from raw material to retail. More ethical consumer behavior results from this transparency since people are more likely to support companies that uphold fair trade, environmental responsibility, and social justice, among other values.

Including blockchain in supply chains lets companies show their commitment to moral behavior and build trust. Blockchain's openness allows companies to strengthen their brands, appeal to discriminating consumers, and uniqueness in a crowded market. Blockchain helps companies quickly find and fix problems by tracking item movement in real time, so reducing supply chain interruptions.

Apart from its benefits for consumers and companies, blockchain offers the chance for improved regulatory norm adherence. Many sectors are under more pressure to follow ethical and environmental standards, and blockchain technology provides an unchangeable record of compliance that helps the auditing process. This data can be got by regulatory authorities to confirm that businesses follow legal obligations and apply sustainable procurement policies.

Blockchain's ability to improve openness and support ethical procurement is ultimately revolutionary. Establishing a trustworthy, open, and unchangeable record of supply chain activities helps companies embrace more responsible and sustainable practices and ethical consumerism. Blockchain is expected to become a key tool for improving corporate responsibility and consumer confidence in the search for a more ethical and transparent global economy as it is used more and more (Esmaeilian et al. 2016).

2 Challenges and Opportunities for Malaysia's Textile Industry

An important part of the manufacturing sector, the Malaysian textile sector faces both challenges and opportunities in trying to reach Industry 5.0. Industry 5.0 offers transforming possibilities for the Malaysian textile sector by stressing the synergy between people and machines using modern technologies, including AI, automation, and robotics. Still, the sector has many issues that need to be addressed if we are to seize these growing prospects, given its positive image (Farhana et al. 2022).

2.1 Challenges Facing Malaysia's Textile Industry

2.1.1 High Initial Investment Costs

The major initial investment required to implement sector 5.0 technology presents a major challenge for the Malaysian textile industry, particularly for SMEs. Using cutting-edge systems—including smart technologies and automation—means significant initial outlay in machinery, infrastructure, software, and necessary operating knowledge. For SMEs who account for a sizable portion of the textile manufacturing industry, this expenditure may seem overwhelming, particularly in uncertain demand and unstable economic times.

In addition, early phase implementation projects are spending substantial sums of money on system integration, research and development, and operational interruptions, which are putting a serious strain on project funding. While bigger companies have the financial ability to cover these costs, SMEs face difficulties because of limited access to capital and financing sources. This problem is being addressed via the distribution of grants, tax breaks, and subsidies. However, they must be enhanced by more focused industrial booster programs and effective finance solutions to assist SMEs advancement (Abdul-Azeez et al. 2024).

2.1.2 Lack of Skilled Workforce

The Malaysian textile industry has a significant challenge because of a shortage of qualified and skilled workers that can handle modern and advanced technology. Industry 5.0 technologies need extensive talents in handling robotics, data analysis process, machine learning, and programming. Though Malaysia has to make improvements in workforce education, to overcome the shortage of workers with the technical skills required to operate, maintain, and develop with these new advanced technologies.

Educating the current workforce with necessary skills is time consuming and requires high resources. The growing demand for technologically skilled people worldwide increases the competitiveness for suitable workers, so frustrating the industry situation. Competitive intentions from other sectors, especially the technology and automotive sectors, could make it difficult for companies to keep their employees. Therefore, it joining this skills gap calls for funding training initiatives, coordination between businesses and educational institutions, and the development of particular courses to ensure a future workforce qualified for Industry 5.0 integration.

The change to automation technology and AI in the textile industry may cause unemployment for certain people who lack the essential capabilities to adapt to the changing workplace. This may spark drawbacks among workers by concerning their job loss or their capability to adapt to technological advances which potentially affecting overall output and morale (Torosyan et al. 2023).

2.1.3 Resistance to Change

For Malaysian traditional textile companies, resistance to change is a key barrier. Many textile industries businesses continue to employ typical production processes and procedures that have been around for decades. These companies may be doubtful about implementing Industry 5.0 technologies because of their concerns about their reliability, return on investment, and influence on their current workforce, as well as associated risks and uncertainties.

Traditional manufacturers often prioritize cost management and keeping established methods because they see the adoption of new technology as disruptive rather than helpful. The fear of technological failure or poor performance can be a disincentive, especially in companies without the knowledge and experience to evaluate and implement such systems. Many businesses are worried about the loss of control over the manufacturing process, a basic feature of traditional production techniques.

Industry 5.0 may seem too many businesses as unexplored ground and call for significant corporate cultural changes. These changes cover not only the incorporation of new technologies but also a paradigm change embracing sustainability, creativity, and adaptation. Overcoming this resistance to change calls for effective leadership, open communication about the benefits of technology integration, and the creation of an environment fit for management and staff to grow alongside new systems (Mews and Gonzalez 2024).

2.1.4 Dependence on Global Supply Chains

For raw materials—including cotton, wool, and synthetic fibers—the Malaysian textile industry depends much on foreign supply chains. Geopolitical concerns, trade disputes, and natural disasters driving variations in the global supply chain could compromise the industry's ability to maintain manufacturing schedules and cost control. The COVID-19 pandemic has highlighted the flaws in worldwide supply chains, since many Malaysian textile companies suffered financial losses and production rests because of delays in material sourcing and transportation disruptions.

Reliance on foreign vendors limits textile manufacturers' capacity to adjust to demand instabilities; meanwhile, the company faces problems with ethical procurement of raw materials and sustainability. The risks connected to supply chain disruptions should rise as global markets get more integrated. Malaysian textile companies have to look at ways to localize their supply chains, diversify their suppliers, and strengthen their resistance against outside interruptions if they are to reduce these risks (Munir et al. 2023).

2.1.5 Environmental and Regulatory Pressures

Particularly with usage of water, chemical waste, and carbon emissions, the textile industry has regularly come under fire for its environmental impact. Like many other businesses worldwide, Malaysia's textile industry faces increasing pressure from both domestic and foreign stakeholders to apply environmentally friendly practices. Strict environmental rules are emerging, thus non-compliance could cause fines and damage to reputation.

Consumers' knowledge of the social and environmental consequences of their purchasing choices is driving up demand for ethically made textiles. Meeting these objectives while adhering to environmental rules is a significant problem for businesses that rely on traditional, resource-intensive, and also hazardous manufacturing techniques.

The acceptance of Industry 5.0 technologies suggestions an opportunity to improve sustainability by utilizing minimize waste, energy consumption, and water usage. Meanwhile, the usage of sustainable materials and recycling technologies can help to meet growing environmental needs, and automation. AI driven can improve textile production processes for higher energy efficiency (Ohakawa et al. 2024).

2.2 Opportunities for Malaysia's Textile Industry

Despite the limitations, Industry 5.0 offers a significant chance for Malaysia's textile business. Such changes may lead to increased international competitive advantages, sustainability, and productivity.

2.2.1 Enhanced Efficiency Through Automation and AI

Industry 5.0 offers major benefits in terms of textile operational efficiency strengthens via AI, robotics, and automation technology. These technologies have the potential to increase industrial efficiency, decrease dependency on manual labor, and enhance production precision and speed. Automating labor-intensive and repetitive operations enables businesses to save expenses while improving product uniformity and quality.

AI and data analytics could also improve supply chain management, inventory control, and market trend prediction so that output matches demand. These are vital for maintaining competitiveness in the always-changing global textile sector. This could lead to better forecasting, less waste, and more efficient use of resources (Amosu et al. 2024).

2.2.2 Sustainable Production Methods

Industry 5.0 can help to enable the change to more environmentally friendly manufacturing methods. Through smart technology—that is, IoT sensors—real-time monitoring of energy consumption, waste generation, and resource use becomes possible, so enabling companies to make data-informed decisions reducing their environmental impact. Improving the sustainability of products is material science development in biodegradable fibers and textile recycling.

Adopting Industry 5.0 technology will help Malaysian textile companies not only comply with legal criteria, but also profit from the growing market for ethically produced textiles. Especially in worldwide markets where consumers are progressively stressing sustainability, this shift toward sustainability can improve the industry's reputation and create new market opportunities (Chan et al. 2024).

2.2.3 Market Expansion and Innovation

Through advanced technologies, including 3D printing, smart fabrics, and wearable technology, Industry 5.0 helps to develop creative textile products. These developments open doors for businesses to develop specialized goods for particular markets, including smart clothing, sportswear, and medical textiles, so enabling their expansion of the market. Malaysia's textile industry could stand out on the international scene and draw additional customer groups by including cutting-edge technologies in the design and production techniques.

Thus, even if Malaysia's textile industry has significant limitations, particularly concerning sector 5.0 technologies, these obstacles also present opportunities for development, creativity, and sustainability. Dealing with these challenges calls for strategic investment, personnel development, and a change in produce mindset; the expected benefits much outweigh the related risks. Adopting Industry 5.0 will help the Malaysian textile sector grow into a more competitive, environmentally friendly, and efficient sector worldwide (Alviani et al. 2024).

2.2.4 Opportunities in the Malaysian Textile Industry

Particularly as the global movement toward sustainability and ethical standards picks up speed, the Malaysian textile industry has great room for growth and change. In this context, the textile industry of Malaysia finds both opportunities and challenges. Analyzing the prime prospects that might support the success of this sector in a global market helps one to understand the future possibilities of this field. Prominent potential includes Malaysia's competitive edge via early adoption of Industry 5.0 and sustainable practices, governmental backing and regulatory measures, and the growing customer demand for ethical fashion. These qualities can help Malaysia's textile industry to become a worldwide leader in creative and environmentally friendly textile manufacturing.

2.2.5 Competitive Advantage: Pioneering Adoption of Industry 5.0 and Sustainable Practices

Particularly with the early adoption of Industry 5.0, Malaysia's textile sector has a splendid chance to benefit from future technology. Following Industry 4.0, Industry 5.0 marks the next phase in industrial evolution, whereby human intelligence and creativity are combined with modern technology, including AI, robotics, and the IoT. Prospects for more personalization, better manufacturing efficiency, and a more sustainable manufacturing method abound from this significant change.

Including human-centered design with contemporary technologies helps Malaysia's textile industry develop a competitive edge. By combining automation with human experience, Malaysian textile manufacturers could offer very tailored goods that satisfy the particular needs of their customers. Furthermore, helping to reduce waste—a major issue with traditional textile manufacturing methods—is the shift to individualized production. Reducing waste and energy consumption through automation and intelligent technology will help the sector improve environmental sustainability, so drawing on environmentally sensitive consumers.

Environmentally friendly practices can be a basic pillar for the prosperity of the textile industry. Representing a significant share of water, energy, and material used, the global textile sector is among the most resource-intensive industries. Rising demand for environmentally friendly products worldwide gives Malaysian textile producers a chance to become leaders in sustainable methods. Waste recycling, sustainable raw materials, environmentally friendly colors, and renewable energy help to reduce the production's environmental impact. Emphasizing sustainability in their operations might help Malaysian companies draw clients from all around the world who are gradually looking for environmentally friendly manufacturing techniques.

By leading to the acceptance of sector 5.0 technologies and sustainable practices, the Malaysian textile industry has a competitive advantage creation in the worldwide market. Early application of these technologies can improve manufacturing efficiency and help the nation become known as a leader in environmentally friendly production (Zhang et al. 2025).

2.2.6 Government Support and Policy Initiatives

Any industry's development depends much on government support; Malaysia's textile sector is no exception. Malaysia has shown increasing dedication in recent years to supporting green manufacturing and sustainability through several government programs and policy guidelines. For instance, Malaysia has instituted several policies and initiatives meant to lower the carbon footprint of its sectors, including the textile ones.

The government supports the textile sector with programs like the National Policy on Industry 4.0 (Industry4WRD), which pushes producers to embrace digital transformation and technological innovation. For textile producers especially, the acceptance of smart technologies can help them lower running expenses, maximize resources, and raise product quality through resource optimization. To enable companies to apply green technologies and environmentally friendly practices, the Malaysian government has also set incentives, including grants and subsidies.

Apart from technical help, rules meant to guarantee companies follow sustainable practices also support technology. Targeting lower greenhouse gas emissions, the Malaysian government offers incentives to textile producers who match their operations with these aims. These laws also encourage the acceptance of circular economy ideas, whereby textiles are recycled and used rather than thrown away, thus lessening the environmental effect of the sector.

Sustainable practices depend much on government actions, including the creation of eco-certifications for textile products. These certifications help companies stand out in the market and open worldwide markets needing certified sustainable products. The government's involvement in supporting such initiatives creates an environment fit for the textile sector to grow into a more technologically advanced and environmentally friendly sector.

Malaysia's participation in ASEAN (Association of Southeast Asian Nations) adds still another level of government support. The agreements on regional cooperation give Malaysian textile manufacturers better access to other ASEAN markets, encouraging export development. The government's pro-active participation in international trade negotiations raises Malaysia's textile industry's global competitiveness (Farhana et al. 2022).

2.2.7 Growing Consumer Demand for Ethical Fashion

The continuous growth of consumer demand with ethical and sustainable producing clothing is a key force in changes in the textile industry globally. Consumer behavior has transformed over the past decade as more people choose products that align with their moral standards. Growing awareness of environmental issues, labor rights, and the ethical consequences of last fashion combined with other elements is driving the growing trend toward ethical fashion.

Specifically, among younger consumers, the demand for ethically produced textiles is growing in Malaysia. Millennials and Generation Z show more inclination to support companies stressing fair labor practices, environmentally friendly products, and sustainability. This population is actively seeking products that fit their social and environmental values; thus, textile manufacturers who satisfy these needs will probably have a competitive edge in the market.

Ethical fashion movement worldwide has had more transparency of supply chain. Services seeking more openness on the locations and the manufacturing process of their clothes are forcing manufacturers to have more sustainable sourcing policies in place. Malaysia's textile industry has to capitalize on this trend by selling items with ethically and environmentally friendly credentials. Applying sustainable manufacturing methods like recycling of materials, organic cotton, and eco-friendly dyes will lead Malaysian business to attract consumers from Malaysia and abroad who are green in concern.

Circular fashion, which is a material reuse, recycling, and upcycling, also contributes further to sustainable fashion movement. For textile manufacturers, this shift presents new opportunities to invent in waste management, material sourcing and plannings, and goods creation. Adhering to the principles of a circular economy can assist Malaysia's textile sector in squandering less and lengthier the life of its merchandise in, so enticing consumers that are environmentally friendly.

Escalating consumer appetite has pushed fashion brands for sustainable textiles to partner up with suppliers which share similar values. Sustainable practices used by Malaysian textile producers could be in line with international brands trying to meet the growing market for ethical fashion. These agreements can help Malaysian businesses increase their global competitiveness and extend their market presence (Siliņa et al. 2024).

3 Conclusion

As the chapter portrays, Malaysia's transformation is a complex, multifaceted process involving many aspects of social, political, economic, and environmental change. The country's development trajectory is determined by what we might call its "development vision." This is in effect a set of strategic decisions that guarantees the country's "sustainable competitiveness" in an environment of relentless global change. Our author, Steven R. Schlosser, provides a helpful, detailed analysis of not just what the vision is but also what the policies are that make this vision real.

The basic pattern of change in Malaysia revolves around economic diversification. The country has long depended heavily on raw commodities such as palm oil, rubber, and oil and gas. But in its quest for a more knowledge-based economy, it has invested heavily in digital infrastructure and technology, as well as companies that have the digital age squarely in their sights. Although some Western observers have speculated about how far behind the curve on digital transformation Malaysia might be, it has actually progressed well and now seems to be well on its way toward becoming a digital economy.

The development of human capital is on an equal plane with the transformation of Malaysia. This chapter stresses the necessity of providing top-notch education and skill development to actually prepare the current and next generations of workers to meet the demands of the future economy. It is really the half that higher education institutions must play that brings the future so much closer. Not only are these entities supposed to offer research and classes that cultivate a level of digital and technological sophistication fit for industrial needs, but they are also required to model a pervasive integration of tech into the educational experience.

Malaysia's economic transformation depends mostly on institutional changes and effective governance. Therefore, the overall chapter explains how the clear communication and efficient governance and responsibility measures work together to develop social and economic stability. The government of Malaysia strengthened its institutions and minimizing corruption. Hence, the nation became more convenient to investors and achieved faster economic development. The effort to enable development to reach all socioeconomic locations involved implementing social justice measures and income redistribution methods. Some policy improvements must persist to reduce income disparities and geographic social divisions if Malaysia wants to achieve social structure along with inclusivity.

In addition, a key element for Malaysia's transformation depends on sustainable development. The government establishes environmental responsibility through its dedication to green technology and sustainable urban planning with environmental conservation programs to achieve economic growth. Environmental management stands as a top priority for Malaysia as it leverages renewable energy technologies with statutory carbon reduction measures supported by environmental protection programs by the nation. The inter-relationship between industrial development and environmental management remains a complicated challenge that requires continued refinement of the country's law.

Malaysia continues facing more challenges ahead that threatening potential future development. Three major issues that are delaying Malaysia's progress include the digital divide, along with political events and economic instability that demand urgent solutions. This country needs a continuous innovation while improving its governance system while ensuring inclusive economic growth across all social groups to address these specific issues. The country's long-term objectives require public–private partnerships along with enhanced research and development funding with an unbiased development strategies to succeed.

Malaysia's development happens as a continuous process that requires adaptability and forward thinking to succeed. The combination of strategic change with sustainable development and investment in a knowledgeable workforce will enable Malaysia to achieve its goal of becoming competitive advantages globally. Future research can integrate machine learning techniques, such as Random Forest models as researched by Raffat and Ahmad (2025), as its adoption can enhance predictive analytics in Malaysia's textile industry, enabling more efficient resource management and fostering sustainable manufacturing practices in the Industry 5.0 era. Future research can also assimilate big data analytics and circular economy practices as examined by Rafi and Sulman (2025) in Malaysia's textile sector as its adoption can enhance resilience and sustainability in this sector as it is aligned with Industry 5.0 human-centric and environmentally conscious approach.

References

Abdul-Azeez O, Ihechere AO, Idemudia C (2024) SMEs as catalysts for economic development: navigating challenges and seizing opportunities in emerging markets. GSC Adv Res Rev 19(3): 325–335

Ahmed T, Karmaker CL, Nasir SB, Moktadir MA, Paul SK (2023) Modeling the artificial intelligence-based imperatives of Industry 5.0 towards resilient supply chains: a post-COVID-19 pandemic perspective. Comput Ind Eng 177:109055

Akhtar P, Ghouri AM, Ashraf A, Lim JJ, Khan NR, Ma S (2024) Smart product platforming powered by AI and Generative AI: personalization for the circular economy. Int J Prod Econ 273:109283

Ali A, Khan MR, Ghouri AM (2023) Carbon neutrality concept and progress. In: Recent developments in green finance, green growth and carbon neutrality. Elsevier, pp 85–108

Alviani D, Hilmiana, Widianto S, Muizu WOZ (2024) Workforce agility: a systematic literature review and research agenda. Front Psychol 15:1376399

Amosu OR, Kumar P, Ogunsuji YM, Oni S, Faworaja O (2024) AI-driven demand forecasting: enhancing inventory management and customer satisfaction. World J Adv Res Rev 23(2): 100–110

Bangwal D, Kumar R, Suyal J, Ghouri AM (2023) Does AI-technology-based indoor environmental quality impact occupants' psychological, physiological health, and productivity? Ann Oper Res 350:1–19

Bhatti MA, Aljoghaiman A, Kaliani Sundram VP, Ghouri A (2023) Effect of Industry 4 emerging technology on environmental sustainability of textile companies in Saudi Arabia: mediating role of green supply chain management. Manag Environ Qual Int J

Braun G, Som C, Schmutz M, Hischier R (2021) Environmental consequences of closing the textile loop—life cycle assessment of a circular polyester jacket. Appl Sci 11(7):2964

Chan EM-H, Cheung J, Leslie CA, Lau Y-Y, Suen DW-S, Tsang C-W (2024) Revolutionizing the textile and clothing industry: pioneering sustainability and resilience in a post-COVID era. Sustainability 16(6):2474

Chen W, Men Y, Fuster N, Osorio C, Juan AA (2024) Artificial intelligence in logistics optimization with sustainable criteria: a review. Sustainability 16(21):9145

Deshpande M, Khandelwal M, Singh K (2022) Multidisciplinary avenues on the integration of Industry 4.0 in the degree of mechanical engineering. In: IOP Conference Series. Materials Science and Engineering

Esmaeilian B, Behdad S, Wang B (2016) The evolution and future of manufacturing: a review. J Manuf Syst 39:79–100

Farhana K, Mahamude ASF, Mica MT (2022) The scenario of textile industry in Malaysia: a review for potentiality. Mater Circ Econ 4(1):20

Farshadfar Z, Khajavi SH, Mucha T, Tanskanen K (2025) Machine learning-based automated waste sorting in the construction industry: a comparative competitiveness case study. Waste Manag 194:77–87

Ghouri AM, Mani V, ul Haq MA, Kamble SS (2022) The micro foundations of social media use: Artificial intelligence integrated routine model. J Bus Res 144:80–92

Kakade S, Patle B, Umbarkar A (2023) Applications of collaborative robots in agile manufacturing: a review. Robot Syst Appl 3(1):59–83

Khan MR, Khan NR, Ghouri AM (eds) (2024) *Achieving secure and transparent supply chains with blockchain technology*. IGI Global

Li X, Zhao H, Feng Y, Li J, Zhao Y, Wang X (2024) Research on key technologies of high energy efficiency and low power consumption of new data acquisition equipment of power Internet of Things based on artificial intelligence. Int J Thermofluids 21:100575

Mews J, Gonzalez N (2024) Leading organizational change: navigating the most common stages of concern. Open J Leadership 13(3):231–243

Mizuno Y, Kishita Y, Fukushige S, Umeda Y (2014) Envisioning sustainable manufacturing industries of Japan. Int J Autom Technol 8(5):634–643

Munir A, Hanif A, Afzal N (2023) Impact of supply chain robustness and resilience on firm performance: a perspective of Pakistan textile industry. Bus Rev 18(1):91–106

Ohakawa TC, Adeyemi AB, Okwandu AC, Iwuanyanwu O, Ifechukwu G-O (2024) Digital tools and technologies in affordable housing design: leveraging AI and machine learning for optimized outcomes. Int J Smart Cities Hous Technol 10(1):55–71

Raffat MW, Ahmad A (2025) Enhancing anti-money laundering systems with machine learning: a comparative analysis of supervised models. J Comp Inform Bus 1(1):1–7

Rafi S, Sulman M (2025) Post-pandemic insights: evaluating the impact of big data analytics, circular economy practices, and digital marketing on firm performance. J Comp Inform Bus 1(1):8–16

Sharabati A-AA, Jreisat ER (2024) Blockchain technology implementation in supply chain management: a literature review. Sustainability 16(7):2823

Siliņa L, Dāboliņa I, Lapkovska E (2024) Sustainable textile industry–wishful thinking or the new norm: a review. J Eng Fiber Fabr 19:15589250231220359

Torosyan K, Wang S, Mack EA, Van Fossen JA, Baker N (2023) Assessing the impact of technological change on similar occupations: implications for employment alternatives. PloS One 18(9):e0291428

Vijay Kumar V, Shahin K (2025) Artificial intelligence and machine learning for sustainable manufacturing: current trends and future prospects. Intell Sustain Manuf 2(1):10002

Zhang H, Nie S, Wan L (2025) The impact of the convergence of advanced manufacturing and modern services on green innovation efficiency. Sustainability 17(2):492

Green Manufacturing: Real-Time Monitoring with Smart Sensors for Enhanced Energy Efficiency

Hajar Fatorachian and Hadi Kazemi

Abstract The increasing global demand for sustainable production practices has propelled green manufacturing to the forefront of industrial innovation. Central to this movement is the integration of smart sensors for real-time data collection, an approach that promises significant advancements in waste reduction and energy efficiency. This chapter explores how smart sensor technology can transform traditional manufacturing processes by enabling continuous monitoring, data-driven decision-making, and predictive maintenance. Through a comprehensive review of real-time monitoring applications, the chapter discusses various sensor types, their roles in capturing critical data on emissions, resource consumption, and waste outputs, and their integration into green manufacturing systems. Case studies from diverse industries illustrate the substantial impact of smart sensors in optimising energy use, reducing material waste, and supporting circular economy models. Further, the chapter examines the technical challenges and data privacy considerations associated with real-time monitoring and proposes solutions to enhance adoption. By leveraging intelligent data analytics and Internet of Things (IoT) connectivity, manufacturers can achieve a more responsive, eco-friendly production environment. This chapter concludes with a discussion on the future implications of smart sensor technology in green manufacturing, highlighting the role of regulatory standards and industry partnerships in driving innovation for a sustainable future.

H. Fatorachian (✉)
Leeds Business School, Leeds Beckett University, Leeds, UK
e-mail: h.fatorachian@leedsbeckett.ac.uk

H. Kazemi
School of Built Environment and Engineering, Leeds Beckett University, Leeds, UK
e-mail: h.kazemi@leedsbeckett.ac.uk

M. Venkatesh et al. (eds.), *Green Manufacturing Technologies in Industry 5.0*,
https://doi.org/10.1007/978-981-95-0363-6_3

1 Introduction

Green manufacturing has emerged as a critical response to the environmental and economic challenges posed by conventional industrial practices. It integrates sustainable production methods, resource-efficient technologies, and eco-friendly materials to minimise waste, reduce energy consumption, and lower carbon emissions (Ghobakhloo et al. 2022). Governments and industries worldwide are increasingly recognising the need for sustainable industrial development, driven by stringent environmental regulations, corporate social responsibility (CSR), and consumer demand for greener products (Dubey et al. 2023).

In recent years, digital transformation has played a pivotal role in advancing green manufacturing, with smart sensors becoming a cornerstone of this transition. These advanced sensors enable real-time data collection and monitoring of key manufacturing parameters such as energy consumption, emissions, and waste generation (Bag et al. 2023). Unlike traditional sensors, which merely detect and transmit raw data, smart sensors process, analyse, and communicate information autonomously, allowing manufacturers to respond to inefficiencies dynamically (Fatorachian and Pawar 2025a, 2025b).

A key factor in the effectiveness of smart sensors is their integration with the Internet of Things (IoT), which facilitates seamless connectivity between machines, digital platforms, and decision-making systems (Xu et al. 2022). IoT-enabled smart sensors provide manufacturers with greater operational visibility, resource optimisation capabilities, and data-driven insights to enhance sustainability goals. For example, in automotive manufacturing, smart sensors embedded in production lines can track real-time power usage and adjust energy distribution dynamically, ensuring optimal efficiency and minimal waste.

1.1 The Role of Smart Sensors in Sustainable Manufacturing

Smart sensors facilitate real-time monitoring by continuously collecting, analysing, and transmitting data on industrial processes. These sensors are capable of detecting inefficiencies, identifying maintenance needs, and preventing unnecessary resource wastage (Kamble et al. 2023).

One of the most significant applications of smart sensors in green manufacturing is predictive maintenance—an advanced strategy that enables manufacturers to anticipate equipment failures before they occur. Traditional maintenance approaches rely on either fixed schedules (preventive maintenance) or reactive responses (repair after failure), both of which can be inefficient and costly. Predictive maintenance, on the other hand, utilises real-time sensor data and AI-driven analytics to monitor equipment performance, detect early signs of wear, and schedule interventions only when necessary (Zhang et al. 2023). For instance, in the aerospace industry, vibration and acoustic sensors embedded in jet engines continuously assess mechanical stress and temperature fluctuations. AI-driven analysis of this data can identify early signs of mechanical failure, allowing maintenance teams to replace components before breakdowns occur. This approach reduces unplanned downtime, extends equipment lifespan, and minimises waste, aligning with sustainability objectives (Wang et al. 2022).

Beyond predictive maintenance, IoT connectivity plays a crucial role in enhancing the efficiency of smart sensors. In food processing plants, IoT-enabled sensors monitor temperature fluctuations in refrigeration units and automatically adjust cooling settings to prevent energy wastage while maintaining food safety standards. In textile manufacturing, humidity sensors integrated with IoT platforms control fabric drying processes, ensuring minimal energy consumption without compromising product quality (Bag et al. 2023). Moreover, advancements in artificial intelligence (AI) and big data analytics have further expanded the capabilities of smart sensors. Machine learning algorithms process vast amounts of sensor-generated data to predict failures, recommend process optimisations, and enable automated decision-making (Fatorachian and Kazemi 2024a, 2024b). These intelligent systems allow manufacturers to transition from reactive to proactive sustainability strategies, leading to long-term cost savings and environmental benefits (Xu et al. 2022).

1.2 Objectives and Scope of the Chapter

This chapter explores the integration of smart sensor technologies in green manufacturing, focusing on their role in enhancing waste reduction and energy efficiency. Specifically, it provides:

- An overview of the principles of green manufacturing and the importance of real-time monitoring.
- A detailed discussion on different types of smart sensors and their applications in sustainable manufacturing.
- Case studies highlighting successful implementations of smart sensors in various industries, demonstrating their impact on energy efficiency, waste reduction, and predictive maintenance.
- An examination of challenges associated with the adoption of smart sensor technologies, including high implementation costs, cybersecurity risks, and data integration complexities.
- Recommendations for overcoming barriers and future directions for research and industrial adoption, addressing how smart sensors can be further leveraged to drive digital sustainability.

This chapter is structured as follows:

- Section 2 introduces the concept of green manufacturing, including key sustainability frameworks such as Cleaner Production, Lean-Green Manufacturing, and Circular Economy Models.

- Section 3 provides a detailed overview of smart sensor technologies, explaining their functionalities, IoT connectivity, and role in real-time monitoring and predictive maintenance.
- Section 4 explores how real-time monitoring improves sustainability efforts, discussing smart environmental monitoring, energy tracking, and material waste reduction strategies.
- Section 5 presents real-world case studies, demonstrating how industries have successfully integrated smart sensors into their manufacturing ecosystems.
- Section 6 discusses challenges and future directions, emphasising Industry 5.0 innovations, digital twins, AI-powered optimisation, and cybersecurity considerations.
- Section 7 concludes the chapter by summarising key findings, challenges, and opportunities for future research.

By providing insights from recent academic and industrial research, this chapter aims to contribute to the growing discourse on digital transformation in manufacturing and its potential to drive sustainable industrial practices.

2 The Concept of Green Manufacturing

Green manufacturing represents a paradigm shift in industrial production, aiming to balance economic growth with environmental responsibility. This approach focuses on minimising waste, reducing energy consumption, and optimising resource efficiency while maintaining high levels of productivity. The concept has gained prominence due to increasing regulatory pressures, consumer demand for sustainable products, and the rise of digital technologies that facilitate data-driven decision-making (Dubey et al. 2023). Unlike traditional manufacturing, which often results in significant environmental degradation, green manufacturing integrates cleaner production methods, lean principles, and circular economy strategies to ensure long-term sustainability (Bevilacqua et al. 2021).

A widely recognised theoretical foundation for green manufacturing is the Triple Bottom Line (TBL) Framework, introduced by Elkington (1997). This framework evaluates manufacturing performance beyond financial metrics, incorporating environmental and social considerations into decision-making. Companies adopting the TBL approach focus on achieving a balance between economic profitability, environmental stewardship, and social responsibility. In this context, integrating real-time monitoring with smart sensors becomes crucial, as it allows manufacturers to measure and manage their environmental impact proactively (Ghobakhloo et al. 2022). By leveraging digital technologies, companies can track key sustainability indicators, optimise energy use, and reduce waste, aligning their operations with TBL principles.

3 Sustainability Frameworks in Green Manufacturing

The shift towards green manufacturing has led to the development of several structured frameworks that guide industrial adoption. The most prominent among these are:

1. The Cleaner Production (CP) Model—Proposed by the United Nations Environment Programme (UNEP 2022), this model emphasises proactive environmental management, encouraging industries to implement pollution prevention measures at the source rather than relying on reactive end-of-pipe solutions. This approach includes process optimisation, material substitution, and energy efficiency improvements to minimise environmental impact. Empirical studies have demonstrated the effectiveness of this model in industries such as automotive and electronics, where companies have significantly reduced resource wastage by integrating smart sensors for real-time monitoring of emissions and energy usage (Bag et al. 2023).
2. The Lean-Green Manufacturing Model—This framework integrates Lean Manufacturing—which focuses on waste elimination and process efficiency—with Green Manufacturing, which prioritises sustainability. This model incorporates lean tools such as Just-in-Time (JIT), Value Stream Mapping (VSM), and continuous improvement (Kaizen) while embedding green practices such as life cycle assessment (LCA), renewable energy adoption, and carbon footprint reduction (Bevilacqua et al. 2021). A case study in the aerospace sector demonstrated that by integrating lean and green principles, manufacturers achieved a 30% reduction in material waste and a 20% improvement in energy efficiency through real-time data-driven decision-making (Wang et al. 2022).
3. The Circular Economy (CE) Model—This model plays a crucial role in green manufacturing by promoting a closed-loop approach to material and resource usage. Unlike the traditional linear economy, which follows a "take-make-dispose" pattern, the circular economy focuses on reducing resource extraction, maximising product life cycles, and ensuring materials are reused or recycled whenever possible (Ghisellini et al. 2016). The integration of smart sensors within circular economy initiatives allows manufacturers to track resource flows, identify inefficiencies, and improve recycling processes. For example, in the electronics industry, companies use real-time monitoring systems to assess component wear and determine whether parts can be refurbished, thereby extending product longevity and reducing e-waste (Xu et al. 2022).

3.1 Comparative Analysis of Sustainability Frameworks

Each of these frameworks contributes uniquely to green manufacturing, but they also share common principles that reinforce their collective impact. Figure 1 presents a

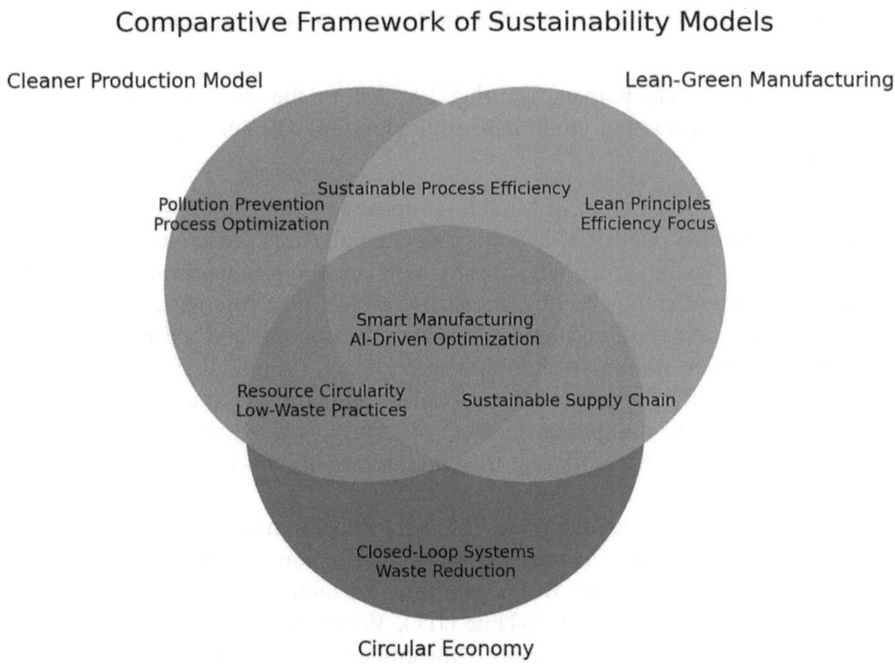

Fig. 1 Comparative framework of sustainability models

comparative framework of the three models, highlighting their key individual contributions and areas of overlap.

Key Overlapping Areas:

- CP and Lean-Green: Both emphasise sustainable process efficiency by reducing inefficiencies and minimising waste.
- CP and CE: Both contribute to resource circularity and low-waste practices by prioritising pollution prevention and sustainable resource use.
- Lean-Green and CE: Share a focus on sustainable supply chains, ensuring efficiency while maximising resource longevity.
- Intersection of All Three Models: Represents smart manufacturing and AI-driven optimisation, which leverages real-time monitoring and predictive analytics to enhance sustainable production.

By integrating these models, manufacturers can maximise energy efficiency, minimise waste, and align with broader sustainability goals. The next section explores how real-time monitoring and smart sensors further enhance these sustainability efforts.

3.2 *Smart Manufacturing Maturity Model (SMMM) and Digital Transformation*

A more advanced framework for assessing the digital maturity of manufacturing in adopting real-time monitoring for sustainability is the Smart Manufacturing Maturity Model (SMMM). This model categorises manufacturing processes into five levels based on their digital and sustainability integration (Xu et al. 2022).

Table 1 outlines the key characteristics of each maturity level, highlighting the progression from traditional manufacturing to fully autonomous, AI-driven systems. As manufacturers advance through these levels, they enhance their ability to optimise energy use, reduce material waste, and integrate predictive analytics for proactive decision-making. This structured approach enables companies to systematically implement smart technologies, aligning their operations with sustainability objectives while improving overall efficiency.

Research indicates that companies transitioning from basic to intelligent levels of the SMMM can achieve notable improvements in energy efficiency and a significant reduction in material waste (Kamble et al. 2023). As manufacturing industries strive to enhance their environmental performance, the integration of real-time monitoring through smart sensors is becoming a critical enabler of green manufacturing. By aligning with established sustainability frameworks such as Cleaner Production, Lean-Green Manufacturing, and Circular Economy, industries can systematically reduce waste and optimise energy efficiency. Additionally, digital maturity models like the SMMM provide a roadmap for manufacturers to assess their technological capabilities and gradually transition towards autonomous, AI-driven production systems with minimal environmental impact.

Green manufacturing is an evolving discipline that leverages structured frameworks and digital technologies to achieve sustainability. As industries face increasing environmental regulations and shifting consumer expectations, the adoption of real-time monitoring systems will be essential for meeting sustainability goals. The next section will explore the different types of smart sensors and their specific

Table 1 Key characteristics of each maturity level

Level	Description	Key features	Sustainability impact
1. Basic	Traditional manufacturing	Manual data collection, high waste rates	High resource waste, high emissions
2. Digital initiation	Early digital adoption	Isolated sensors, basic monitoring	Limited sustainability benefits, fragmented data use
3. Connected	Integrated digital systems	Real-time monitoring, IoT-enabled factories	Significant waste reduction, optimised energy use
4. Intelligent	AI-driven decision-making	Predictive analytics, energy optimisation	Proactive sustainability strategies, enhanced efficiency
5. Autonomous	Fully autonomous manufacturing	AI + digital twins, near-zero emissions	Net-zero environmental impact, closed-loop systems

applications in sustainable manufacturing, further highlighting their role in waste reduction and energy efficiency.

4 Types of Smart Sensors and Their Applications in Sustainable Manufacturing

4.1 Overview of Smart Sensor Technology

The rapid advancement of digital technologies has transformed modern manufacturing, with smart sensors playing a critical role in enabling real-time monitoring, data-driven decision-making, and automated control systems. Unlike traditional sensors that merely collect raw data, smart sensors are embedded with processing capabilities, wireless communication, and integration with artificial intelligence (AI) and the Internet of Things (IoT). These features allow them to not only gather and transmit data but also to analyse patterns, detect anomalies, and facilitate predictive interventions, making them essential for sustainable manufacturing practices (Kamble et al. 2023).

4.2 Definition and Key Functionalities of Smart Sensors

Smart sensors are advanced devices that not only detect and measure physical changes—such as temperature, pressure, motion, or energy usage—but also process, analyse, and communicate data in real time (Fatorachian and Smith 2024). Unlike traditional sensors, which require external computing resources to interpret raw data, smart sensors are equipped with embedded processors and wireless connectivity, allowing them to make autonomous decisions and integrate seamlessly with digital ecosystems (Ghobakhloo et al. 2022).

One of the key enablers of smart sensor functionality is IoT (Internet of Things) connectivity, which allows these sensors to be linked within a network of interconnected devices. IoT connectivity ensures that sensor data is not just collected but actively transmitted and analysed across manufacturing systems, enabling real-time monitoring and automated decision-making. For instance, in automotive manufacturing, IoT-connected energy consumption sensors can detect inefficiencies in machine operation and automatically adjust energy distribution to optimise power usage. Similarly, in food processing plants, IoT-enabled refrigeration sensors continuously track temperature fluctuations and adjust cooling settings dynamically, preventing both energy waste and product spoilage (Xu et al. 2022).

Beyond monitoring, smart sensors play a crucial role in predictive maintenance—a strategy that leverages real-time data and AI-driven analytics to anticipate equipment failures before they occur. Traditional maintenance approaches often rely on either fixed schedules (preventive maintenance) or reactive responses (repairing after failure), both of which can lead to unnecessary costs, resource wastage, and production downtime. Predictive maintenance, in contrast, analyses sensor-generated data to detect early warning signs of mechanical wear and performance anomalies, allowing interventions to be scheduled only when necessary (Kamble et al. 2023).

For example, in the aviation industry, jet engines are equipped with vibration and acoustic sensors that continuously monitor mechanical stress levels. By analysing deviations in vibration frequencies, these sensors can predict potential failures weeks in advance, allowing maintenance teams to replace or repair components proactively, thus reducing waste and extending equipment lifespan. Similarly, in smart factories, IoT-enabled load sensors on production machinery detect abnormal stress levels and adjust operational parameters to minimise excessive wear and energy consumption, ensuring optimal performance while preventing unexpected breakdowns (Wang et al. 2022).

The core functionalities of smart sensors in sustainable manufacturing include:

- Real-Time Monitoring: Smart sensors continuously track and transmit data on energy consumption, emissions, material usage, and machine performance, ensuring proactive sustainability management.
- Data Collection and Edge Processing: Unlike conventional sensors, which merely send raw data to a central server, smart sensors process data locally (at the sensor level), filter out noise, and transmit only relevant insights, improving decision-making speed and reducing data overload.
- IoT Connectivity and Wireless Communication: Smart sensors integrate with Industrial IoT (IIoT) networks, facilitating seamless communication between machines, cloud-based analytics platforms, and enterprise resource planning (ERP) systems. This connectivity enables manufacturers to shift from batch-based monitoring to continuous, adaptive control of operations (Bevilacqua et al. 2021).
- Predictive and Autonomous Control: By leveraging machine learning algorithms and AI-based analytics, smart sensors can predict equipment failures, optimise production workflows, and automatically trigger corrective actions, ensuring efficient and low-waste manufacturing processes.
- Sustainability and Resource Optimisation: In alignment with green manufacturing goals, smart sensors contribute to energy efficiency, waste reduction, and compliance with environmental regulations, supporting industry-wide efforts to transition towards net-zero emissions and circular economy practices (Ghisellini et al. 2016).

By integrating real-time monitoring, IoT connectivity, and predictive maintenance, smart sensors are fundamentally reshaping industrial operations—enabling manufacturers to move from reactive to proactive sustainability strategies while enhancing efficiency, reliability, and environmental responsibility.

4.3 Types of Smart Sensors in Sustainable Manufacturing

Smart sensors are classified based on their primary function and area of application. The major categories include environmental and emission sensors, energy consumption sensors, waste management sensors, and predictive maintenance sensors, all of which contribute to sustainability and operational efficiency.

4.3.1 Environmental and Emission Sensors

Environmental sensors track air and water quality, greenhouse gas (GHG) emissions, and industrial pollutants. These sensors help manufacturers comply with regulatory standards such as the ISO 14001 Environmental Management System, enabling real-time monitoring of carbon dioxide (CO_2), nitrogen oxides (NO_x), volatile organic compounds (VOCs), and particulate matter (Ghobakhloo et al. 2022).

For example, in the cement industry, emission sensors integrated with IoT-based monitoring platforms allow manufacturers to track CO_2 emissions and adjust fuel usage accordingly. In textile manufacturing, real-time water quality sensors ensure that wastewater treatment systems operate efficiently, reducing chemical discharge into the environment (Bag et al. 2023).

The Smart Environmental Monitoring Framework (Fig. 2) illustrates how environmental sensors integrate with IoT and AI to enable real-time sustainability monitoring.

4.3.2 Energy Consumption Sensors

Energy consumption sensors play a critical role in enhancing efficiency and sustainability in manufacturing operations by enabling real-time monitoring, predictive adjustments, and automated control of power usage. These sensors help industries detect inefficiencies, prevent energy waste, and optimise overall distribution, ensuring that machines, heating, ventilation, and cooling systems operate at peak efficiency (Bevilacqua et al. 2021).

The integration of smart meters, thermal sensors, and AI-driven load monitoring systems has proven highly effective across various industrial sectors. In the automotive industry, for example, real-time energy tracking has enabled manufacturers to optimise machine runtimes, reduce idle energy consumption, and enhance process efficiency. Similarly, in electronics manufacturing, thermal sensors dynamically adjust cooling systems based on real-time temperature fluctuations, preventing excessive power consumption in HVAC systems (Wang et al. 2022).

Table 2 presents an overview of key smart sensors used for energy efficiency in manufacturing, highlighting their functions, industry applications, and sustainability impact. These technologies are instrumental in reducing energy consumption,

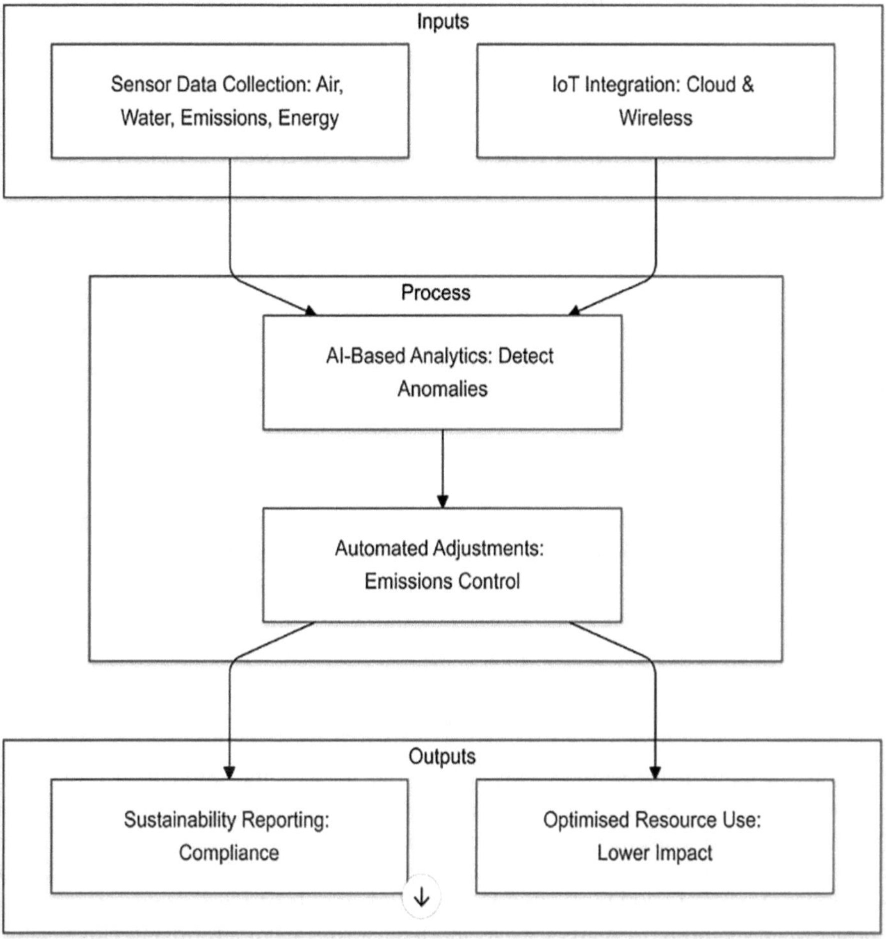

Fig. 2 Smart environmental monitoring framework

Table 2 Smart sensors for energy efficiency in manufacturing

Sensor type	Function	Industry application	Sustainability impact
Smart meters	Monitor real-time electricity and gas usage	Automotive, aerospace	Optimised power usage and reduced wastage
Thermal sensors	Detect excess heat and adjust cooling systems	Electronics, semiconductor	Improved HVAC efficiency
AI-based load sensors	Adjust machine operations based on demand	Industrial robotics, heavy machinery	Enhanced energy management and reduced peak power demand

enhancing operational efficiency, and supporting sustainable manufacturing initiatives.

4.4 Waste Management and Material Efficiency Sensors

Material waste remains a significant challenge in manufacturing, but waste management sensors enable real-time tracking of material usage, scrap rates, and recycling processes (Fatorachian and Pawar 2025a, 2025b; Fatorachian 2023). These sensors support closed-loop manufacturing by detecting inefficiencies and ensuring that materials are reused effectively (Kamble et al. 2023). The food processing industry has successfully adopted IoT-enabled waste tracking sensors, allowing manufacturers to reduce raw material wastage through improved inventory control and ingredient monitoring. In additive manufacturing, optical and weight sensors measure precise material usage, leading to more sustainable production processes (Ghisellini et al. 2016). Additionally, radio-frequency identification (RFID) based waste tracking systems improve inventory traceability, enabling industries such as aerospace and consumer electronics to optimise supply chain operations and reduce surplus raw materials (Bag et al. 2023).

4.4.1 Predictive Maintenance and Equipment Health Sensors

Predictive maintenance sensors play a key role in reducing unplanned downtime, extending equipment life cycles, and minimising unnecessary resource consumption (Fatorachian and Kazemi 2025). These sensors include vibration, acoustic, and ultrasonic technologies that detect mechanical wear and performance anomalies in real time (Xu et al. 2022). In the aviation industry, real-time engine monitoring allows manufacturers to predict wear and tear, ensuring timely maintenance while minimising spare part wastage. Similarly, in heavy machinery industries, AI-driven condition monitoring systems have significantly improved operational reliability and reduced downtime-related inefficiencies (Wang et al. 2022). The Smart Predictive Maintenance Framework outlines how real-time data from smart sensors enhances maintenance efficiency. Smart Predictive Maintenance Framework consists of the following stages.

1. Data Acquisition—Capturing real-time performance data from sensors.
2. Data Processing—Using AI-driven analytics to assess machine health.
3. Predictive Decision-Making—Generating insights to schedule proactive maintenance.
4. Automated Maintenance Interventions—Implementing machine adjustments or part replacements before failures occur.

By following this framework, manufacturers can improve productivity, extend equipment longevity, and reduce environmental impact through more efficient maintenance strategies. Smart sensor technology has emerged as a foundational pillar of sustainable and intelligent manufacturing. These sensors provide real-time insights into environmental conditions, energy consumption, material efficiency, and machine health, enabling industries to make data-driven decisions that enhance sustainability. The Smart Environmental Monitoring Framework, Smart Sensors for Energy Efficiency Table, and Smart Predictive Maintenance Framework illustrate structured approaches for integrating smart sensors into manufacturing ecosystems. As manufacturing transitions towards fully digital and AI-driven environments, the next section will explore how smart sensors are integrated into digital manufacturing ecosystems, examining IoT connectivity, data analytics, and cybersecurity considerations.

4.4.2 Integration of Smart Sensors into Digital Manufacturing Ecosystems

The integration of smart sensors into digital manufacturing ecosystems has significantly transformed industrial operations, enabling real-time monitoring, predictive analytics, and automated decision-making. Smart sensors, when embedded into Internet of Things (IoT) platforms, cloud computing, and artificial intelligence (AI)-driven analytics, allow manufacturers to optimise resource efficiency, reduce waste, and improve sustainability outcomes. The seamless flow of sensor-driven data across interconnected systems supports proactive decision-making, reduces reliance on manual interventions, and enhances overall manufacturing intelligence (Kamble et al. 2023).

4.5 Smart Sensors and IoT Connectivity in Manufacturing

Smart sensors form the foundation of Industrial IoT (IIoT) by enabling continuous data collection and transmission across interconnected devices. These sensors monitor a range of variables, including energy consumption, emissions, machine performance, and material efficiency, and relay data in real time to cloud-based storage systems for analysis (Xu et al. 2022). IoT-enabled sensors allow manufacturers to shift from traditional batch monitoring to continuous, adaptive process control, ensuring greater efficiency in resource utilisation and waste management.

A typical IoT-enabled smart manufacturing framework involves:

1. Data Acquisition—Smart sensors collect real-time data on various manufacturing parameters.
2. Data Transmission—The data is sent via wireless or wired networks to centralised IoT platforms.
3. Cloud Computing and Edge Processing—Advanced analytics tools process and interpret the data.

4. Automated Responses and AI Integration—AI-driven decision-making systems trigger responses, such as adjusting machine settings or sending maintenance alerts.
5. Data Visualisation and Decision Support—Results are presented via dashboards, enabling human operators to make informed decisions.

This interconnected structure ensures that real-time sensor data flows seamlessly across the manufacturing ecosystem, improving process optimisation and energy efficiency.

4.6 Role of Big Data Analytics in Sensor Integration

The vast amount of data generated by smart sensors presents an opportunity for manufacturers to extract meaningful insights through big data analytics. Advanced machine learning algorithms analyse historical and real-time sensor data to identify patterns, detect inefficiencies, and optimise production processes. Predictive analytics, powered by AI and digital twins, enables manufacturers to anticipate maintenance needs, adjust energy consumption dynamically, and enhance operational sustainability (Bevilacqua et al. 2021). For example, in automotive manufacturing, AI-driven analytics applied to smart sensor data has facilitated:

- Dynamic energy adjustments to optimise machine power usage.
- Predictive maintenance scheduling to prevent equipment failures.
- Automated process optimisations to minimise material waste.

4.7 Cybersecurity Considerations in Sensor-Driven Manufacturing

With the growing reliance on IoT-based smart sensors, cybersecurity threats have become a major concern. Cyberattacks targeting manufacturing networks can lead to production disruptions, data breaches, and loss of sensitive intellectual property (Ghobakhloo et al. 2022). Ensuring robust cybersecurity protocols in sensor-integrated manufacturing systems is essential to mitigate risks and maintain operational integrity (Khan et al. 2024).

Key cybersecurity measures include:

- End-to-End Encryption—Securing sensor data transmission through encryption protocols.
- Access Control Mechanisms—Implementing authentication measures to restrict unauthorised access.
- Intrusion Detection Systems (IDS)—Deploying real-time monitoring tools to detect cyber threats.

- Blockchain for Data Security—Using decentralised ledger technology to ensure data integrity.

These security frameworks provide manufacturers with resilience against cyber vulnerabilities, ensuring that sensor-driven data remains secure, reliable, and tamper-proof.

5 Challenges in Implementing Smart Sensors in Digital Manufacturing

Despite the advantages of integrating smart sensors into manufacturing, several challenges hinder widespread adoption. These include:

- High Implementation Costs—Upfront investment in sensor deployment, cloud storage, and AI-driven analytics.
- Data Overload and Processing Complexity—Managing and interpreting vast amounts of real-time data.
- Integration with Legacy Systems—Many traditional factories still rely on outdated technologies, making seamless integration difficult.
- Skill Gaps in Digital Manufacturing—Workforce upskilling is required to manage and leverage smart sensor data effectively.

To address these challenges, manufacturers need to develop clear digital transformation strategies, invest in IoT-compatible infrastructure, and ensure effective training programmes to enhance digital literacy among employees. Therefore, the integration of smart sensors into digital manufacturing ecosystems represents a transformative shift towards data-driven, sustainable, and intelligent industrial operations. By leveraging IoT connectivity, AI-powered analytics, and real-time process monitoring, manufacturers can significantly improve energy efficiency, reduce waste, and enhance predictive maintenance capabilities. However, cybersecurity risks, integration complexities, and data management challenges must be carefully addressed to ensure successful implementation. As manufacturing systems continue evolving towards fully autonomous and AI-driven production, future research should explore advanced cybersecurity solutions, AI-based anomaly detection, and blockchain-integrated IoT frameworks to further strengthen digital sustainability.

6 Real-Time Monitoring in Green Manufacturing

The application of real-time monitoring in green manufacturing has become a pivotal strategy for enhancing sustainability, optimising energy consumption, and reducing waste. Real-time data collection, processing, and automated interventions enable manufacturers to move beyond reactive decision-making to proactive and predictive management of resources. By integrating smart sensors, Internet of Things (IoT) networks, and artificial intelligence (AI)-powered analytics, manufacturing systems can continuously track key environmental and operational metrics, ensuring greater efficiency and regulatory compliance (Ghobakhloo et al. 2022).

6.1 Role of Real-Time Data in Sustainable Manufacturing

Traditional manufacturing models rely on periodic or batch monitoring, which limits the ability to detect inefficiencies as they occur. Real-time monitoring systems, powered by smart sensors, enable instantaneous data collection and analysis, allowing manufacturers to identify anomalies and address sustainability issues dynamically (Kamble et al. 2023).

The key advantages of real-time monitoring in green manufacturing include:

- Immediate Waste Detection—Sensors can identify inefficiencies in raw material usage, reducing unnecessary waste.
- Energy Efficiency Optimisation—Real-time energy tracking allows for dynamic adjustments in power consumption, preventing excessive usage.
- Emission Control and Environmental Compliance—Continuous monitoring of greenhouse gas emissions ensures that manufacturing processes remain within regulatory limits.
- Predictive Maintenance for Resource Efficiency—Sensors detect early signs of machine failure, preventing unnecessary replacements and downtime.

To illustrate these capabilities, the Real-Time Green Manufacturing Monitoring Framework (Fig. 3) presents a structured approach to integrating smart sensors for sustainable industrial operations.

- Data Collection Layer—Smart sensors track energy consumption, emissions, waste generation, and equipment health.
- IoT and Cloud Processing Layer—Sensor data is transmitted and analysed in real time using AI and big data platforms.
- Decision-Making and Automation Layer—AI-powered insights trigger automated system responses, such as adjusting machine settings, initiating maintenance, or optimising energy use.
- Visualisation and Reporting Layer—Data is presented via dashboards for operators, enabling human oversight and regulatory compliance tracking.

This framework provides a systematic approach to real-time monitoring, ensuring that sustainability measures are embedded into manufacturing processes at all levels.

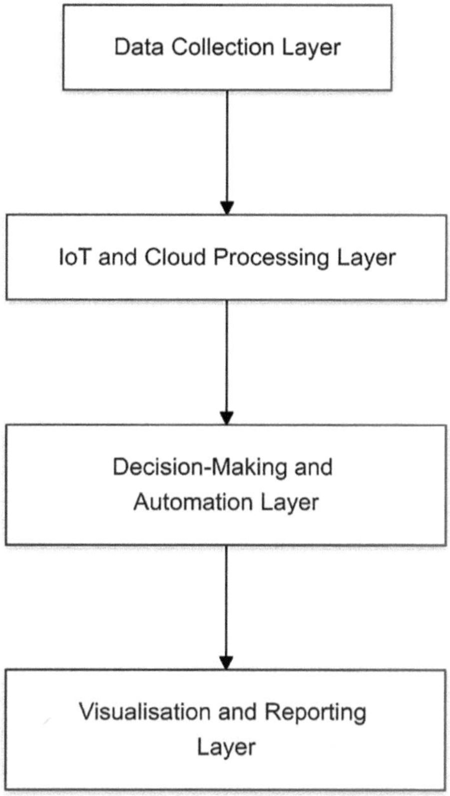

Fig. 3 Real-time green manufacturing monitoring framework

6.2 Applications of Real-Time Monitoring in Green Manufacturing

The adoption of real-time monitoring has been successfully implemented in various industries, delivering significant sustainability benefits. Below are key applications of real-time monitoring across waste reduction, energy efficiency, and predictive maintenance.

Waste Reduction Through Real-Time Material Tracking
One of the most critical applications of real-time monitoring is in material efficiency, ensuring that raw materials are optimally used, recycled, or repurposed rather than wasted. IoT-enabled waste tracking sensors can monitor production lines, detect material inefficiencies, and enable closed-loop manufacturing (Ghisellini et al. 2016). For example, in food processing, real-time monitoring of raw material usage ensures that ingredients are used efficiently, minimising spoilage. In automotive manufacturing, waste management sensors have enabled manufacturers to track scrap rates and improve material reusability by identifying points of inefficiency in the production process (Bevilacqua et al. 2021). The ability to track real-time

Table 3 Real-time monitoring for waste reduction in manufacturing

Manufacturing sector	Monitoring technology used	Sustainability outcome
Food processing	IoT waste tracking sensors	Reduced ingredient wastage and optimised supply chain
Automotive	Smart scrap detection sensors	Improved raw material efficiency and recycling rates
Textile production	Water and dye usage monitoring	Minimised chemical waste and enhanced wastewater recycling

material waste and optimise reuse strategies positions smart sensors as a key enabler of circular economy models in manufacturing. A structured comparison of waste reduction strategies using real-time monitoring is presented in Table 3.

6.3 *Energy Efficiency and Real-Time Monitoring*

Smart sensors play a vital role in tracking and optimising energy consumption in manufacturing plants. Energy meters, thermal sensors, and AI-driven load monitoring systems ensure that power usage is optimised, preventing unnecessary energy waste (Wang et al. 2022). For example, in electronics manufacturing, real-time monitoring of power usage in semiconductor production has led to a reduction in excessive cooling system energy consumption through AI-driven thermal regulation. Similarly, in steel manufacturing, real-time energy tracking has enabled dynamic adjustments in furnace operations, leading to optimised fuel consumption (Bag et al. 2023). By continuously tracking and analysing energy consumption, manufacturers can make dynamic process adjustments, reducing carbon footprints while maintaining production efficiency (Fatorachian 2024b; Fatorachian and Kazemi 2024a, 2024b).

6.4 *Predictive Maintenance and Equipment Health Monitoring*

Traditional reactive maintenance approaches lead to excessive downtime, inefficient resource use, and increased environmental impact. With real-time monitoring, predictive maintenance strategies ensure that machine failures are detected before they occur, enabling proactive interventions (Xu et al. 2022). In the aviation sector, smart sensors in jet engines track vibrations, pressure levels, and operational efficiency in real time. By predicting component failures, airlines reduce unnecessary engine part replacements, lowering waste generation. In heavy machinery industries, real-time vibration sensors monitor the wear and tear of critical equipment, ensuring timely maintenance while reducing overall resource consumption (Ghobakhloo et al. 2022).

6.5 Challenges and Future Prospects of Real-Time Monitoring

Despite the numerous advantages, real-time monitoring in manufacturing faces several challenges that must be addressed for widespread adoption.

- High Initial Investment Costs—Deploying IoT-enabled sensors and integrating cloud-based monitoring systems require significant financial investment.
- Data Overload and Interpretation Complexity—Processing vast amounts of real-time data requires advanced AI-driven analytics and skilled personnel.
- Cybersecurity Risks—As real-time monitoring relies on IoT connectivity, securing sensor data from cyber threats is crucial.
- Integration with Legacy Systems—Many factories operate on traditional infrastructure, making it challenging to fully integrate real-time digital monitoring solutions.

However, the future of real-time monitoring in green manufacturing is promising, with advancements in AI, blockchain security, and edge computing expected to further improve efficiency, security, and scalability. In conclusion, real-time monitoring is a cornerstone of green manufacturing, enabling industries to move from reactive to predictive and autonomous decision-making. By leveraging IoT-enabled smart sensors, AI-driven analytics, and real-time process optimisation, manufacturers can significantly improve waste reduction, energy efficiency, and predictive maintenance strategies. The Real-Time Green Manufacturing Monitoring Framework, alongside case studies in food processing, automotive, electronics, and aviation, highlights the transformative potential of real-time monitoring in achieving sustainability goals.

7 Applications of Smart Sensors in Industry: Case Studies

The adoption of smart sensors in green manufacturing has led to substantial improvements in waste reduction, energy efficiency, and predictive maintenance across various industries. By leveraging real-time monitoring, IoT connectivity, and AI-driven analytics, companies are transforming their manufacturing processes to align with sustainability goals. This section explores case studies from key industrial sectors, demonstrating how smart sensors contribute to improved environmental performance and operational efficiency.

7.1 Smart Sensors in the Automotive Industry

The automotive industry has been at the forefront of integrating smart sensors for sustainable production and resource efficiency. Automakers have adopted

IoT-enabled energy meters, thermal sensors, and AI-driven quality control systems to optimise manufacturing processes and reduce environmental impact. A leading automotive manufacturer implemented real-time monitoring of energy consumption across multiple production plants. By integrating smart meters and thermal sensors, the company detected inefficiencies in paint shops, welding lines, and assembly stations, leading to dynamic energy adjustments. This initiative resulted in a significant reduction in electricity and gas consumption, lowering the plant's overall carbon footprint. Furthermore, AI-powered vision sensors have been deployed for real-time defect detection during vehicle assembly. By minimising material waste associated with defective components, the company optimised raw material usage while improving product quality. This case study highlights how real-time energy monitoring and AI-driven waste reduction strategies can enhance both sustainability and operational efficiency in the automotive sector (Ghobakhloo et al. 2022).

7.2 *Electronics Manufacturing: Optimising Energy and Material Use*

In the electronics industry, manufacturers have incorporated AI-driven energy sensors, waste tracking systems, and predictive maintenance technologies to improve sustainability. Given the sector's high energy demands and material-intensive production, real-time monitoring has been crucial in optimising resources. One electronics manufacturer adopted smart thermal sensors to monitor heat dissipation in semiconductor fabrication. These sensors provided continuous feedback to AI-driven cooling systems, ensuring that energy consumption was dynamically adjusted based on real-time production loads. This approach reduced overall energy usage while preventing overheating and extending equipment lifespan (Bag et al. 2023). Additionally, waste tracking sensors have been instrumental in optimising material usage. Smart optical sensors monitored the application of solder paste on circuit boards, detecting excess material deposits and adjusting dispensing rates accordingly. By preventing unnecessary material use, the company reduced waste generation in PCB assembly lines, improving resource efficiency and cost savings.

7.3 *Food and Beverage Industry: Minimising Waste and Enhancing Supply Chain Efficiency*

The food and beverage industry face sustainability challenges due to perishable goods, high water consumption, and packaging waste. To address these issues, leading companies have integrated IoT-based waste tracking, water monitoring sensors, and predictive analytics for supply chain management. A major dairy producer deployed real-time tracking sensors to monitor raw ingredient use in yogurt and cheese production. These sensors measured ingredient flow rates, detected inefficiencies in mixing processes, and provided automated adjustments to prevent excess material usage. This implementation resulted in a measurable reduction in ingredient waste, improving supply chain sustainability (Bevilacqua et al. 2021).

In addition, water sensors have been used to optimise cleaning and sanitation processes in beverage bottling plants. By continuously measuring water consumption in cleaning cycles, manufacturers have reduced overall water usage while ensuring compliance with hygiene standards.

7.4 Textile and Apparel Industry: Reducing Water and Chemical Waste

The textile industry is one of the most resource-intensive sectors, consuming large quantities of water, chemicals, and energy. The introduction of smart water monitoring sensors, dye usage optimisation systems, and energy-efficient textile drying technologies has helped improve sustainability. A textile manufacturer adopted real-time water quality monitoring sensors to track dye effluent discharge. These sensors measured chemical concentrations in wastewater, ensuring compliance with environmental regulations and wastewater treatment protocols (Ghisellini et al. 2016). By adjusting dye formulations based on real-time sensor feedback, the company reduced overall chemical waste and improved water recycling rates (Fatorachian and Smith 2022). Moreover, energy-efficient drying systems equipped with thermal sensors and AI-driven optimisation algorithms have helped reduce energy consumption in fabric drying processes. These systems dynamically adjust temperature and airflow based on moisture sensor readings, leading to improved efficiency in textile production (Xu et al. 2022).

7.5 Aerospace Industry: Predictive Maintenance and Fuel Efficiency

The aerospace industry has adopted predictive maintenance sensors and AI-driven diagnostics to enhance operational efficiency and reduce environmental impact. Aircraft manufacturers use real-time engine monitoring systems to detect component wear and predict potential failures before they occur. For example, an aerospace company integrated vibration and acoustic sensors into jet engines to continuously monitor performance metrics. AI-driven analytics processed sensor data, identifying early warning signs of mechanical stress. This predictive approach allowed maintenance teams to schedule repairs before major failures occurred, reducing aircraft downtime and minimising resource wastage (Wang et al. 2022). Additionally, fuel efficiency monitoring sensors have helped airlines optimise fuel consumption during

Table 4 Smart sensor applications in different industries

Industry	Smart sensor application	Sustainability impact
Automotive	IoT energy meters and AI defect detection	Reduced energy consumption and material waste
Electronics	Smart thermal sensors and waste tracking	Optimised cooling systems and reduced material waste
Food & Beverage	Ingredient flow monitoring and water sensors	Lower ingredient waste and improved water efficiency
Textile & Apparel	Water quality sensors and AI drying optimisation	Reduced dye pollution and lower energy use
Aerospace	Predictive maintenance and fuel monitoring	Lower aircraft downtime and improved fuel efficiency

flights. These sensors analyse engine performance, altitude, and aerodynamic conditions, providing pilots with real-time adjustments to improve fuel efficiency. By leveraging big data analytics and AI-based optimisation, the aerospace industry continues to drive sustainability improvements in aviation operations. The successful implementation of real-time monitoring and smart sensors in various industries demonstrates their potential to drive sustainability improvements, reduce waste, and optimise energy consumption. Table 4 summarises key smart sensor applications and their sustainability outcomes across different industrial sectors.

7.6 Challenges in Scaling Smart Sensor Applications

Despite the demonstrated success of smart sensors in manufacturing, certain challenges hinder their scalability and widespread adoption.

- High Implementation Costs—Many industries face capital investment barriers in deploying large-scale sensor networks.
- Data Integration Complexity—Legacy manufacturing systems often lack the infrastructure to integrate IoT-based real-time monitoring solutions.
- Cybersecurity Concerns—As manufacturing ecosystems become more digitised, protecting sensor-driven data from cyber threats is a growing challenge.
- Workforce Skill Gaps—Implementing AI and predictive analytics requires workforce upskilling in data interpretation and digital systems management.

Addressing these challenges will require strategic investments in IoT infrastructure, cybersecurity frameworks, and workforce training, ensuring that sensor-driven manufacturing remains secure, scalable, and sustainable.

8 Future Directions for Smart Sensors in Green Manufacturing

The continued integration of smart sensor technology into green manufacturing is set to revolutionise sustainable production, energy efficiency, and waste management. As industries transition towards Industry 5.0, AI-driven automation, and hyper-connected IoT ecosystems, smart sensors will play an even greater role in enhancing decision-making, reducing carbon footprints, and optimising circular economy models. This section explores emerging technologies, policy developments, and industrial trends that will shape the future of smart sensor applications in sustainable manufacturing.

8.1 Advancements in AI and Machine Learning for Sensor Analytics

The future of smart sensor technology is closely linked with advancements in AI and machine learning (ML). As sensors generate increasingly large volumes of real-time data, AI-driven analytics will enable manufacturers to move beyond traditional reactive strategies towards fully predictive and autonomous decision-making (Ghobakhloo et al. 2022).

Future AI-driven sensor applications will include:

- Self-Learning Manufacturing Systems—AI-integrated sensors will continuously learn from operational data, enabling production lines to self-optimise and reduce energy consumption.
- Advanced Anomaly Detection—Machine learning models will be capable of identifying micro-level inefficiencies in real time, allowing for ultra-precise waste reduction and energy efficiency strategies (Bevilacqua et al. 2021).
- Automated Supply Chain Optimisation—AI will analyse sensor data from supply chains, optimising logistics, material flows, and inventory management to minimise emissions and waste.

As AI capabilities improve, the integration of digital twins—virtual replicas of physical manufacturing systems—will further enhance predictive maintenance, resource allocation, and sustainability forecasting.

8.2 Next-Generation IoT and Edge Computing in Manufacturing

The rapid evolution of IoT connectivity and edge computing will further refine how smart sensors are deployed and managed in green manufacturing. Traditional cloud-based data processing presents challenges such as latency issues and high data transfer costs. To address this, the shift towards edge computing will enable real-time data processing at the sensor level, reducing the need for centralised computing power (Fatorachian et al. 2013; Xu et al. 2022).

Key future IoT and edge computing advancements include:

- 5G-Enabled Smart Factories—High-speed 5G networks will enhance real-time data exchange between smart sensors and industrial control systems, allowing for faster responses to energy fluctuations and process inefficiencies.
- Decentralised Edge AI—AI-driven edge sensors will process environmental, energy, and machine health data locally, enabling real-time process optimisation without cloud dependency.
- Blockchain-Integrated IoT—To improve data security and transparency, blockchain networks will be used to validate and store sensor data, ensuring tamper-proof sustainability monitoring.

By reducing reliance on centralised cloud computing, manufacturers will achieve lower operational costs, improved data privacy, and increased sensor responsiveness in green manufacturing applications.

8.3 Smart Sensors in the Circular Economy and Zero-Waste Manufacturing

The transition towards a circular economy—where materials are continuously reused, repurposed, or recycled—will be significantly enhanced by real-time smart sensor data. Future sensors will play a pivotal role in monitoring material flows, product lifecycles, and supply chain sustainability, ensuring that waste is minimised at every stage of production (Ghisellini et al. 2016).

Key circular economy advancements driven by smart sensors include:

- Automated Waste Sorting and Recycling Sensors—AI-powered waste tracking sensors will detect recyclable materials and classify waste streams, enabling fully automated recycling processes.
- Extended Product Life Cycle Monitoring—Embedded sensors in products will track wear and tear, allowing manufacturers to design remanufacturable and upgradable products, reducing landfill waste.
- Reverse Logistics Optimisation—IoT-enabled tracking of returned products and used components will help manufacturers implement cost-effective remanufacturing and material recovery strategies.

As industries adopt net-zero waste models, sensor-driven circular economy solutions will be critical to achieving sustainability objectives.

9 Policy and Regulatory Implications for Future Smart Sensor Adoption

With the increasing adoption of sensor technologies in green manufacturing, policymakers are expected to implement stricter environmental reporting requirements and incentivise smart technology adoption. Governments and international organisations will likely introduce new data-driven sustainability regulations that mandate real-time monitoring for carbon emissions, energy usage, and waste production (Kamble et al. 2023).

Future regulatory developments may include:

- Mandatory Smart Sensor Deployment—Industries may be required to integrate IoT-based emissions and energy monitoring sensors to comply with global sustainability targets such as the Paris Agreement and Net-Zero Carbon Goals.
- Standardised Sustainability Data Reporting—A global framework for real-time sustainability data collection will drive greater transparency in corporate environmental impact reporting.
- Government Incentives for Green Technology Investments—Policies encouraging industries to adopt AI-driven energy management sensors and waste reduction technologies will accelerate the adoption of sensor-based manufacturing ecosystems.

By aligning technological advancements with policy frameworks, smart sensors will become a central tool for global sustainability compliance and industrial decarbonisation efforts. Hence, the future of smart sensors in sustainable manufacturing is defined by greater AI integration, decentralised IoT processing, circular economy adoption, and regulatory advancements. As sensor accuracy, connectivity, and AI-driven decision-making continue to evolve, manufacturers will unlock new opportunities for waste reduction, energy efficiency, and emissions control. With the emergence of Industry 5.0, where human–AI collaboration enhances industrial intelligence, sensor-driven sustainability solutions will be at the forefront of achieving long-term environmental and economic goals (Fatorachian 2024a; Fatorachian and Smith 2022). As businesses, researchers, and policymakers continue advancing data-driven green manufacturing strategies, the role of smart sensors will only become more critical in shaping a sustainable and efficient industrial future.

10 Conclusion

The integration of smart sensors in green manufacturing has fundamentally transformed how industries monitor, optimise, and manage their production processes in real time. By leveraging Internet of Things (IoT) connectivity, AI-driven analytics, and real-time data processing, manufacturers have successfully reduced waste, improved energy efficiency, and enhanced sustainability compliance. Smart sensors have evolved from simple data collection tools to intelligent, self-learning systems that drive predictive and autonomous decision-making.

This chapter has explored the different types of smart sensors, including environmental sensors, energy consumption sensors, waste tracking sensors, and predictive maintenance systems. Their impact has been demonstrated through real-world applications in automotive, electronics, food processing, textiles, and aerospace industries, highlighting their role in waste minimisation, energy efficiency, and circular economy models. Real-time monitoring has been a critical enabler of green manufacturing, shifting industries from reactive to proactive sustainability strategies. Structured frameworks, such as the Real-Time Green Manufacturing Monitoring Framework and Smart Predictive Maintenance Model, have provided a foundation for integrating sensor-driven decision-making into industrial operations. These technological advancements have contributed to reduced emissions, optimised resource utilisation, and alignment with global sustainability targets.

However, despite these benefits, challenges remain in scaling smart sensor adoption. High implementation costs, cybersecurity vulnerabilities, data integration complexities, and workforce skill gaps continue to hinder widespread deployment. Overcoming these barriers requires advancements in AI, 5G connectivity, blockchain security, and edge computing, which will make sensor technologies more scalable, secure, and cost-effective.

Future Research Directions: Integrating Smart Sensors with Industry 5.0 and Digital Twins Looking ahead, Industry 5.0 and AI-driven automation will further enhance smart sensor capabilities, paving the way for self-optimising manufacturing systems that predict inefficiencies, adapt dynamically, and optimise processes in real time. Decentralised AI and blockchain-integrated IoT ecosystems will be crucial in ensuring that sensor-generated data remains secure, transparent, and regulatory-compliant, allowing industries to meet increasingly stringent environmental policies and global sustainability commitments.

Additionally, the integration of smart sensors with digital twins—virtual models that simulate real-world manufacturing environments—will provide new opportunities for predictive analytics, process optimisation, and sustainability forecasting. By creating real-time digital replicas of physical production systems, manufacturers can test and refine sustainability strategies before implementing them, further reducing waste, downtime, and resource inefficiencies.

As green manufacturing continues to evolve, smart sensors will become indispensable in achieving energy-efficient, zero-waste, and net-zero carbon production models. The future of sustainable manufacturing will be driven by real-time automation, predictive decision-making, and intelligent digital ecosystems, with Industry 5.0 and digital twins playing a pivotal role in shaping next-generation industrial sustainability.

References

Bag S, Dhamija P, Mishra A, Gupta S (2023) Smart manufacturing and sustainability: exploring the role of digital technologies. J Clean Prod 385:135654

Bevilacqua M, Ciarapica FE, Mazzuto G (2021) Integrating lean and green manufacturing: a structured framework for sustainability. J Clean Prod 295:126470

Dubey R, Bryde DJ, Gunasekaran A (2023) Green manufacturing in the era of Industry 4.0: a review and future research agenda. Resour Conserv Recycl 191:106795

Elkington J (1997) Cannibals with forks: the triple bottom line of 21st century business. Capstone, Oxford

Fatorachian H (2023) The significance of Industry 5.0 in the globalization of supply chain management. Eur Econ Lett (EEL) 13(5):872–885

Fatorachian H (2024a) Sustainable supply chain management and Industry 5.0. In: Eco-innovation and sustainable development in Industry 5.0. IGI Global, pp 172–197

Fatorachian H (2024b) Leveraging artificial intelligence for optimizing logistics performance: a comprehensive review. Glob J Bus Soc Sci Rev 12(3):146–160

Fatorachian H, Kazemi H (2024a) AI-enhanced fault-tolerant control and security in transportation and logistics systems: addressing physical and cyber threats. Complex Eng Syst 4:1–18

Fatorachian H, Kazemi H (2024b) Integrating learning-based solutions in intelligent transportation systems: a conceptual framework and case studies validation. Cogent Eng 11(1):2427235

Fatorachian H, Kazemi H (2025) Sustainable optimization strategies for on-demand transportation systems: enhancing efficiency and reducing energy use. Sustain Environ 11(1):2464388

Fatorachian H, Pawar K (2025a) Waste efficiency in cold supply chains through Industry 4.0-enabled digitalisation. Int J Sustain Eng 18(1):2461564

Fatorachian H, Pawar K (2025b) Sustainable cold chain management: an evaluation of predictive waste management models. Appl Sci 15(2):770

Fatorachian H, Smith C (2022) Impact of CPS on enhancing supply chain resilience, with a focus on solutions to pandemic challenges. In: Cyber-physical systems. CRC Press, pp 109–125

Fatorachian H, Smith C (2024) A nexus of digital entrepreneurship and Industry 5.0. In: Alternative Finance. Routledge, pp 29–44

Fatorachian H, Shahidan M, Kazemi H (2013) Role of internet in supply chain integration: empirical evidence from manufacturing SMEs within the UK. ECMLG2013-proceedings for the 9th European Conference on Management & Leadership

Ghisellini P, Cialani C, Ulgiati S (2016) A review on circular economy: the transition to sustainable production and consumption. J Clean Prod 114:11–32

Ghobakhloo M, Fathi M, Iranmanesh M (2022) Smart manufacturing and green production: how Industry 4.0 technologies impact environmental sustainability. Comput Ind 140:103683

Kamble SS, Gunasekaran A, Sharma R (2023) IoT and AI in sustainable manufacturing: a review of applications, challenges, and future research. Sustain Prod Consum 35:468–482

Khan MR, Masood F, Gul I, Khan MR (2024) Blockchain technology in supply chain management: opportunities and challenges. In: Convergence of Industry 4.0 and supply chain sustainability. IGI Global, pp 275–295

UNEP (2022) Cleaner production: strategies for a sustainable future. United Nations Environment Programme

Wang Y, Xu X, Liu Z (2022) Leveraging smart sensors for real-time waste monitoring in manufacturing: a data-driven approach. J Manuf Syst 62:331–345

Xu LD, He W, Li S (2022) Internet of things in industries: a survey on security, applications, and future directions. Futur Gener Comput Syst 125:24–39

Zhang C, Huang GQ, Luo H (2023) Machine learning-driven energy efficiency improvement in smart factories. Energy Rep 9:211–225

Green Product Eco-Design and Life Cycle Assessment

B. R. Manoj Kumar, Naveed R. Khan, N. Thilaka, R. Banumathi, and Blessy Doe

Abstract Eco-design has emerged as a central theme in contemporary production, solving serious environmental issues using sustainable product creation. This chapter explores the planned integration of the principles of eco-design throughout the life cycle of a product through phases like market analysis, conceptualization, component selection, and validation of designs. A specific emphasis is put on Life Cycle Assessment (LCA), supplying objective measures with which to estimate and improve ecological performance. The dialogue focuses on the utilization of recyclable and renewable materials, modular design to simplify repair, and design methods in favor of longevity and minimization of waste. Implementation includes sustainable manufacturing planning, ethical purchasing, and consumer engagement through explicit communication and education on environmental advantages. Post-launch evaluation, stakeholder input, and adaptive redesign mechanisms ensure continuous improvement. Examples from the furniture and fashion industries demonstrate how these principles can be applied by companies to achieve success. Essentially, eco-design is a strategic bridge between innovation, regulatory compliance, and long-term environmental responsibility, providing both ecological and economic benefits in an increasingly sustainability-focused market.

B. R. Manoj Kumar (✉)
Xavier Institute of Management and Entrepreneurship, Bangalore, India
e-mail: phd.brmk@gmail.com

N. R. Khan
UCSI University, Kuala Lumpur, Malaysia
e-mail: naveed.r.khan@gmail.com

N. Thilaka · R. Banumathi
Acharya Bangalore B-School, Bangalore, India
e-mail: narayanthilaka@gmail.com; banuscc3@gmail.com

B. Doe
Sri Ramakrishna Engineering College, Coimbatore, India
e-mail: blessydoem@gmail.com

M. Venkatesh et al. (eds.), *Green Manufacturing Technologies in Industry 5.0*,
https://doi.org/10.1007/978-981-95-0363-6_4

1 Introduction

Eco-design is an important approach in contemporary production that responds to the immediate need for sustainability, addressing a complex and pressing environmental issue (Kishita et al. 2021). Such an approach aims to reduce the ecological footprint of a product during its entire life cycle, supporting the global effort for sustainable development as described in the United Nations SDGs. There is an increasing proportion of consumers (77%) that would pay a premium for an eco-friendly product, making eco-design not only an opportunity but also a driver for manufacturers to meet tighter regulation and work on resource efficiency and saving costs. There is a promising market for sustainable products that is projected to be $150 billion by 2027. Life cycle assessment, durability, recyclability, and user centric design are some of the core principles of eco-design and are very important for innovation, and as a tool, to foster the change to a circular economy (Ng and Tang 2022). In the end, eco-design increases product value, facilitates sustainability, and earns businesses a long-term sustainable profit (Kishita et al. 2021).

1.1 Definition

Eco-design refers to sustainable design practice and is the process of developing products, services, and systems that have an attributable low environmental cost throughout their life cycle (Kishita et al. 2021). In this method, environmental factors are taken into account in the design process in order to promote sustainability, waste minimization, and resource conservation. It highlights the importance of product development taking into account the whole life cycle of products, within raw material extraction, production, use, and ultimate disposal (Faud-Luke 2009).

One of the primary factors is that eco-design promotes circular economy principles aimed at reducing waste and recycling and re-fabricating or remanufacturing the material (Choudhury and Hashmi 2020). To contribute to this, life cycle assessment happens to be an important tool; this enables designers to see the environmental performance of a product along the whole life cycle right from its manufacture right through to disposal for encouraging more sustainable options (Faud-Luke 2009).

Eco-design also aims at reconciling the expectations of consumers with environmental sustainability, that is designing products that satisfy the expectation of users and at the same time reduces the environmental impact (Kishita et al. 2021). It can be considered, for example, a strategic tool for companies to innovate, improve the brand image, and to comply with the law, whereas at the same time, customer needs in terms of sustainability are taken into account. In general, eco-design represents a multi-structured approach that goes beyond the simple approach of product design to include the production and distribution systems and processes, leading to sustainability throughout the chain (Ng and Tang 2022).

1.2 Eco-Design Process

The eco-design method is a structured methodology that imbeds environmental considerations into the design and production of products. Its aim is to mitigate environmental adverse impact and promote sustainability in the production cycle by production lifecycle (Faud-Luke 2009).

2 Research and Analysis

Market analysis is an important element of the eco-design framework because it can shed light on both consumer behaviors and the competitive environment in the sustainability field.

2.1 Identify Consumer Needs and Preferences Related to Sustainability

Consumer needs can be determined for designing products which meet the needs of targeted groups. Research shows that the consumer is increasingly weighing sustainability when purchasing (Kishita et al. 2021). The best proxies indicate discerning consumer preferences, including green materials, ethically procured raw material supply chain transparency (Choudhury and Hashmi 2020). Survey, focus group, and social media shaming can all be applied in efforts to gather both qualitative and quantitative information on how consumers feel about sustainability.

2.2 Analyze Competitor Products and Trends in Eco-Design

Based on the analysis of competing products, one gains very specific information on market trends and good practices regarding eco-design. This includes an assessment of how competitors manage sustainability in their products, material, and application of marketing (Kishita et al. 2021). For instance, companies like Patagonia and IKEA have made sustainability an integral element at the core of their business models, setting the lead. Thus, with the analysis of these competitors, companies learn about the available market opportunities, where opportunities might arise for innovation, and what really makes them stand out.

2.3 Environmental Impact Assessment

An essential process workflow of eco-design is an environmental impact assessment, which is a tool providing the designers with the understanding of ecological effects that the products they are designing may induce. This implies the assessment of parameters such as energy consumption, emissions, generation of waste, and resource use. There exist tools like environmental footprint calculators that allow for obtaining a first concept of the ecological significance of a product. An alternative is the life cycle assessment (LCA) technique, which analyzes the product impact from the raw material extraction to final disposal (Choudhury and Hashmi 2020). The data-based reconstructions with system/process analysis, as required by ISO 14040 (2006) (LCA) allow designers to choose more-optimal materials, manufacturing processes, and waste management strategies (Klüppel 2005 & Lewandowska and Matuszak-Flejszman 2014). In the long terms (e.g., ultimate purpose) it helps companies to pinpoint where improvements are possible and to focus on more sustainable products.

2.4 Regulatory Review

Compliance with the right regulations is crucial for the proper compliance process, as well as in providing recommendations for sustainable design practices. This regulatory evaluation comprises reviewing environmental regulations at local, national, and international levels, which affect design of products (Ng and Tang 2022). It includes material regulation, emission regulation, waste regulation, and safety regulation. For example, the Registration, Evaluation, Authorisation and Restriction of Chemicals (REACH) directive of the European Union, and the Waste Electrical and Electronic Equipment (WEEE) Directive have set very strict requirements about the use of hazardous substances and waste related to electronic equipment. By keeping themselves abreast of such rules, the businesses can avert costly problems and allow their green products to comply with the requirement standards that will increase their marketability and customer confidence (de Figueiredo and Guillén 2011).

2.5 Concept Development

Development phase plays a substantive role in the eco-design process such that the first brainstorming ideas are refined and become viable product concept. At this stage, it is all about cooperation and innovation, where sustainability is their contribution.

2.5.1 Brainstorming Sessions

Idea generation session is crucial for the nurturing of creative thoughts in the arena of sustainable product features and materials. Such teamwork meetings promote creativity and innovation and open the way for new possibilities that can lead to remarkable ideas. Methods, such as mind mapping and role play, can help participants to imagine how a product might contribute to the sustainability targets. E.g., such discussions may range from how energy efficiency can be enhanced, less waste or environmental-friendly packaging materials can be employed. The study analysis clearly demonstrates that the heterogeneous groups' brainstorming yield creative solutions more frequently than heterogeneous groups because multimodal inputs add value in communications and result in more integrative products (Kishita et al. 2021).

2.5.2 Collaboration with Interdisciplinary Teams

Interdisciplinary team work is a priority in successful eco-design. Engaging designers, engineers, environmental scientists, and marketers ensures that every aspect in product development is taken into account. If not, this diversity, which stimulates creativity, can also lead to a more refined appreciation of sustainability problems. Collaborations allow teams to recognize possible problem(s) in the early stages of design process, e.g. material feasibility and reality of sustainable features (Ng and Tang 2022). Regular cross-departmental meetings are able to make communication better and prevent sustainability from being at the edge of product development processes (Kishita et al. 2021).

2.6 *Material Selection*

2.6.1 Identifying Sustainable Materials

This is a very important part of an eco-design regarding the selection of the most ideal materials. And there are ongoing efforts toward the discovery of sustainable alternatives such as recyclable, renewable, and biodegradable. Ashby and Johnson (2002) observe that the materials resulting from the minimum environmental damage—such as bamboo or recycled plastics—present a considerable step forward. The life-cycle consequences of a chosen material should also enter into the designer's consideration; for this reason, tools like the Material Passport may be adopted to evaluate material choices in relation to sustainability criteria, thus acquiring a quantitative input regarding the amount of environmental impact made by different kinds of materials (Choudhury and Hashmi 2020).

2.6.2 Evaluating Environmental Impacts

Evaluation of environmental impacts of potential materials is a crucial stage where environmental impacts of the materials are reviewed (Choudhury and Hashmi 2020). This may be achieved through an analysis of the environmental impacts of materials from raw to disposal using Life Cycle Assessment (LCA), namely, the environmental impacts of materials from their extraction to final disposal. It is through such learning that designers are able to make informed choices on the energy consumption, emissions, and resource consumption of each material (Chekima and Gbadamosi 2016). For instance, measuring the carbon footprint of virgin and recycled materials may help designers identify the benefits of choosing the greener options (Tezer and Bodur 2021).

2.7 Design for Longevity

2.7.1 Enhancing Product Durability

Longevity focus is an important consideration within eco-design. The integration of features extending a product's life greatly helps to reduce the need for replacement and hence contribute to reducing waste. This may, for instance, mean the implication of robust materials, improved manufacturing processes, and applying rigorous quality control in achieving product robustness (Ng and Tang 2022). Maintenance-ease strategies, such as user-replaceable parts and easy-to-follow maintenance instructions, can also be used to extend product life. In product development literature, durability products benefit the Earth; at the same time, both environmentally conscious consumers who would love to save the world and keep their emissions lower and the good old-fashioned consumerists who prefer the impact they bring on earth reduced through purchasing the durable good with longer longevity get the opportunity of owning a similar product (Faud-Luke 2009).

2.7.2 Exploring Modular Designs

Modular design is yet another great and effective solution toward eco-design because of easy repairing or replacing product. The structural and architectural improvements in the designs of products contribute to making reparability easier, thereby reducing overall replacement. Thus, this system creates a new economy of loops, keeping commodities in circulation as long as possible by adapting them according to changing conditions (Ng and Tang 2022). For example, modular "smartphones" allow for part replacement without whole phone replacement.

3 Design and Prototyping

In the eco-design process, the design and prototyping stage is very crucial as it transforms ideas into real products. This phase emphasizes the importance of sustainable design considerations, iterative evolution, and user input.

3.1 Sketching and Modeling

3.1.1 Creating Initial Sketches and Digital Models

The design process starts with sketching and modeling, both of which are very important in visualizing the product concept. Sketching has the benefit to quickly explore and iterate on many design aspects, especially in relation to sustainability features, for example, materials and energy. Images, as well as communication tools, are tools for solving problems between design, as a way to speed up the refinement of thought by designers (Ng and Tang 2022). Digital modeling has improved drafting by making designs more accurate. Computer-Aided Design (CAD) software permits designers to specify and design intricate 3D models that have specific sustainable attributes, such as modularity or energy efficiency. These models enable early visualization of product use, interaction, or ease of adjustments in alignment with principles of design or considerations related to sustainability.

3.1.2 Utilizing Design Software

Sophisticated design software serves as a critical input toward enhancing product concepts. The designers can be equipped with tools such as SolidWorks, Rhino, and Autodesk Fusion 360 to simulate product behavior in different environments and analyze the performance of materials while quantifying environmental effects. Then, through simulation, the designers can identify potential flaws or impasses before the actual prototype development phase begins. The ability to use this software provides for a more complete evaluation of how sustainable features can be integrated effectively into product design (Kishita et al. 2021).

3.2 Prototyping

3.2.1 Developing Physical Prototypes

Prototyping for the product involves making physical mockups of the product so that the design and functionality can be tested. At this stage it is critically important to assess the effectiveness of the sustainable functions within real-world scenarios.

Prototypes can be made of any material so that one can test which of the options is working most effectively in terms of performance and of environmental factors. As part of the prototyping loop, low-fidelity prototypes, usually in the form of paper models or 3D-printed parts, are introduced to allow for rapid evaluation of shape and fit.

3.2.2 Ensuring Prototypes Incorporate Eco-Design Principles

It is of high importance to embed eco-design principles into the prototyping process. This includes the development of products that are easy to take apart, and that can be recycled and reused after their use is over. Methods such as use of snap-fits or of adhesive residues can be used to create products that are easy to disassemble and sort into recyclable components. Not only does this aid sustainability, but it is also in accordance with the principles of circular economy (CE) as it encourages the efficient use of resources (Faud-Luke 2009).

3.3 User Testing

3.3.1 Conducting Usability Testing with Target Users

After the designing of prototypes, usability testing is an integral part of experience wherein real users take part in giving feedback. In this stage, the behavior of the users with the product is tracked, thus allowing one to obtain excellent information about the usability and performance of the same. Through user testing, design errors and fields that require improvement may be discovered, allowing designers to hone the product before mass production. Feedback from the user can lead to higher acceptance of the product and satisfaction of the customers.

3.3.2 Assessing Consumer Perceptions of Sustainability and Functionality

Aside from usability, consumer perceptions of product sustainability and functionality should be measured too. This may affect marketing strategies and guide future product design based on comments made on eco-design features. Surveys, interviews, and focus groups are vehicles providing qualitative information relating to consumer values and preference toward sustainability. Knowledge of consumer's perception of the sustainable elements of a product allows the designers to adjust features with a view to meet market needs and increase the acceptance of the product as a whole (Sayal and Pant 2023).

4 Evaluation Using Life Cycle Assessment (LCA)

4.1 Life Cycle Assessment (LCA)

Life Cycle Assessment (LCA) emerged with the growing concerns of environmental pollution and nonrenewability of energy and resources. LCAs can be applied to define the environmental impact and resource utilization of products through the entire life cycle—from extraction of raw materials, production, use, and waste disposal. Methods for LCA have undergone a dramatic change since they were first introduced in the 1960s and have been used quite frequently during the last decade (Baumann and Tillman 2004).

LCA is a means of evaluating the environmental impacts of products during their entire life cycle. It has found to be an accepted field, as well-established practices exist for carrying out LCA and guidelines for its use in applications such as eco-design, environmental labeling, and carbon emission reductions are available (Baumann and Tillman 2004).

The methodological basis for LCA is described in the ISO 14040 series, and other ISO standards provide support and guidelines on various applications, including eco-design (ISO 14062, ISO 14006), communication of environmental performance (ISO 14020 series), or greenhouse gas emissions reporting (Klüppel 2005).

4.2 LCA Framework

The ISO standard breaks down LCA into four key phases:

4.2.1 Goal and Scope Definition

In this first stage, for this preliminary phase, the objectives of the study are set and the limits of analysis are defined. It lies in the choice of the functional unit (a quantitative surrogate of the function of the product), processes in the product system, and impact assessment parameters. The scope covers establishment of geographical and temporal limits, definition of a point of view for the study (consequential or attributional), and proper critical review of the study, if it has the potential to make public claims (Baumann and Tillman 2004).

4.2.2 Inventory Analysis

At this point, data of inputs to the product system, emissions, waste, and outputs of the product system are collected. This is made possible, for instance, through scaling flows with the functional unit and encompasses utilization of information taken from the databases currently available. These include production databases of materials, waste disposal or other databases. Outputs are a life cycle inventory, which can account for the physical movements associated with the product or the service.

4.2.3 Impact Assessment

The next phase converts these physical flows into environmental impacts using models from environmental science. The assessment involves selecting impact categories, classifying the inventory flows, and characterizing their environmental effects (Baumann and Tillman 2004). The mandatory elements of this phase include:

- *Selection of Impact Categories:* Identifying categories and models to assess the impacts.
- *Classification:* Assigning flows to these categories.
- *Characterization:* Quantifying impacts based on the assigned flows, resulting in a total impact score for each category.

The additional elements are:

1. *Normalization:* Comparing the characterized impacts to common reference impacts.
2. *Grouping or Weighting:* Ranking or weighting impact categories to facilitate decision-making, leading to an overall environmental impact score.

This structured approach helps businesses make informed decisions about the environmental impact of their products and improve sustainability practices.

4.2.4 Interpretation

The results of an eco-design and life cycle assessment (LCA) study are analyzed to answer the questions defined at the start. This analysis includes information regarding inventory analysis as well as an impact assessment, including the definition of environmental impacts and information normalization or weighting. Such assessment should correspond to the purpose and the scope of the study, but should take into account, for example, geographical, temporal, or technical limitations (Faud-Luke 2009). Sensitivity and uncertainty analyses are performed in order to assess the reliability of the results as well as the highlights and tips for research to delineate the improvements of outputs.

5 Application of LCA

LCA has recently and broadly been applied in guiding decisions for environmental sustainability efforts, a cost sharing effort involving many different groups—government, industry, and consumers.

5.1 Government Perspective

LCA is used by governments to develop and implement, monitor, and evaluate policies and environmental impacts. For example, the European Commission's Integrated Product Policy (IPP) aspires to minimize environmental effects by assessing the entire lifecycle of products, from manufacture to end of life. LCA can support the decision-making process in terms of introducing new technologies, for example, biofuels and electric vehicle, and the selection of waste management systems with the main focus on sustainability and human health safety (Baumann and Tillman 2004).

5.2 Industry Perspective

Life cycle assessment is an integrative approach used by industry to assess the environmental performance of a product, process, or service throughout its life cycle, from cradle to grave. The overall objective is to identify ways in which the various negative environmental effects can be minimized and how sustainability can be achieved. LCA plays different roles in various industries (Baumann and Tillman 2004).

5.2.1 Product and Process Development

LCA plays an important role in product and process development because it allows companies to know about environmental problems and improve their designs of a product. Since the life cycle analysis of a product's whole life cycle involves raw material extraction, production, use, and final disposal, LCA helps companies recognize the opportunities in order to make environmental savings. For instance, on the example, LCA could show that the production phase of a product is very energy-intensive or wasteful. Businesses could improve manufacturing methods, select more sustainable materials, or redesign products for better recyclability. This preventive mindset to mitigate environmental impacts promotes innovations and contributes to sustainability measures.

5.2.2 Marketing

In competitive markets, LCA aids in marketing activities through the provision of trustworthy, data-added information about the ecological efficiency of a product. A high percentage of businesses rely on the outcome of LCA to attain eco-certifications and eco-labels, for example, Energy Star or Cradle to Cradle, in order to convince consumers of their environmental responsibilities (McDonough and Braungart 2002). These labels serve as a proof of concept for sustainability and are used to brand companies in the market. Furthermore, LCA helps companies to verify environmental claims for products, such as carbon footprint reduction or minimization of water consumption, hence increasing transparency and trust in the brand by the consumer (Tezer and Bodur 2021).

5.2.3 Monitoring

LCAs are the key monitoring tools that allow companies to monitor environmental performance over time. Companies can also measure the impact of their products and processes lifelong through use of metrics such as carbon footprint, water consumption, and energy. This continuous monitoring contributes to the identification of trends and to monitoring of the achievement of sustainability targets and compliance with environmental regulations or standards. It also enables companies to take strategic decision, to optimize their strategy, and to choose the actions which will improve their environmental performance (Tezer and Bodur 2021).

5.2.4 Supplier Selection

In modern global supply chains, LCAs are critical in order to achieve sustainability in sourcing. With the consideration of different suppliers through the environmental impact, companies may choose to deal with partners who share their schedule on sustainability goals. For example, LCA can help a company determine the CO2 footprint, water consumption, and waste production of the various suppliers' products and choose the supplier that has the smallest environmental footprint. This ensures that sustainability is not just built at the end product level, but it is put into the whole supply chain, and, in return, leads to improve the environmental strategy of the company at large (Tezer and Bodur 2021; Khan et al. 2023).

5.2.5 Strategic Planning

LCA is, above all, supportive of strategic planning to enable companies to define and attain sustainable targets. Uniquely, it was well-known companies, like Unilever, that applied LCA to inform their steps toward effectively reducing environmental footprint along both product lines and supply-chain. For instance, Unilever plans for an environmental impact reduction by 2030 of 50%. LCA means that companies can measure their performance; monitor the success level of the sustainability actions taken; and with data-driven information, best change the approach taken (Baumann and Tillman 2004). With this strategic use of LCA, environmental objectives are integrated into the long-term vision of the company and are systematically targeted.

5.2.6 Challenges for Smaller Enterprises

Small enterprises have problems in LCA implementation despite the many advantages that LCA can provide. Due to the significant cost associated with performing an extensive LCA, companies with restricted capital can have considerable difficulties with performing LCAs. Secondly, LCA requires specific expertise, and smaller firms may not be equipped with the expertise to understand intricate data and to take the right decision. Further, the procurement of reliable life cycle data from suppliers and other contributors to the supply chain can be difficult and time-consuming for small firms (Wimmer et al. 2004). Even with those challenges, nevertheless, some small businesses are devising strategies to address those challenges in the form of simplifying LCA, professional collaboration to obtain data on environmental impact, or using low-cost online LCA tools to obtain necessary data on an agency's environmental impact (Pati 2023).

5.2.7 Citizen Perspective

The results of an LCA also influence consumer choice behavior, i.e., eco-labels and supportive information, which enable the choice of more environmentally friendly products. Moreover, LCA results may influence residents in case of government applications to infrastructure choices, for example waste management.

6 Using LCA Results for Improvement

The output of the LCA can effectively guide design choices for greater product sustainability. If only some of the lifecycle phases exhibit strong influences, then designers can target these aspects for enhancement, e.g. for selecting a different material or for a good manufacturing process (Choudhury and Hashmi 2020). By integrating LCA results into the design process, manufacturers can guarantee product compliance with the market requirements while meeting with sustainability requirements (Ng and Tang 2022).

6.1 Sustainability Criteria

The sustainability evaluation of products is considered an important step toward compliance of products with the environment and responsive to the dynamic demand of consumers. The approach provides transparency for the designer through the cradle-to-grave overview of all environmental performances of their products—including usage of materials, energy consumption, and generation of wastes. By a systematic assessment of these factors, designers can identify points for improvement and take smart decisions for the sustainability of the product. This does not only assist business houses to achieve environmental regulations but also generates an atmosphere of trustworthiness and loyalty among environment-conscious consumers which in turn goes on to build a better, more sustainable world (Kishita et al. 2021).

6.2 Evaluating Against Sustainability Criteria

Sustainability criteria have prompted the design and manufacture of eco-friendly goods (Kishita et al. 2021). These standards emphasize key factors, such as energy efficiency, waste reduction, and responsible sourcing of materials, diminishing a product's environmental impact and encouraging lasting sustainability (Choudhury and Hashmi 2020).

6.2.1 Energy Efficiency—Lifecycle Energy Analysis

Lifecycle Energy Analysis includes assessing the energy of a product during its entire lifecycle—from the raw material mine through production, shipment, use, and end. During this process, those process stages that are using the most energy are pinpointed so business owners can direct their improvement efforts toward specific areas that need to be made. For instance, energy-greedy operations can be optimized at the manufacture stage by incorporating more energy-efficient technologies, or energy-consuming operations can be optimized at the utilization stage by designing products for lower energy consumption, which can effectively minimize total environmental impact. Thus, by understanding energy use at each step, designers are in a position to take informed decisions on how to reduce the energy use and related emission (Tezer and Bodur 2021).

6.2.2 Operational Efficiency

Operational efficiency is the design of products that consume the least energy in operation. This means devices such as LED lights that use less power than ordinary incandescent bulbs, as well as energy-saving appliances like refrigerators or washing machines. Operational efficiency is especially relevant in high energy demand fields such as transportation (electric vehicles), electronics and construction (energy efficiency buildings). The amount of energy used in the operational stage of the product can be limited to give businesses a way to reduce the emissions of the company and reduce the environmental impacts of the product, all the while providing savings to consumers from lowered energy costs.

6.2.3 Carbon Footprint Reduction

Minimizing a product's carbon footprint is one of the key ways to tackle climate change. Carbon emissions can also be reduced by saving energy at every stage of a product's life cycle. For instance, energy-saving products, such as LED lights and electric vehicles, lower the level of greenhouse gases emitted, as they rely on less fossil-based energy. In this regard, industrial emissions can be minimized by applying energy-saving technologies, such as using alternative energy sources or enhancing the efficiency of energy used in the production process. Reducing energy consumption in both the production phase and end-of-life processes also helps lower carbon footprints, mitigating global warming and reducing reliance on fossil fuels (Tezer and Bodur 2021).

6.2.4 Consumer Transparency

Consumer transparency relates to the provision of readily understandable information regarding a product's energy efficiency and its environmental footprint. Energy labeling programs, like Energy Star or European Union energy label, help consumers make informed purchasing decisions based on the energy demand of products. Manufacturers provide consumers with the tools to choose energy-demanding and energy-efficient products through the clear presentation of a product's energy usage. With this level of transparency, there is an incentive for a market-based impetus for sustainability in consumption, whereby consumers are increasingly demanding eco-friendly goods consistent with their environmental sustainability values (Chekima and Gbadamosi 2016).

6.3 Waste Reduction

6.3.1 Manufacturing Waste

Manufacturing waste refers to the waste products, materials, and scrap generated during the manufacturing process (Nematchoua et al. 2022). Landfill waste can take many forms, such as unused raw materials, rejected goods, or packaging materials

(Baskar et al. 2022). An integral part of waste management in manufacturing is the design of production processes. These processes should be designed to generate the least amount of waste possible. Strategies like lean manufacturing focus on increasing efficiency while eliminating unnecessary, waste-generating steps. In closed-loop systems, materials are reintegrated into the production flow, which helps reduce the amount of waste sent to landfills or simply discarded. Additionally, the introduction of precision manufacturing methods and quality control automation helps reduce the number of defective products, thereby decreasing waste. Companies are also investing in technology to recycle materials, which reduces dependence on new raw materials and further decreases landfill waste (Baskar et al. 2022).

6.3.2 Sustainable Packaging

Sustainable packaging defines the packaging as low footprint, recyclable or even using biodegradable or renewable materials. The traditional packaging, especially plastic, is not considered sustainable, as it may take up to hundreds of years to decompose. This has influenced firms to shift away from the traditional bulky packages to much minimalist packaging packages using minimal materials, however still assuring the product's integrity. Such recycling schemes as Loop, where packaging is designed to be deposited for refilling, rather than to be thrown away, are proofs of this concept. Additionally, companies use biodegradable plastics or compostable packaging that biodegrade and do not persist as pollutants. Sustainable packaging seeks to reduce the environmental footprint of product packaging, yet simultaneously address consumer's desire for an environmentally sustainable choice (Kishita et al. 2021).

6.3.3 End-of-Life (EOL) Management

EOL management refers to processes used in handling consumer goods which have completed their lifecycle. Thus, as products are recyclable, it is easy, and also repaired and demountable, with the aim that less waste winds up in a landfill. E.g., modular phablets could give users not to replace a whole system by replacing some particular parts and extended the lifecycle of their device and electronics waste. Designing to disassemble pertains to the process of producing products that are easily disassembled in order to allow the reuse or recycling of materials. An important part of EOL management is ensuring that the products are integrated into recycling processes, which helps close the loop while disposing waste. Many companies are now envisaging ways of providing repair services or take-back initiatives, where customers can return items for remanufacture, reutilization, or proper waste disposal instead of being dumped in a landfill (Baskar et al. 2022).

6.3.4 Circular Economy Practices

The circular economy is different from the more common linear mode of economic activity, where material is used and then discarded. In a circular economy, materials are circulated longer by means of reuse, repair, recycling, and remanufacturing (Keivanpour 2023). This aims to inflate the service life of items and decrease resource consumption and waste. Businesses are becoming more familiar with take-back schemes that motivate consumers to submit used goods to recycling, repair, or refurbishment processes. This approach keeps waste streams out of landfills and waste material recovered for reuse (Baskar et al. 2022). Additional companies are designing products with easy-to-replace or upgradeable modular components which extend product lifetimes and minimize replacement to new units. By embracing circular economy principles, businesses contribute to reducing waste and creating a more sustainable, resource-efficient economy (Fukushige et al. 2025).

7 Material Sourcing

7.1 Sustainably Sourced Materials

Originating and acquired renewable and sustainably sourced materials are extracted or produced by methods that minimize environmental degradation, promote ethical behavior, and protect natural environments (Kishita et al. 2021). Verified materials, for example, FSC-certified wood or Fairtrade-certified cotton, ensure that the raw materials are produced responsibly, taking into account both the environment and the communities producing them. Sustainable sourcing activities include responsible forest management, meaning forest management for regeneration, as well as responsible farming that does not use harmful pesticides and that enhances soil health. With the help of these tools, companies are able to mitigate the damage caused to the environment, conserve biodiversity, and enhance good work practices. Consumers have started to find more importance in sustainable sourcing because of increased awareness among them toward the environmental and social impacts their consumption decisions trigger (Tezer and Bodur 2021). Such third-party certification through either Forest Stewardship Council or Rainforest Alliance ensures that materials meet tight criteria for sustainability.

7.2 Recycled and Renewable Materials

The use of recycled and renewable materials is an important aspect in reducing the usage of scarce resources and creation of wastes. Use of recycled materials, such as plastics from ocean litter, post-consumer litter, or industrial waste, can reduce virgin virgin raw material requirements and minimize the environmental footprint of extraction activity. This could involve making plastics, metals, or glass available for recycling. The manufacturer could then play a role in reducing waste sent to landfills and contribute to reduction of carbon footprints as a result of creating new material from scratch. It can then be substituted with renewable materials: bamboo, hemp, and bioplastics. It is therefore noted that the materials are renewable, meaning that they can be regrown or refilled on a timescale much less than a million years as contrasted with the non-renewable resources, which take up times greater than a million years to form. A good example of such material is bamboo, a fast-growing plant used in flooring and clothing, and it has a significantly lower impact on the environment compared to traditional hardwood (Chen et al. 2023).

7.3 Material Transparency

Material Transparency is one of the conditions where companies shall disclose information and traceability so that the people know the material source, how it is developed, and possible environmental impacts because of the introduction of such new materials. Some innovative tools or technologies, in this case Material Passports, or Blockchain technology may help consumers find standardized pieces of information over the ingredients comprising a product. This transparency allows consumers to make informed decisions, and that promotes the products that align with their environmental and ethical values. Material transparency also compels companies to become accountable and begin to use more sustainable sourcing options and reduce the environmental footprint of their products. Moreover, transparency enhances a brand image as today's consumers require products for which they can be confirmed to be transparent and verifiable in the supply/material origins and work related to sustainability (Kishita et al. 2021). Companies demonstrate their commitments to sustainability and gain the trust of environmental concerned shoppers, by adopting transparency and accountable practices.

7.4 Local Sourcing

Local sourcing means acquiring raw materials from local suppliers, which means reducing the distance of transportation on the environment, and also strengthening the local economy. The decrease in the materials' distance traveled from generating to manufacturing plants enables corporations to minimize their carbon footprint, fuel consumption, and greenhouse gas emissions of logistics and distribution chains (Tezer and Bodur 2021). Second, local sourcing will also enable to retrieve the regional economy and help out a local business, enterprise for social sustainability to be attained. Moreover, in general, the material sourced locally is relatively fresh and of good quality, mainly because it is circumventing the process of long shipping or storage. Companies seeking local sourcing can strengthen their links with their local communities and improve the ability of their supply chain to cope with disruptions by reducing their dependence on global supply chains. Moreover, this approach also improves the traceability of the materials, as companies have a better understanding of the conditions in which the materials are produced.

7.5 Innovative Alternatives

Innovative alternatives refer to new materials and technologies devised to minimize negative impacts on the environment, achieve optimized product sustainability, and substitute harmful or limited resources. Among these advancements have been biodegradable materials—such as bioplastics—bioplastics break down under natural conditions without causing lasting pollution. For example, Danimer Scientific has served as a pace-setter by demonstrating how bioplastics with a "renewable from plant sources sustainable" origin are possible and in this manner is how greener biodegradable equivalents to petro-based plastics exist. Plant leather alternatives include non-(dust) and totally biodegradable mushroom or pineapple case alternatives to animal casings. These materials increase the independence from limited resources while providing solutions that fit with changing consumer values on sustainable and ethically made goods. Emerging innovative replacements are also being put into use in construction, where hempcrete and mycelium (material produced by fungi) are being investigated as replacers for the traditional concrete and insulation materials. In using these new materials, businesses can push the envelope on sustainability while simultaneously providing consumers with eco-friendly alternatives that contribute to a circular economy (Fukushige et al. 2025).

7.6 Holistic Approach to Sustainability

Energy efficiency and waste reduction will be integrated in the product's lifecycle, to bring about sustainability across every stages, from design and production to usage and disposal. Actions taken here, stimulate innovation, while enhancing the reputation of a brand, supporting sustainable goals across geographies and permitting companies to take healthy and sustainable nature (Nuryakin and Maryati 2020).

7.7 Adjusting Design Based on Assessment Findings

Designers might have to alter product design according to the sustainability assessment of criteria. This is an iterative process that is crucial to success in finding optimal sustainable results. For instance, if a product is identified as energy-heavy, designs may look into alternative technology or feature that can increase energy efficiency, for example by including smart functionality that optimizes energy use (Faud-Luke 2009). Alterations may also include variations of materials, modification of processes in manufacturing, and an elongation of the life of the product through extensions of durability.

8 Implementation

8.1 Production Planning

Production planning forms an essential part of the eco-design process, which allows for sustainable manufactured product production. It involves creation of detailed blueprints of the production of such products that include resources and their allocation, timelines, acquisition of materials, energy usage, and treatment of waste (Baskar et al. 2022). An aim is to develop a system that is effective, minimizes ecological cost, and is consistent with the sustainable goals.

An important factor in the productivity of a production planner is getting the most from the resources provided. This means the strategic sourcing of sustainably sourced raw materials and waste reduction during manufacture. For instance, companies can decide to use renewable or recycled products, use energy-efficient ways of producing, and introduce technologies to reduce material waste. Taking into account resources efficiency allows companies to reduce their footprint and simultaneously keep their products of optimum quality and cost in control (Kishita et al. 2021). It serves as a structured production plan but also has the potential to optimize operations and reduce energy. Methods like lean manufacturing and just-in-time (JIT) production manage energy consumption only when it is demanded, and they produce with high efficiency and speed (Tezer and Bodur 2021). This reduces wasteful energy use and carbon release, hence becoming a more environmentally responsible production system. Minimization of waste is perhaps the most significant element of sustainable production planning. Solutions for minimizing waste and environmental pollution should form a part of planning. Companies reduce and manage their waste by applying techniques such as recycling, repurposing, and cleaner production approaches [1]. Waste minimization not only helps organizations reduce their waste and their ecological footprint but also meets and boosts productivity and reduces costs, simultaneously (Baskar et al. 2022).

Production planning also needs to consider the whole supply chain. It is also essential to ensure that suppliers are aligned with the company's sustainability goals. This can be in terms of the suppliers who source materials that are sustainable, who conduct business in an ethical manner, or who have taken steps to reduce their environmental impact. By integrating sustainability into the supply chain, companies are working toward making their eco-design activities extend beyond the boundaries of manufacturing (Faud-Luke 2009). Effective production planning involves thinking about a product's end-of-life. Producing products that are easy to disassemble, recycle, or repurpose minimizes their ecological footprint at the end of their life cycle. These factors need to be incorporated into the manufacturing process and, thereby, can help organizations minimize the environmental footprint of their products long after they have left the factory gates.

8.2 *Developing a Sustainable Production Plan*

Developing a production strategy that focuses on sustainability calls for an analysis of the available manufacturing technologies to begin with. Sustainable manufacturing is the manner in which goods are manufactured to avoid harmful effects on the environment yet remain economically viable, consuming fewer amounts of energy and natural resources. This may incorporate lean manufacturing strategies in the belief of waste minimization, as well as cleaner production strategies toward minimization of resource use and emissions (Kishita et al. 2021). This is through integration of renewable energy, meaning the carbon tailpipe of production can be reduced a lot. For instance, corporations can put up solar panels or wind energy in order to provide energy for their facilities, and thereby decrease their reliance on fossil fuels. A good production plan should be equipped with measurable environmental performance and metrics that support continuous sustainability improvements.

8.3 *Ensuring Suppliers Align with Sustainability Goals*

The selection of dedicated sustainable suppliers through sustainable production planning is of great importance. Indeed, many experts have widely suggested that working with suppliers could bring in new innovative sustainable materials and processes. In addition, the companies must have selection criteria for suppliers, focused at a first stage on sustainability, which should be based on actual certifications (e.g., ISO 14001), evidence of robust environmental management. Routine auditing of suppliers would keep compliance with metrics on sustainability and ensure constant improvement (Kishita et al. 2021).

8.4 *Marketing Strategy*

A good marketing plan is the foundation for a successful transfer of the eco-friendly attributes of a product to the consumer. Branding can now focus on sustainability as a way of gaining competitive advantage.

8.4.1 Creating a Marketing Plan That Highlights Eco-Friendly Features

Creating a marketing plan that highlights eco-friendly features and effective marketing campaigns, needs to highlight the unique environmental attributes of the product, which include the use of recyclable materials, saving energy, and an environmentally friendly production process. As Kotler and Keller put it, "clear communication regarding sustainability" needs to be done to attract consumers who prefer the product over its alternatives.

The storytelling technique can help consumers connect emotionally to the product because it enhances a marketing strategy about the product being developed. By writing stories relating to their sustainability material supply chain and its repercussions in the immediate local community, the brands increase brand loyalty toward the consumers because of the cause they support; this is much more than signaling the product being environmentally friendly.

8.4.2 Educating Consumers on Sustainable Design Benefits

The most significant outcomes of a consumer education program on sustainable design and the product lifecycle are awareness and informed purchasing. For many consumers, there is an awareness gap concerning the environmental impact of their decisions, as identified by Peattie and Peattie (2009). Clear and understandable information regarding the benefits of sustainable products—such as low energy consumption and waste reduction—empowers consumers to make better decisions (Tezer and Bodur 2021). There must be multiple learning tools used in the corporation. The corporation is provided with the right platform through which sustainable design practices and a product life cycle can be communicated. Workshops and webinars can be done in better involving the consumer and breeding the community to become environmentally aware (Kishita et al. 2021).

8.4.3 Product Launch

The product launch in the eco-design process has many lead roles to play and is not just the release of a new product—it is also a platform to present the associated sustainable design embedded in the design. During this stage it is possible to highlight the deliberate decisions on selection of materials, the supply chain, and the whole product life cycle intended to reduce their impact on the environment. These sustainability features can be demonstrated by companies through successful advertising about their sustainability efforts to environmentally aware customers. This is an opportunity to tell the story of the product and how its design can contribute to achieving wider ecological benefits, such as decreasing emissions of carbon, decreasing waste, and encouraging recycling (Faud-Luke 2009). Aside from the benefits of connecting with customers in bigger, more meaningful ways, product launch also becomes an opportunity for discussions on the value of sustainable living. Such a link increases brand loyalty and makes the product appear as a responsible selection in a market saturated by competition. Executed to perfection, the launch can create trust, enhance positive brand image, and support growth that is happening along with sustainable consumerism.

8.4.4 Introducing the Product to the Market

The importance of revealing the eco-design attributes of a new product to the target consumer group is critical. This can be achieved through the use of different types of marketing strategies, for example, ad campaigns based on sustainability. Brands will be looking to emphasize sustainable features in their communications to facilitate consumers' better appreciation of environmental value in the purchases they make (Kishita et al. 2021).

Besides the general marketing strategies, social media can be used by the product while promotion involving influential people can contribute to the successfulness of it. Product launches that highlight product lifecycles from conception toward production can draw customers and eventually build a community for the brand.

8.4.5 Monitoring Consumer Responses and Gathering Feedback

It is crucial that when the product is launched, continuous collection of feedback from customers is carried out to assess the effectiveness of the product and point out places for further improvement. Surveys, social media, and sales data collecting bring a useful information about how the customer sees the sustainability features of the product.

Analysis of such feedback could allow companies to modify their marketing strategies and even design their products accordingly to fit consumer needs. Improvement is sustainable if driven by the insight from consumers; therefore, such a mechanism would not only assure innovation in products but also prove to the customer that the company stands for sustainability and development (Nuryakin and Maryati 2020).

9 Post-Launch Review

9.1 Performance Monitoring

One of the critical steps of the eco-design process is performance evaluation, which assesses the effectiveness of green innovative solutions as well as the product's contribution to the market and environment. Continuous assessment ensures that corporations are kept up to speed with their environmental goals, thus in a position to respond to shifting market demands.

9.2 Tracking Product Performance and Environmental Impact

For monitoring performance effectively, the enterprises should determine the key performance indicators reflecting the market success and the environmental ones. This would include sales volume, market share, customer interaction metrics, and environmental metrics, such as carbon footprint, resource use, and waste production. LCA-type tools are useful in assessing what the environmental impact of a product is from start to finish in its complete lifecycle. In this approach, what one knows about where to look can supply businesses with what they require in order to make intelligent decisions which will ensure their sustainability. Also, performance monitoring can expose trends in consumer wants and needs that can be used to modify products so they stay in step with market demand.

9.3 Collecting Customer Satisfaction Data and Tracking Sustainability Claims

Another important thing is to gather information concerning the satisfaction of customers so that the degree to which product based on consumer expectation satisfies the sustainability attribute of the product can be established. Some of the approaches that may be used are surveying, focus groups, and feedback from social media, which can help gather considerable viewpoints of the customer on products (Kishita et al. 2021).

Besides, sustainability claims shall be monitored and thus verified of their actuality in the light of transparency. Companies shall be careful enough to have their marketing communication truthful with proper portrayal of product performance and environmental footprint. Transparency in reporting creates the basis for trust among consumers, which is the need of the hour in this marketplace where ethical consumption is gaining momentum (Tezer and Bodur 2021).

9.4 Feedback Loop

One of the major needs for continuous advancement of eco-design projects is an active feedback loop, which involves continuous gathering of insights from customers, suppliers, and internal staff.

9.5 Creating a System for Ongoing Feedback

An organized feedback mechanism can be created to support product development and innovation. Periodic questionnaires, testing with actual users, and discussions through forums or public community meetings may be part of the system. Companies can identify their needs, wants, and proclivities by creating an environment for open customer communication so that this will be considered in product directions for the future. Involving suppliers and other stakeholders in providing input allows for an entire view of product performance and sustainability. Platforms can be designed and implemented to allow the collaborative exchange of ideas, thereby helping to contribute to innovation in more sustainable ways (Nuryakin and Maryati 2020).

9.6 Using Insights to Guide Future Eco-Design Projects

To drive current and future projects of eco-design, outputs should be understood from the feedback loop. Trend identification in customer behavior together with opportunities for improving is the one that helps an organization alter its strategies and improve existing products in a manner that better aids them achieve sustainability aims.

This is an iterative process that improves product design and underpins the sustainability commitment of the company. Continuous improvements on feedback from the user can promote innovation and build brand loyalty in green-conscious consumers (Chan 2011; Khan et al. 2023).

9.7 Addressing End-of-Life Solutions

The end-of-life phase of a product is very fundamental, especially for eco-design. It affects sustainability directly. Proper options of the end-of-pipe can minimize waste and make recovery much easier.

9.8 Strategies for Product Take-Back or Recycling Programs

One effective strategy is the product return programs where consumers can be educated to return their used items to recycling centers or properly disposed. These activities consummate the product life cycle as well as minimize environmental degradation that waste poses to the environment (Haq et al. 2016). Take-back schemes have already been employed successfully by companies like Dell and HP with the added benefit of enhancing recycling and encouraging sound consumption habits leading to customer loyalty (Tezer and Bodur 2021).

As a result of collaboration with such recycling NGOs to enhance effective return flows with the recovery of materials and then reuse them, new products get ready to be fabricated, which forms a circular economy (Keivanpour 2023).

9.9 Exploring Refurbishment and Remanufacturing Options

Refurbishment and remanufacturing offer additional opportunities for sustainable end-of-life solutions. Restoration and improvement of product life are termed as rebuilding, while remanufacturing refers to the dismantling of a product for recovery of its parts and materials. These methods can reduce waste and resource consumption drastically by lessening demands on new materials. For example, Companies like Caterpillar have integrated remanufacturing into the business model and this has resulted in significant economic and environmental benefits. By making these decisions, companies can enhance their sustainability and live up to the principles of a circular economy by keeping their products functional for a longer period (Fukushige et al. 2025). This process is iterative and incorporates several stakeholders for the development of the product. When this paradigm is applied in a systematic approach, products conceptualized and engineered by designers and organizations can be made to satisfy consumer wants as well as facilitate environmental sustainability. This approach indicates the need to embed sustainability in all stages ranging from the concept generation to the post-launch analysis and encourages the lifestyle of continuing design evolution.

10 Advantages and Disadvantages of Eco-Design

10.1 Advantages

10.1.1 Environmental Benefits

Eco-design relates to resource optimization, energy saving, and waste reduction (Haq et al. 2016). It assesses a product's environmental impact at every phase from acquiring raw materials to dispose or recycling of the product. For instance, beverage companies such as Coca-Cola have begun supplying bottles of plant-based material and more lightweight bottle designs, both of which have the advantage to reduce plastic waste and also reduce carbon emissions. Similarly, LED lamps use less power compared to normal lamps and thus reduce greenhouse gas emission. Adopting eco-design techniques, companies address global warming, pollution, and consumption of natural resources. Over the years, these resultant implications enhance world sustainability and heal the earth to a better health.

10.1.2 Market Differentiation

With a state of green society, observing eco-design creates an edge over the other organizations. In today's time, green products appeal to environmental-friendly consumers (de Figueiredo and Guillén 2011). For instance, outdoor-apparel brand Patagonia has expanded with the help of sustainable materials and responsible practices to appeal to eco-conscious customers. Specifically, in terms of environmental concern, an investment in renewable energy and recyclable material has enhanced IKEA's reputation among environment-conscious consumers (IKEA Sustainability Report 2022). This can help businesses enhance their brand image, win customer loyalty, and expand in new market niches by making a successful adoption of eco-design principles to their business practices.

10.1.3 Regulatory Compliance

As governments enforce stricter environmental regulations, eco-design helps businesses stay ahead of these changes. This approach reduces the risk of fines or penalties while ensuring compliance with legal standards. For example, the European Union's Eco-design Directive requires energy-related products to meet specific efficiency and sustainability criteria. Similarly, the Restriction of Hazardous Substances (RoHS) directive compels electronics manufacturers to limit the use of harmful materials like lead and mercury. By aligning with these regulations, companies avoid legal issues and strengthen their credibility with regulators and stakeholders.

10.1.4 Cost Savings

Although there may be an up-front cost to eco-design, long-term savings are highly probable. Savings through enhanced use of resources and energy efficiency savings lead to overall savings in cost. For instance, Apple reduced the size and weight of its packing, therefore reducing transport costs and environmental impact for the company. For instance, Unilever's water-saving production lines enable it to save on operating costs. In the long term, such savings can be plowed into cutting-edge projects or passed on to customers, giving companies a competitive advantage and driving their business to sustainability.

10.1.5 Innovation

Eco-design encourages innovation and innovative thinking, and pushes companies to open up new material, new techniques, and new technologies, toward achieving the best possible solutions (Nuryakin and Maryati 2020). It is often found that focusing on sustainability helps companies make excellent outcomes. For example, electric vehicles helped Tesla change the contours of the automobile sector by reducing dependence on fossil fuels (Maradin et al. 2022). Similarly, Danimer Scientific is engineering biodegradable polymers from plant-based raw materials to mitigate global issues of waste. While accomplishing this focus on eco-design both environments are served and technologies such as new market opportunities are created and the overall value of products is improved.

10.2 Disadvantages of Eco-Design

10.2.1 Initial Costs

Eco-design typically involves a high initial investment. Companies will likely have to invest in research, green materials, and changes in the manufacturing process. For example, organic fibers and biodegradable plastics are typically more expensive than their traditional counterparts. Companies may also need specialized equipment or employee training to carry out sustainable activities. Such expenses could be too daunting for smaller companies with fewer financial resources.

10.2.2 Complexity

Eco-design can be a complex flow, requiring the skill of specialists in fields like engineering and environmental science. It involves measurement of the complete life cycle of a product, starting from procurement of raw materials until it is discarded. This means very careful prioritization and coordination is required, though this can be challenging for organizations if they don't have in-house sustainability professionals (Pati 2023). This can even cause the product to miss the market because of longer development timescales due to increased complexity.

10.2.3 Market Acceptance

Not all consumers are ready to embrace eco-designed products. Many people prefer to spend less or save time rather than being environmentally friendly. For instance, reusable bags, although environmentally friendly, are often offered at a discount in favor of single-use plastic bags because they are cheaper or easier to use. Similarly, eco-friendly products are, generally, more expensive, which reduces their price competitiveness compared to cheaper alternatives. Getting consumers to choose sustainability is still a significant challenge (Foschi et al. 2020).

10.2.4 Limited Availability of Materials

Finding sustainably sourced materials is often challenging. Such materials cannot be found in the market easily, especially in certain regions of the world. In such a case, costs will be increased, and design possibilities will be limited too. For example, bamboo is widely available but poses a supply chain or quality problem. These obstacles may not allow for the creation of products with functional and aesthetic requirements that are combined with sustainability (Chen et al. 2023).

10.2.5 Performance Trade-offs

There is a possibility that eco-designed products can at times perform less than conventional counterparts, thus leading to customer dissatisfaction. For instance, biodegradable plastics may not be as tough as ordinary plastics, thus limiting their uses (Foschi et al. 2020). Products from recycled materials may also not be as durable or have the same finish as their new counterparts. Ideally, trade-offs may discourage consumers, especially during performance-critical industries.

11 The Importance of Eco-Design for the Future

11.1 Environmental Sustainability

Eco-design favors the inclusion of renewable materials, as well as reduction of waste, which can assist in preserving ecosystems, and thereby decrease environmental impact. It supports the development of a circular economy through recycling and reuse of materials. For example, Adidas has made shoes out of plastic waste recovered from the oceans, thereby converting waste into a product. This method not only saves resources but also reduces the impacts on marine ecosystems, so showing concrete environmental advantages of eco-design (Fukushige et al. 2025).

11.2 Climate Change Mitigation

Eco-design is one of the factors through which climate change can be reduced with minimized greenhouse gas emissions. For instance, companies like Tesla produce electric cars that significantly reduce the carbon emissions produced by such cars compared to the normal gasoline-burning automobiles (Maradin et al. 2022). Similarly, energy-saving appliances from companies like LG and Samsung minimize energy usage, thus being active participants in complying with the global action against climate change. Through these innovations, products not only meet the demands of consumers but also respond profoundly to critical environmental problems.

11.3 Economic Benefits

Although adopting eco-design may be an investment in the short run, it provides considerable economic benefits in the long run. Sustainable practices usually reduce production costs and increase efficiency. For instance, IKEA's focus on recyclable and renewable materials usage in the collection of furniture products allows it to cut costs on expenses while also winning people with an ethical agenda (IKEA Sustainability Report 2022). Again, Unilever has introduced water-saving technology into its production process, thus saving on costs and increasing resource utilization. Examples as above show that eco-design begets new market opportunities and boosts economic activities.

11.4 Regulatory Compliance

Eco-design helps organizations to stay alive with the increasing environmental legislation and thus reduces legal risk potential and thereby improves the stakeholder's view of the organization (Subramanian et al. 2023). The Ecodesign Directive of the European Union requires to meet both energy efficiency and environmental requirements which promotes innovation among companies while keeping them strictly under control of the regulation (Nuryakin and Maryati 2020). Philips, for example, has already developed energy-saving lights that meet these standards and is a forerunner in the sustainable design field. Anticipatory implementation of eco-design will enable companies to avoid and escalate the cost of environmental sanctions and improve environmental market position (Wimmer et al. 2004).

11.5 Social Responsibility

It is highly encouraged by a very eco-design-commitment that a company's image becomes good and it takes social and ethical responsibility. This is well realized in Patagonia, since it specializes on sustainable apparel with waste reduction initiative to extend products' life so as to establish sustainable use of them. Since this is its affinity for sustainability design, they align with customers who are green-conscious, establishing trust and will eventually lead the long-term loyal customers (Kishita et al. 2021). Focused on sustainable companies end up fulfilling more than social responsibility; they ensure a competitive edge in the market due to sustainable-conscious customers.

11.6 Innovation and Competitiveness

Eco-design inspires creativity by driving companies to seek new materials, technologies, and processes. For instance, companies like Danimer Scientific have developed biodegradable packaging options that help meet global demand for substitutes to single-use plastics (Foschi et al. 2020). These inventions, besides environmental benefits, also help businesses change market dynamics and maintain profitable operation in the long run (Nuryakin and Maryati 2020).

11.7 Global Responsibility

Eco-design contributes toward the attainment of global sustainability goals, such as the United Nations Sustainable Development Goals. That is, by putting responsibility into production and consumption processes. For example, Nestlé has applied eco-design for packaging to increase waste reduction and recyclability, which positively impacts the environment. By adopting the concept of eco-design, businesses contribute to the international efforts to address environmental problems and produce common goods for society and nature (Tezer and Bodur 2021).

Case Study-1

IKEA's "Sofa for Life" Initiative

Background: In the furniture retail market, IKEA has now introduced the "Sofa for Life" (SFL) concept to design a sustainable sofa featuring comfort, durability, and eco-friendliness. This project aims to increase waste reduction and increase circular economy by the use of renewable and recyclable materials and by designs focussing on durability.

Key Features

Materials: The green sofa is designed to be sustainable right from the outset. It is constructed using sustainably sourced wood in such a way that wood derived is responsibly forested to take care of environmental protection. Further, fabric material is recycled polyester, thus preventing waste and cutting down the dependency on virgin material. Through renewable and recycled material, the sofa reduces its environment impact without reducing durability or aesthetic appeal.

Modular Structure: The sofa is designed with a modular structure, so different parts can be easily replaced or repaired. This design not only extends the life of the product but also follows circular economy approaches by promoting maintenance instead of discarding the product. Regardless of whether a change to cushion cover or structural repair is required, the modular design allows users to repair either of these components without replacing the entire sofa. This method saves waste, and it is a relatively low-cost option for the consumer and has a benefit in terms of sustainability.

Sustainable Disposal Options: To further sharpen its vision of sustainability, IKEA provides a take-back scheme as the end-of-life solution for its products. At the end of their service life, customers can send the sofa to IKEA, and it will be disposed of appropriately. The returned sofas are recycled, materials are salvaged for future use, or repurposed in new products. This tool ensures that the life cycle of sofas matches environmental goals by not depositing waste into landfills, but instead by promoting a circular economy (IKEA Sustainability Report 2022).

Energy Efficiency: It has an energy efficiency-based production process. By utilizing renewable resources (as in, wind and sun power), IKEA significantly reduces its carbon footprint from the manufacturing process (IKEA Sustainability Report 2022). This dedication to clean energy not only reduces the environmental footprint of the sofa but also sets a standard for sustainable production processes in furniture making (Çınar 2005).

The "Sofa for Life" initiative exemplifies eco-design in action, illustrating how a major corporation can integrate sustainability into its product offerings. Analyzing this case study helps stakeholders understand the complexities and benefits of eco-design, as well as the strategies required for successful implementation.

Questions for Discussion:

1. How does the use of recycled materials in the "Sofa for Life" initiative help reduce the product's overall environmental footprint? What challenges might IKEA encounter in sourcing sustainable materials?
2. In what ways could consumer perceptions of sustainability affect their purchasing decisions regarding the "Sofa for Life"?
3. What specific design strategies are used in the "Sofa for Life" to ensure it is modular and repairable?
4. What competitive advantages could IKEA gain from leading in eco-design initiatives?
5. What limitations might the "Sofa for Life" face regarding scalability and market reach?

Case Study 2

Implementing Life Cycle Assessment (LCA) and Sustainable Practices in the Fashion Industry

Background

Software packages have been designed and commercially merchandised to help companies perform Life Cycle Assessment. Such tools can be applied by companies to estimate a manufacturing process's environmental footprint, providing important information as to where investments may be directed toward developing sustainability. That is the case now, when a lot of companies use these resources to learn about what the ecological impact of their activities have on us and put the sustainability in practice during manufacturing. LCA is an important tool for facilitating Eco Design through the provision of the information that designers are able to use to make informed, sustainable design decisions early in the product development process (Cimatti et al. 2017).

The Evolution of Sustainable Manufacturing

The 3R principle, which has been the basis for eco-based manufacturing processes for a long time, is Reduce, Reuse, and Recycle. However, this paradigm has been transformed by the 6R extension to include Reduce, Reuse, Recover, Redesign, Remanufacture, and Recycle. This extended framework offers an integrative perspective on sustainability, enabling moving from a linear, one-way life cycle to a closed loop, multi-life cycle system. The 6R framework assures that in each step of the production and consumption, sustainability is introduced, making way for a cyclic economy in which resources are continuously recuperated.

Recycling in Fashion Production

Recycling is a simple option to enhance the sustainability of fashion production (Cimatti et al. 2017). Recycling refers to the procedure in which materials that would normally be thrown away are collected to produce new materials. Waste, or more specifically, surplus fabrics, leather, and other materials, creates a specific problem that piles up in the fashion industry (Liscio and Sospiro 2023). Through recycling, the materials can be transformed into new clothes, thus reducing the dependency on virgin material and subsequently the environmental footprint. This way, the use of resources and reduction of waste makes this practice possible in an environmentally friendly fashion industry (Cimatti et al. 2017).

Social Perception and Consumer Demand

This aspect of sustainability in fashion covers social issues as well, with the way customers perceive sustainable practices as the most prominent. Consumer demand and purchasing habits are critical components that determine the effectiveness of recycling or other environmentally conscious practices (Liscio and Sospiro 2023). This is more particularly the case for the luxury fashion market, whose purchasing decision drivers are typically nonproduct-related, meaning they involve perceptions of brand, status, and exclusivity rather than the intrinsic value or utility of the product. Thus, for these houses to continue doing well in the market, they have to find a way to fit their approaches toward sustainability in tandem with consumers' preferences and attitudes (Chen et al. 2023).

Eco Design in Fashion

Eco Design is a concept which takes environmental consequences of products at all its life stages into account. In the context of fashion, this includes the use of sustainable materials such as organic cotton or wool instead of synthetic fibers, and manufacturing processes that cause less harm to the environment (Cimatti et al. 2017). For example, if toxic chemicals can be replaced by natural compounds, then the environmental impact of fashion articles can be significantly reduced (Liscio and Sospiro 2023). Applying Eco Design approaches, manufacturers can propose products which are both green and attractive to environmentally conscious consumers.

Life Cycle Assessment (LCA) as a Tool for Sustainability

LCA is an essential method for assessing the environmental impacts and resource consumption associated with a product throughout its entire life cycle, from the extraction of raw materials to disposal. Through assessment along every step of the product life cycle, LCA facilitates companies with scope for development to optimize the productive operations, taking informed decisions to enhance its sustainability. It is due to such reasons many software packages have also been developed supporting companies to work through LCA for better appreciation of and for diminishing the load that production generates upon the environment. These instruments are widely utilized in various industries, including the fashion industry, to track indicators of sustainability and make strategic decisions (Liscio and Sospiro 2023).

Conclusion

Using the 6R concept in coordination with LCA tools can be considered one of the significant steps toward attaining more sustainable practice, especially in an area where both environmental and social impacts need to be taken into consideration, like the fashion industry (Liscio and Sospiro 2023). Businesses can reduce their environmental footprint while fulfilling consumer needs by using sustainable design, manufacturing, and consumption methods. There is the role of that concept where brand image and prestige play a major role in consumer's purchase decisions, which is especially pronounced in the luxury fashion industry (Cimatti et al. 2017). If the fashion businesses apply carefully the Eco Design and sustainability methods, it is possible to reduce their environmental impact and thereby have followers among the growing number of eco-aware customers.

Questions

1. How can Life Cycle Assessment (LCA) influence decision-making in fashion production to promote sustainability?
2. Why is it important for luxury fashion brands to align their sustainability practices with consumer desires and perception.
3. How do the principles of Eco Design and LCA work together to reduce the environmental impact of fashion products?

References

Ashby M, Johnson K (2002). Materials and design: The art and science of material selection in product design.

Baskar C, Ramakrishna S, Baskar S, Sharma R, Chinnappan A, Sehrawat R (2022) Handbook of solid waste management: sustainability through circular economy. Springer. https://doi.org/10.1007/978-981-16-4230-2

Baumann H, Tillman AM (2004) Introduction to life cycle assessment. Kluwer Academic Publishers

Chan HK (2011) Green process and product design in practice. Procedia Soc Behav Sci 25:398–402. https://doi.org/10.1016/j.sbspro.2012.02.050

Chekima B, Gbadamosi A (2016) Consumer values and green products consumption in Malaysia: a structural equation modelling approach. In: Handbook of research on consumerism and buying behavior in developing nations. IGI Global

Chen X, Rahman MK, Rana MS, Gazi MAI, Rahaman MA, Nawi NC (2023) Predicting consumer green product purchase attitudes and behavioral intention during COVID-19 pandemic. Front Psychol 12:760051. https://doi.org/10.3389/fpsyg.2021.760051

Chen X, Chen F, Yang Q, Gong W, Wang J, Li Y, Wang G (2023) An environmental food packaging material part I: a case study of life-cycle assessment (LCA) for bamboo fiber environmental tableware. Ind Crop Prod 194:116279. https://doi.org/10.1016/j.indcrop.2023.116279

Choudhury IA, Hashmi MSJ (2020) Encyclopedia of renewable and sustainable materials. Elsevier

Cimatti B, Campana G, Carluccio L (2017) Eco design and sustainable manufacturing in fashion: a case study in the luxury personal accessories industry. Procedia Manuf 8:393–400. https://doi.org/10.1016/j.promfg.2017.02.050

Çınar H (2005) Eco-design and furniture: Environmental impacts of wood-based panels, surface and edge finishes. For Prod J, 55(11):27–33

de Figueiredo JN, Guillén MF (2011) Green products: perspectives on innovation and adoption. CRC Press. https://doi.org/10.1201/b12784

Faud-Luke A (2009) Eco-design: the sourcebook. Thames & Hudson

Foschi E, Zanni S, Bonoli A (2020) Combining eco-design and LCA as decision-making process to prevent plastics in packaging application. Sustainability (Switzerland) 12(22):9738. https://doi.org/10.3390/su12229738

Fukushige S et al (2025) EcoDesign for circular value creation. Springer Nature, Singapore

Haq MA, Khan NR, Parkash R, Jabeen A (2016) Impact of JIT, waste minimization, and flow management on operational performance of manufacturing companies. Qual-Access Success 17(153)

IKEA Sustainability Report 2022: IKEA Group (2022) People & planet positive: IKEA sustainability report. IKEA Sustainability

Keivanpour S (2023) Circular economy in engineering design and production: concepts, methods, and applications. Synthesis lectures on sustainable development. Springer, Cham. https://doi.org/10.1007/978-3-031-44652-8

Khan MR, Khan NRK, Tufail M, Ali L (2023) A path towards a greener future: fostering green supply chain, green marketing, and environmental sustainability. LogForum 19(1):1–14

Kishita Y, Matsumoto M, Inoue M, Fukushige S (eds) (2021) EcoDesign and sustainability II: social perspectives and sustainability assessment. Sustainable production, life cycle engineering and management. Springer, Singapore. https://doi.org/10.1007/978-981-15-6775-9

Klüppel HJ (2005) The revision of ISO standards 14040-3 – ISO 14040: environmental management—life cycle assessment—principles and framework – ISO 14044: environmental management—life cycle assessment—requirements and guidelines. Int J Life Cycle Assess 10:165

Lewandowska A, Matuszak-Flejszman A (2014) Eco-design as a normative element of environmental management systems—the context of the revised ISO 14001:2015. Int J Life Cycle Assess 19:1794–1798. https://doi.org/10.1007/s11367-014-0787-1

Liscio MC, Sospiro P (2023) Life cycle assessment on fashion industry: four case studies. International Conference on Environmental Science and Technology Proceedings, 18. https://doi.org/10.30955/gnc2023.00283

Maradin D, Malnar A, Kaštelan A (2022) Sustainable and clean energy: the case of tesla company. J Econ Financ Manag Stud 05(12):3531–3542. https://doi.org/10.47191/jefms/v5-i12-10

McDonough W, Braungart M (2002) Cradle to cradle: remaking the way we make things. North Point Press

Nematchoua MK, Orosa JA, Reiter S (2022) Life cycle assessment of waste products of a green-neighbourhood. Waste Biomass Valori 13(8):3639–3651. https://doi.org/10.1007/s12649-022-01749-2

Ng CY, Tang WC (2022) Evaluation of design options for green product development: a combined cuckoo search and life cycle assessment approach. Int J Life Cycle Assess 27(5):665–679. https://doi.org/10.1007/s11367-022-02056-7

Nuryakin N, Maryati T (2020) Green product competitiveness and green product success. Why and how does mediating affect green innovation performance? Entrep Sustain Issues 7(4):3061. https://doi.org/10.9770/jesi.2020.7.4(33)

Pati SN (2023) Life cycle assessment: future challenges. Taylor & Francis

Peattie K, Peattie S (2009) Social marketing: A pathway to consumption reduction? J Bus Res, 62(2):260–268. https://doi.org/10.1016/j.jbusres.2008.01.033

Sayal A, Pant M (2023) Examining the moderating effect of green product knowledge on green product advertising and green product purchase intention: a study using SmartPLS SEM approach. Int J Asian Bus Inf Manag 13(1):1–16. https://doi.org/10.4018/IJABIM.316175

Subramanian V, Peijnenburg WJGM, Vijver MG, Blanco CF, Cucurachi S, Guinée JB (2023) Approaches to implement safe by design in early product design through combining risk assessment and Life Cycle Assessment. Chemosphere, 311(Part 1), 137080. https://doi.org/10.1016/j.chemosphere.2022.137080

Tezer A, Bodur HO (2021) The green consumption effect: how using green products improves consumption experience. J Consum Res 47(1):25–39. https://doi.org/10.1093/JCR/UCZ045

Wimmer W, Züst R, Lee K-M (2004) ECODESIGN implementation: a systematic guidance on integrating environmental considerations into product development, Alliance for global sustainability bookseries. Springer, Dordrecht. https://doi.org/10.1007/978-1-4020-3071-0

Sustainable Resilient Strategies to Tackle Disruptions in the Era of Industry 5.0

Mohammad Yavari

Abstract The emergence and evolution of Industry 5.0 has coincided with the increase in disruptions. These disruptions can be caused by human actions, such as terror attacks and labor strikes, or by natural events like floods and pandemics. In recent years, the COVID-19 pandemic has revealed major weaknesses in production systems and supply chains in the presence of disruptions. The occurrence of disruptions not only affects the economic aspect of the supply chain but also has negative impacts on the environmental and social dimensions of sustainability. To deal with disruptions, several traditional resilience strategies have been developed and used in different industries. Some of these resilience strategies have sustainability implications. Furthermore, focusing solely on sustainability makes production systems and supply chains more vulnerable to disruptions. Therefore, the interplay between sustainability and resilience must be properly addressed. New technologies of Industry 4.0 and its subsequent evolution into Industry 5.0 have provided significant potential to contribute to the resilience and sustainability of supply chains simultaneously. The pillars of Industry 5.0 rely on increased resilience, sustainability, and a human-centric approach. In this regard, this chapter first explores various aspects of resilience and sustainability issues in manufacturing supply chains. We will review the evolution of sustainability, proactive, and reactive resilience strategies, and the main approaches to balancing resilience and sustainability. Next, this chapter develops a framework for using the capabilities of new technologies such as blockchain to develop new resilience strategies, on the one hand, and to foster human-machine collaboration, on the other hand, to create a dynamic real-time optimization decision-making environment to ensure the sustainability and resilience of manufacturing supply chains. The proposed framework uses real-time disruption probability data to continuously update decisions and resilience strategies and addresses disruptions in a dynamic environment where the disruption probability of facilities and their interconnected links changes over time.

M. Yavari (✉)
Department of Quantitative Studies, University Canada West, Vancouver, BC, Canada
e-mail: mohammad.yavari@ucanwest.ca

M. Venkatesh et al. (eds.), *Green Manufacturing Technologies in Industry 5.0*,
https://doi.org/10.1007/978-981-95-0363-6_5

1 Introduction

Global supply chains, despite their benefits, have made manufacturing and logistics more vulnerable to disruptions. These disruptions can arise from man-made events and natural disasters. Major events such as natural disasters, geopolitical tensions, and the COVID-19 pandemic have increased the frequency of disruptions in manufacturing supply chains. Covid-19, in particular, showed major shortcomings in supply chains and manufacturing systems (Khan et al. 2024a). Natural disasters have also contributed to the increase in disruptions. For example, in 2020, the Atlantic hurricane season set a record with 30 named storms, causing over $20 billion in damages. Such events are becoming more frequent and severe due to climate change, further compounding supply chain vulnerabilities. More recently, in January 2025, California experienced the costliest wildfire in its modern history, which is also likely the most expensive wildfire in U.S. history. A preliminary estimate from AccuWeather puts the damage and economic losses at between $135 billion and $150 billion.

Being sustainable is crucial for manufacturing systems while remaining resilient against disruptions. Resilience and sustainability are interconnected, with both having mutual impacts on each other (Yavari et al. 2024). Today, they are viewed as complementary, with businesses focused on creating supply chains that are adaptable to disruptions while minimizing environmental impact and promoting social equity (Crane et al. 2014). While scholars have proposed various resilient strategies to cope with disruptions, there are notable shortcomings, such as potential negative impacts on sustainability in certain cases and limited effectiveness in handling severe and frequent disruptions.

New digital technologies of the Industry 4.0 revolution have provided new capabilities to enhance the performance of the manufacturing supply chains particularly under disruptions. Industry 4.0 emphasizes the interconnection of physical systems and digital systems. It realizes a high degree of automation and intelligence in manufacturing through technologies such as digital twins, the Internet of Things, and data analysis (Yang et al. 2024). In this context, some existing resilient strategies have been adjusted to leverage new capabilities, while new strategies have been developed using technologies such as blockchain, digital twins, and AI. One of the key questions surrounding Industry 4.0 is the role of humans in this evolving landscape. Future manufacturing systems should make a proper human–machine collaboration to ensure that the manufacturing system will get the most out of resources to make a resilient-sustainable manufacturing system. Industry 5.0, however, takes the smart factory a step further and emphasizes close cooperation between humans and machines, embracing human-centric, resilience, and sustainability as core values (Xun et al. 2021; Rame et al. 2024).

Industry 5.0 refers to robots and smart machines working alongside people with added resilience and sustainability goals included. Where Industry 4.0 focuses on technologies such as the Internet of Things and big data, Industry 5.0 seeks to add human, environmental, and social aspects back into the equation. There is a major question. Does humanity need to be reintegrated into the equation, or can it truly help systems tackle resiliency and sustainability challenges in new ways? How do human and machine technologies complement each other through collaboration? It is crucial to properly define the roles of humans and machines to ensure their long-term coexistence in Industry 5.0.

To this end, this chapter focuses on how to design a human-centric decision-making manufacturing system using the capabilities of new technologies. This chapter presents a framework designed to enhance human–machine collaboration by leveraging real-time data to address disruptions in a dynamic environment, where disruption probabilities evolve over time. The framework has three key components. It incorporates disruptive technologies like blockchain and IoT to develop sustainable-resilient strategies, machine-based disruption predictions, and optimization models to adapt to disruptions. This real-time decision-making aims to enhance both sustainability and resilience.

2 Sustainability: From Forestry to Integration with Resilience

The word sustainability comes from the Latin verb sustinēre, "to maintain, sustain, support, endure," made from the roots sub, "up from below," and tenēre, "to hold." The German equivalent of Nachhaltigkeit was first used in the forestry book *Sylvicultura Oeconomica* by Hans Karl von Karlowitz in 1713 (Robertson 2017). In 1987, the definition of sustainability in the Brundtland Report by the World Commission on Environment and Development explicitly referred to the link between the environment, economy, and equity for the first time. In the report, sustainable development was defined as "development that meets the needs of the present without compromising the ability of future generations to meet their own needs." This highlighted the integration of economic, environmental, and social aspects of sustainability (Dabija 2021). The economic dimension of sustainability focuses on optimizing the total profit obtained by the system that is financially viable over the long term without harming the other two conflicting dimensions. The environmental dimension aims to ensure the availability of natural resources for future generations by mitigating pollution, reducing waste, promoting the circular economy, and minimizing the ecological footprint. Finally, addressing social expectations such as equality, labor rights, and so on has become an important part of sustainability. The multidimensional and complex nature of sustainability has increased the need for integration across sectors, systems thinking, and balance between sustainability pillars to ensure long-lasting positive outcomes.

Disruptions to manufacturing systems and supply chains, especially in global supply chains, have been exacerbated over the past decade by multiple factors such as the COVID-19 pandemic, natural disasters, geopolitical tensions, cyberattacks, and climate change. These frequent and severe events have prompted businesses to revise their preparedness and resilience strategies to cope with future disruptions. An interconnected global economy is more vulnerable to major disruptions like COVID-19. The COVID-19 pandemic impacted both demand sides of the supply chain, exacerbating their impacts on supply chain performance. Moreover, according to USAFACTS, of the 10 years with the most natural disasters, nine were in the last decade which shows that climate change has been increasing natural disasters. Besides, geopolitical tensions such as Brexit, tariff wars, the US-China trade war, and the Ukraine war have significantly caused disruptions that affect global trade flows with ripple effects that go beyond affecting local affected regions. Recently, by expanding the use of digital technologies, every point in the supply chain potentially is vulnerable to cyberattacks. The Ponemon Institute reported that the average cost of a data breach jumped to USD 4.88 million from USD 4.45 million in 2023, in addition to the continued increase in the number of attacks per year. In response to the increasing frequency and severity of disruption, most companies focused on risk management strategies.

Disruption can affect the manufacturing supply chain in a variety of ways, including disruptions to the flow of information, goods, and even cash flow, leading to sustainability impacts. In addition, the operation of some facilities and resources may be affected by these events. In many cases, when a disruption occurs, to quickly remediate the affected areas, some investments and budgets for sustainability are temporarily limited to focus on short-term survival rather than long-term sustainability goals. In addition, supply-side disruptions such as resource shortages, or demand-side disruptions such as overconsumption, force companies to use alternative facilities and transportation routes, which are likely to increase carbon emissions during the disruptions. Furthermore, disruptions can downplay long-term strategic planning, sideline innovation, and increase waste, as companies may prioritize quick fixes over resource optimization. To manage these challenges, resilient strategies are presenting themselves as an effective solution. To this end, a technology-based circular economy that relies on a human-centered approach can provide the holistic approach needed to ensure sustainability in the face of disruption. This approach helps design and implement resilient strategies consistent with sustainability in the face of various disruptions in the future.

New revolutionary technologies such as blockchain, IoT, smart contracts, digital twin, and AI have a great potential to upgrade traditional resilient strategies and even help to develop new ones that diminish their sustainability impacts as well. Furthermore, integration between these technologies, such as integrating smart contracts and blockchain with IoT and AI, creates synergies and simultaneously enhances the sustainability and resilience of supply chains by providing visibility, secure data sharing, and real-time data (Ipek et al. 2023; Srhir et al. 2023). Also, these integrations mitigate the risk of human error and fraud, thereby increasing operational resilience (Patidar et al. 2023).

Industry 5.0 demands an overarching data and value-driven framework encompassing production resilience, sustainability, and human-centeredness (Brauner and Ziefle 2022). This transition from Industry 4.0 to Industry 5.0 broadens the definition of sustainability in industrial practices, focusing on long-term

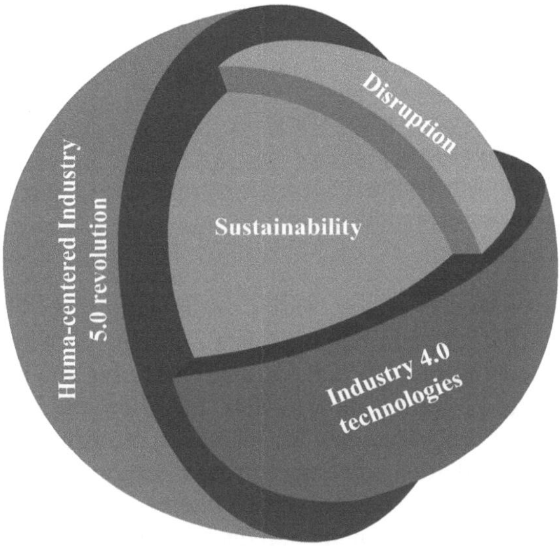

Fig. 1 Sustainability at the core of the Industry 5.0 revolution

environmental, social, and economic viability (Rame et al. 2024). The principles of Industry 5.0, sustainability, resilience, and human-centricity, align closely with sustainability goals, particularly through the use of advanced technologies like IoT, Blockchain, and AI, which can significantly contribute to achieving sustainable development goals and fostering a greener, more equitable world (Rame et al. 2024). Hence sustainability is at the core of Industry 5.0 as shown in Fig. 1. As we move further into the era of Industry 5.0, the convergence of sustainability and resilience will be key to creating manufacturing systems that are both adaptive and sustainable, capable of thriving in an uncertain, rapidly changing world.

3 Resilient Strategies to Cope with Disruptions

Manufacturing supply chains can adopt three key strategies in response to disruptions. First, they can adapt their decision-making processes by considering various scenarios to effectively manage disruptions. Second, resilience can be proactively built into the supply chain to anticipate and prepare for unforeseen events. Finally, recovery strategies can be integrated into network designs to ensure a swift return to normal operations following disruptions. As noted, adapting decision-making processes by considering various scenarios is an effective way to consider disruptions. However, the main challenge with this strategy lies in the number of scenarios, which is equal to 2^n, where n refers to the number of potentially affected links and facilities. This number of scenarios can grow rapidly, even with a limited number of links and facilities at risk of disruption. Hence, solving optimization models based on this strategy becomes difficult by increasing the number of scenarios. As a result, solving optimization models becomes increasingly difficult as the number of

Table 1 The most popular resilient strategies

Proactive strategies	Reactive strategies
Emergency stock/buffer inventory	Lateral transshipment
Extra capacity/capacity flexibility	Adjusting routing plans
Multiple sourcing	Direct shipment
Facility fortification	Dynamic pricing
Collaboration with suppliers	Coalition/collaboration with suppliers

scenarios increases. This issue is particularly challenging when the optimization model relies on real-time data to update decisions, as it significantly impacts both model complexity and run-time. This issue will be addressed in this chapter by using a set of optimization models that allow the incorporation of newly arrived data, particularly regarding the disruption probability of links or facilities.

In addition to scenario-based decision-making processes, both proactive and reactive strategies are crucial but they serve different roles. Proactive strategies are focused on preventing or reducing the impact of disruptions by anticipating potential risks. On the other hand, reactive strategies are about responding effectively once a crisis has already occurred. The most resilient manufacturing systems use a combination of both, staying prepared for the unexpected while being flexible enough to adjust when challenges arise. The most popular proactive and reactive strategies are summarized in Table 1.

Emergency stock, extra capacity, multiple sourcing, and facility fortification are all proactive resilient strategies that adapt systems to disruptions and uncertainties. By maintaining emergency stock at vulnerable facilities throughout the supply chain, the system can continue to meet demand even during severe disruptions. Additionally, extra capacity at select facilities prepares the supply chain for demand fluctuations or disruptions in the flow of products from other parts of the supply chain. Furthermore, the risk of relying on a single supplier can be reduced by securing multiple sources for key components to enhance the resilience of the supply chain. Finally, by facility fortification, as a traditional proactive strategy, businesses expect to reduce the probability of disruptions and their severity in facilities.

On the other side, reactive strategies like lateral transshipment, adjusting routing plans, direct shipment, and dynamic pricing are designed to recover the system and help it to come back to its normal condition. During disruptions, the lateral transshipment between facilities, with the same echelon, allows them to balance inventory levels and helps continue the operation of the system. During disruptions, readiness for modifying some operational decisions like adjusting routing and scheduling decisions and direct shipping, through bypassing distribution channels, can help to respond to some disruptions ensuring deliveries properly. In cases of disruption, supply, demand (e.g., customer behavior and market changes), or both sides can be affected by the disruption. Dynamic pricing acts as a reactive strategy in these situations by balancing the supply and demand sides of the supply chain. To create a resilient supply chain, early research such as Yavari et al. (2020) applied dynamic pricing to manage demands along with other decisions such as routing.

Jafari-Nadouşhan et al. (2024) showed that simultaneously managing disruption risks and uncertainties can improve sustainability objectives. They used a joint pricing strategy to increase supply chain sustainability in the event of disruption. Furthermore, Alptekinoğlu et al. (2024) used two demand levers, one of which was dynamic pricing in the face of supply disruption, and compared it with a product diversification strategy. They showed that dynamic pricing may reduce the required level of safety stock.

Furthermore, building long-term relationships and collaborations with suppliers enhances transparency, facilitates information sharing, promotes flexibility, and encourages joint problem-solving, leading to cost reductions, improved efficiency, increased agility, and adaptation to market changes. Thus, it serves as a resilient strategy that aids the supply chain in maintaining its sustainability, both proactively and reactively, during disruptions. For instance, as a kind of coordination with infrastructure providers, Yavari and Zaker (2019), Yavari and Zaker (2020), and Yavari and Bohreghi (2025) implemented an integrated decision-making model for a supply chain layer and a disrupted power network layer. They investigated the effectiveness of integrated decision-making along with several traditional resilient strategies. Results showed that this integration of decision-making has a great impact on keeping the environmental and economic excellence of the supply chain. Besides traditional resilient strategies, new technologies have a high impact on enhancing the resiliency of the manufacturing supply chain. New disruptive technologies like blockchain and smart contracts enable supply chains to be transparent and share secure real-time data that help the members to collaborate quickly to minimize the sustainability and operational impact of disruptions. By using these technologies all parts are accountable for the sustainability and resilience of the supply chain through real-time monitoring of related metrics.

Resilience and sustainability are mutually interrelated. Therefore, it is very important to maintain their balance. From the perspective of resilience, resilient strategies may leave their mark on sustainability, especially the environmental dimension. Therefore, resilient strategies should be in harmony with sustainability (Carissimi et al. 2021; Yavari and Ajalli 2021; Safari et al. 2024; Mirzaee et al. 2024). While expanding production capacity has profitability benefits, this additional capacity may conflict with sustainability goals. Similarly, the emergency stockpiling strategy releases carbon emissions in terms of energy consumption to hold inventory, which negatively impacts the environment (Wang et al. 2024).

Some resilience strategies may not have any direct impact on sustainability, but if not properly optimized, they will have negative impacts. For example, alternative transportation and multi-channel distribution methods may also increase carbon emissions if transportation decisions are not made optimally (Suryadi and Rau 2023). As another example, a multi-sourcing strategy can improve resiliency and sustainability while spreading risks if local suppliers are optimally selected to maximize both sustainability and supply chain resilience. Also, the alternative sources provided by the backup supplier strategy ensure the flow of products during disruptions. This strategy strengthens resilience and can also support sustainability if backup suppliers are selected based on environmental and social criteria. It is notable that the dynamic pricing strategy can support economic sustainability by optimizing revenue, through managing demand and supply during disruptions, but may not directly impact environmental or social sustainability.

In summary, the current situation has three key shortcomings: (i) the challenge of applying scenario-based decision-making to large-scale real-world problems, particularly in real-time environments during disruptions, (ii) the need to integrate new technologies into systems to effectively address disruptions, and (iii) the oversight of the human role in the process. The first issue will be addressed in this chapter by proposing a framework and using a set of optimization models that allow the incorporation of newly arrived data, particularly regarding the disruption probability of links or facilities. Furthermore, blockchain, as a key technology of Industry 4.0 and evolved further in Industry 5.0, has the potential to enhance resilience while boosting system sustainability. This chapter proposes a framework that integrates different levels of blockchain technology with traditional proactive resilience strategies (e.g., emergency stock) to enable both proactive and reactive strategies within the system.

4 Approaches to Balancing Sustainability and Resilience

Several approaches can help balance sustainability goals with the need for resilience during disruptions. The primary approaches explored thus far include Multi-Objective Optimization Models, and the integration of new disruptive technologies. This section highlights how these approaches contribute to sustainability in the face of disruptions. Notably, other approaches, such as the development of Reverse Logistics, have also been reported as useful strategies as shown in recent studies (Cheramin et al. 2021; Govindan and Gholizadeh 2021; Torshizi et al. 2024; Rajabzadeh et al. 2024).

4.1 Multi-objective Optimization Models

Some scholars have incorporated sustainability through integrating environmental and social objectives along with economic objectives. In this approach, various objectives have been proposed for dimensions of sustainability and resilience in different industries to balance sustainability and resilience. These objectives aim to ensure the long-term sustainability, efficiency, and adaptability of supply chains while maintaining the capacity to recover quickly from disruptions.

Effective cost management and profit maximization are crucial for staying competitive in the manufacturing industry. Economic objectives generally seek to minimize the total cost, maximize the use of available production throughput for factories, and maximize the total profit and financial return while incorporating environmental considerations. From a sustainability point of view, environmental objectives help to ensure a long-term vision to meet future generations' needs. Environmental objectives can be formulated in different ways to cover various aspects of environmental protection and footprint reduction. Examples of environmental objectives include minimizing total greenhouse gas emissions, mitigating waste generation, optimizing water usage, and reducing energy consumption. The overarching aim is to create more resource-efficient and environmentally friendly manufacturing processes. Social goals often focus on maximizing job opportunities, ensuring fair labor practices in the supply chain network, and minimizing unmet demand for specific products. By promoting job creation and supporting local communities, manufacturing supply chains can contribute to regional economic growth and social equity. In addition, these goals consider the broader social impacts of supply chain decisions, such as improving living standards, supporting education, and addressing poverty and inequality through ethical sourcing and community development programs. Furthermore, given the potential disruptions caused by natural and man-made disasters, resilience goals are critical for manufacturing supply chains. Minimizing vulnerabilities, such as dependence on single suppliers or inefficient logistics, is essential to maintaining operational stability. Ensuring the resilience of critical nodes in the supply chain, such as key suppliers or manufacturing facilities, is also crucial. Simplifying the supply chain structure and introducing redundancies where necessary can increase resilience, enabling the system to sustain and recover from unexpected shocks and ensure continued production. Responsiveness objectives focus on improving the agility and flexibility of manufacturing supply chains. This includes minimizing lead times and reducing lead times in manufacturing processes to ensure products reach customers quickly. Maximizing responsiveness also means increasing the supply chain's ability to adapt quickly to changes in demand, market conditions, or disruptions. In today's competitive environment, adapting quickly to changing customer needs and external conditions is essential to staying competitive. By incorporating these diverse objectives into multi-objective mathematical models, scholars and practitioners aim to optimize the manufacturing supply chain balance between sustainability and resilience. These models seek to find the best trade-offs between cost, environmental impact, social benefits, resilience, and responsiveness.

For instance, Sazvar et al. (2021) studied a sustainable-resilient supply chain network by developing a multi-objective mathematical model that minimizes the total cost and environmental pollution and maximizes responsibilities to the social community and level of resilience of the influenza vaccine supply chain. Mostaghim et al. (2024) introduced a resilient model to enhance the sustainability and resilience of the broiler supply chain in the face of demand uncertainty and unforeseen disruptions by incorporating backup facilities and multiple sourcing strategies. The proposed MILP model aims to minimize carbon emissions from transportation while maximizing overall supply chain profit. Moadab et al. (2023) developed a multi-objective model to minimize costs, environmental impact, and negative social impact caused by shortages for a sustainable, resilient, and responsive supply chain. Sharifi et al. (2024) considered resilience and responsive objectives along with sustainability objectives. They formulated the node criticality, complexity, and customer dissatisfaction to optimize resiliency and they minimized the total delivery time for responsiveness. Mohammed et al. (2023) developed a hybrid integrated BWM-CoCoSo-multi-objective programming model to minimize total cost and environmental impact and the maximization of social and resilience paradigms in an agri-food supply chain. Moreover, Tsao et al. (2024) developed a fuzzy MILP model for a flexible and sustainable multilayer SC semiconductor network design problem with four objective functions: minimizing total costs, retailer dissatisfaction level, carbon emissions, and maximizing the total number of job opportunities created. These studies demonstrate the applicability and generality of the multi-objective approach to optimally balance cost, environmental impacts, social benefits, resilience, and responsiveness.

4.2 Integration of New Revolutionary Technologies

In supply chains, blockchain technology facilitates data storage and sharing among all stakeholders within the network (Masood et al. 2024). In the coming years, blockchain is predicted to revolutionize supply chain management (Verma et al., 2025). The capabilities of blockchain as a revolutionary technology have been widely used in diverse supply chains to enhance sustainability and resilience in the face of disruptions like COVID-19.

Several studies investigated utilizing the blockchain specifically in response to the COVID-19 pandemic. To adapt to the COVID-19 pandemic, Hervani et al. (2022) integrated a resource-based view with blockchain technology in a framework to improve the sustainability and resilience of the supply chain in the face of COVID-19. In the same year, de Farias et al. (2022) proposed a model using Industry 4.0 technologies like Digital Twin, IoT, and blockchain to boost supply chain resilience, focusing on improving visibility and developing disruption-mitigation strategies, particularly during the COVID-19 pandemic. As the pandemic continued to impact global supply chains, Moosavi et al. (2022) conducted a comprehensive analysis using bibliometric, network, and thematic methods to evaluate disruption management research. Their findings highlighted resilience, sustainability, and the role of technologies like AI, IoT, and blockchain, suggesting strategies for maintaining critical product supplies, such as PPE and vaccines, during crises. Recognizing the specific challenges faced by healthcare supply chains during the pandemic, Sathiya et al. (2023) identified key areas for improving healthcare supply chains during the pandemic, including resilience, localization, reverse logistics, visibility, and digitalization. To create a smart supply chain and monitor healthcare products, they proposed a Chain of Things technology through the integration of Blockchain and IoT. Recently, Bai et al. (2024) showed that companies with blockchain capabilities experienced less negative stock price reactions during COVID-19 suggesting blockchain helps manage uncertainty and provides a competitive edge.

Additionally, Erol et al. (2022) proposed a decision framework to study critical functions in sustainable supply chains such as sourcing, delivery, transformation, product recovery, customers, and product use for blockchain applicability. In the context of manufacturing and supply chain management, Lupi et al. (2023) introduced smart contracts and blockchain technology for sharing workload and resources under unexpected disruptions in the supply chain to increase the adaptivity of distributed resources using additive manufacturing (AM) technology. Their proposed mechanism provides more resilience through decentralized, transparent, and trusted operations.

Further research by Ekinci et al. (2024) studied the adoption of Blockchain technology to evaluate the resilience of the retail sector. They employed A hybrid BWM Fuzzy TODIM method to ensure the development of a sustainable, resilient supply chain across diverse retail categories, particularly in uncertain situations. Additionally, Lotfi et al. (2024) recommended blockchain technology to enhance agility in disruptions by facilitating quick transactions between components and supporting risk-averse decision-making. They pursued sustainability requirements by minimizing CO2 emissions and energy consumption.

In the energy sector, Abadi et al. (2024) developed a blockchain-based decision-making framework to improve energy management by establishing supply chain resiliency in a distributed smart grid. They used this framework with a dynamic energy pricing model to manage energy distribution, especially during energy crises. In another work by Abdul-Wadud et al. (2025), the practical advantages of blockchain technology in energy trading are discussed.

Blockchain technology has emerged as a vital enabler of sustainability and resilience in diverse supply chains, including manufacturing, healthcare, retail, and energy. With the increasing rate and impact of disruption, the integration of blockchain with other emerging technologies, like the IoT and machine learning, is expected to enhance resilience and promote long-term sustainability.

More recent studies have underlined the abilities of other disruptive technologies, like Digital Twin (DT), to magnify sustainability and resilience. For example, Singh et al. (2024) pointed up that the DT in manufacturing supply chains for resilient-sustainable systems will improve visibility. Cimino et al. (2024) presented a simulation-based cyclical methodology to utilize the supply Chain Digital Twin Paradigm, to enhance the sustainability and resilience of the chains in the agri-food sector. Corsini et al. (2024) utilized the digital twin architecture employing data analytics, machine learning, simulation modeling, and a metaheuristic to optimize replenishment decisions through a learning process. Bakopoulos et al. (2024) integrated DTs with Industry 4.0 technologies to assess resilience in steel manufacturing. Greis et al. (2022) combined augmented reality with DTs for adaptive scheduling. de Farias et al. (2022) proposed a model using DTs, IoT, and blockchain to improve supply chain resilience, informed by the COVID-19 pandemic. Burgos and Ivanov (2021) studied the pandemic's impact on Germany's food retail sector, focusing on resilient inventory management. Finally, Aron and Sgarbossa (2023) reviewed the integration of resilience in the physical Internet and identified future research opportunities.

Besides the potential benefits of new revolutionary technologies, it is necessary to consider the implementation challenges which is emphasized by Khan et al. (2024b) and promoting collaboration and compliance among supply chain members is suggested.

5 Human-Centric Real-Time Sustainable-Resilient Manufacturing Framework

This section proposes a framework that facilitates collaboration between machines and humans leveraging real-time data to cope with disruptions in a dynamic environment. In this environment, the probability of disruptions in facilities and links between them changes over time dynamically. The proposed framework has three folds with a centric role of human. It incorporates disruptive technologies such as Blockchain and IoT to develop sustainable-resilient strategies, facilitates human–machine collaboration for predicting disruption occurrences, and includes optimization models to update decisions for managing disruptions (see Fig. 2). The proposed framework leverages real-time data, blockchain technology, and AI systems, with a

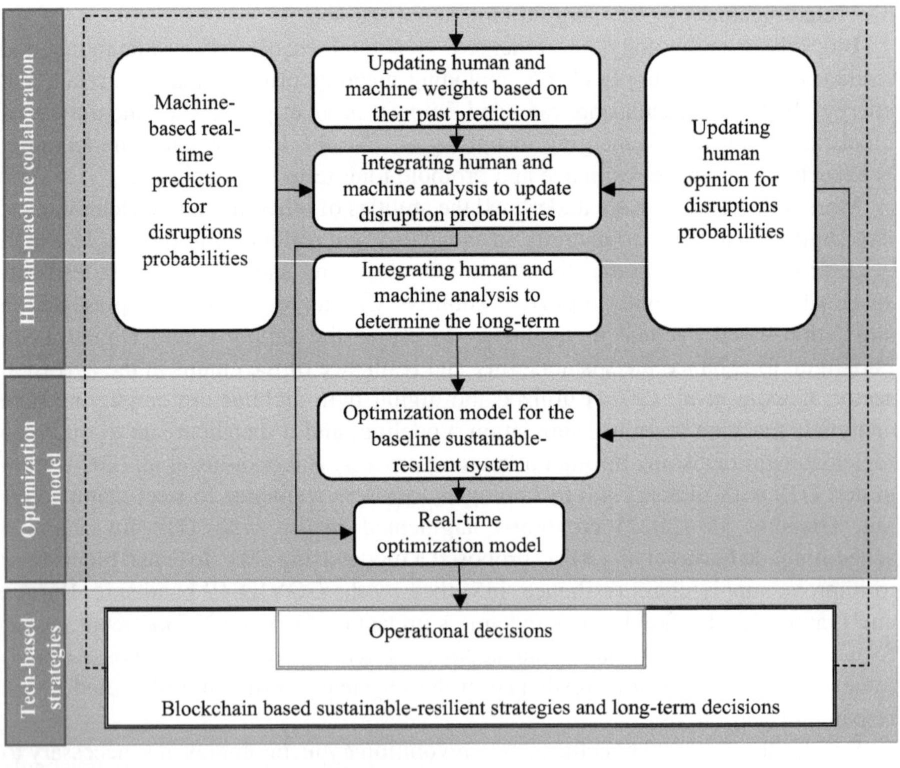

Fig. 2 Human–machine collaborative framework

human-centric approach, to enable real-time decision-making that optimizes both the resiliency and sustainability of the manufacturing system.

As shown in Fig. 2, at the first step of the proposed framework, a base sustainable-resilient strategy for the manufacturing supply chain is developed using long-term disruption probabilities based on experts' opinions. To this end, experts provide the probability and significance of each link and facility. Then, Model I will be solved to determine how to impel proactive and reactive resilient strategies, for example, emergency stock and blockchain technology. Emergency stock at retailers, in the event of a disruption, helps to meet customer demand. This type of inventory can only be used when there is no available path to transfer products to the retailer. The other strategy, blockchain technology, serves as both a proactive and reactive strategy and can be implemented at three different levels. The first level of blockchain provides traceability, ensuring that the root cause of the disruption can be quickly identified (e.g., problematic raw material batch). With the basic blockchain level in a facility, the system will be able to detect the quality disruption faster and can identify the root cause. It helps to keep the facility active and only a portion of the facility capacity will be out of service. On the other hand, the facility cannot continue operations without using blockchain technology in the event of a disruption. At the second level, blockchain evolves from data recording to automating actions via smart contracts, preventing the distribution of defective drugs to facilities and retailers. When a quality issue arises, a smart contract quarantines the affected batch and halts its distribution. It also sends recall notifications to other facilities to remove the product and stops any ongoing production using the faulty raw material. This system effectively controls the social impact of defective products (e.g., medicine). Level 3 of the blockchain technology can decrease the probability of quality disruption in the facility when integrated with IoT devices, as it provides real-time data and allows for continuous monitoring.

After applying Model I, once a new forecast of the probability of disruptions arrives, provided by either human or machine, Model II will be run to choose the most impactful disruptions. In the case where there is a change in the list of the most important disruptions, the framework recommends solving Model III with the updated disruptions to determine new disruption scenarios and adjust decisions related to resilient strategies. The proposed framework integrates data from both human opinions and AI methods to support optimization models for making and updating decisions on resilience strategies and manufacturing supply chain management. The next section explains how these models function and contribute to real-time optimization in the era of Industry 5.0.

6 Optimization model

This section discusses the optimization models in the proposed framework and Fig. 3, starting with the notations presented in Table 2. Three optimization models are then formulated and detailed in Sects. 6.1, 6.2, 6.3, and 6.4.

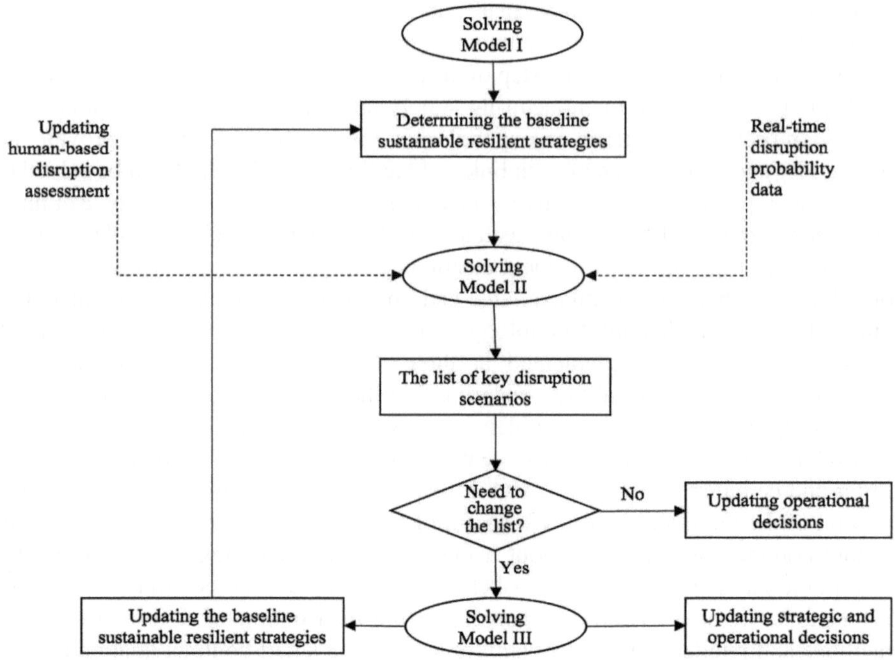

Fig. 3 Real-time optimization to update decisions using disruption probability data

6.1 Optimization Model for Long-Term Resilient and Sustainable System

Model I defines the foundational sustainable and resilient strategies for the manufacturing system, including long-term decisions like supplier selection. To do this, it combines machine learning and human predictions to estimate the likelihood of disruptions across various scenarios. Only the most probable disruption scenarios are considered to simplify the mathematical model. The system operates based on the outputs of Model I until new disruption probability data is available. Once real-time data is received, the system must determine how to respond. It begins by identifying the disruptions in facilities and supply chain links that would have the greatest impact on system performance. Model II is responsible for this analysis. After identifying the most impactful disruption scenarios, Model III updates the system's decisions based on the revised disruption probabilities.

Table 2 Notations

Set	Description
S = {1, 2, …, |S|}	Set of suppliers
R = {1,2,…,|R|}	Set of retailers
M = {1,2,…,|M|}	Set of manufacturing sites
P = {1,2, …, |P|}	Set of products
L = {1,2, …, |L|}	Set of disruption scenarios
I = {1,2, …, |I|}	Set of blockchain technology level
Parameters	
Pr_{sm}	The probability of disruption at the link between supplier s and manufacturing site m.
Pr_{mr}	The probability of disruption at the link between the manufacturing site m and retailer r.
Pr_s	The probability of disruption at supplier s.
Pr_m	The probability of disruption at manufacturing site m.
Pr_l	The probability of scenario l.
$Pr'_s = \gamma \, Pr_s$	The probability of disruption at the supplier s if it is equipped with blockchain level 3, where γ is a positive coefficient less than 1.
$Pr'_m = \gamma Pr_m$	The probability of disruption at manufacturing site m if it is equipped with blockchain level 3.
Pr'_l	The probability of scenario l occurring with implementing blockchain level 3 in some facilities within scenario l.
U_r^l	Equals 0 if retailer r is unable to receive the product in scenario l; otherwise, it equals 1.
λ_{sm}^l	Equals 0 if the link s-m is disrupted in scenario l; otherwise 1.
μ_{mr}^l	Equals 0 if the link m-r is disrupted in scenario l; otherwise 1.
O_s^l	Equals 0 if the supplier s is disrupted in scenario l; otherwise 1.
A_m^l	Equals 0 if the manufacturing site m is disrupted in scenario l; otherwise 1.
D_{rp}	Demand for product p at retailer r.
FS_s	Fixed cost for selecting supplier s.
CS_{smp}	Transportation cost per product from supplier s to manufacturing site m.
CM_{mrp}	Transportation cost per product from manufacturing site m to retailer r.
h_{rp}	Inventory holding cost per product p at retailer r.
PBS_s^i	A binary parameter equals 1 if supplier s uses the i-th level of blockchain in the current state.
PBM_m^i	A binary parameter equals 1 if manufacturing site m uses the i-th level of blockchain in the current state.
PY_s	A binary parameter equals 1 if supplier s works with manufacturing sites in the current state.
FBS_s^i	Fixed cost for enabling supplier s to the blockchain technology level i.
FBM_m^i	Fixed cost for enabling manufacturing site m to the blockchain technology level i.

(continued)

Table 2 (continued)

v_p	Blockchain transaction cost per product
CEB_{sm}	Carbon emitted per product transferred from s to m
CEB_i	Carbon emitted per product for using the i-the level of blockchain
KS_s	Capacity of supplier s
KM_m	Capacity of manufacturing site m
α	The percentage of the supplier's active capacity after a quality disruption, when the supplier is enabled with blockchain level 1 or higher.
β	The percentage of the manufacturer's active capacity after a quality disruption when the manufacturer is enabled with the first level of blockchain or higher.
γ	Coefficient for the disruption probability at a facility when using the highest level of blockchain.
TC_{rp}	The cost of reducing the emergency stock level of product p at retailer r.
W_s	Relative probability of disruption at supplier s, as determined by experts.
W_m	Relative probability of disruption at manufacturer m, as determined by experts.
W_{sm}	Relative probability of disruption at link s-m, as determined by experts.
W_{mr}	Relative probability of disruption at link m-r, as determined by experts.
Variables	
X^l_{smp}	The flow of product p between supplier s and manufacturer m in scenario l.
V^l_{mrp}	The flow of product p from the manufacturer m to retailer r in scenario l.
Y_s	Takes the value 1 if supplier s is selected; otherwise 0.
BS^i_s	Takes the value 1 if supplier s is equipped with blockchain technology level i; otherwise 0. Where $i \in \{0, 1, 2, 3\}$ and i = 0 means the facility is not equipped with blockchain technology.
BM^i_m	Takes the value 1 if manufacturing site m is equipped with level i of the blockchain technology; otherwise 0. Where $i \in \{0, 1, 2, 3\}$ and i = 0 means the facility is not equipped with blockchain technology.
Q_{rp}	Level of emergency stock for product p at retailer r.
Q'_{rp}	The quantity of emergency stock for product p at retailer r, which can only be used when retailer r is disrupted.
SCS_s	A binary variable equals 1 if supplier s is enabled with level 2 or higher of blockchain technology.
SCM_m	A binary variable equals 1 if manufacturing site m is enabled with level 2 or higher of blockchain technology.
N_{sm}	A binary variable equals 1 if supplier s sends product to manufacturer m, otherwise 0.
E^t_l	A binary variable equals 1 if the number of facilities enabled with blockchain level 3 is t, otherwise 0.
DQ_{rp}	The quantity of reduction in the emergency stock level of product p at retailer r.

In Model I, human experts' opinions on the long-term probability of disruption can be gathered and combined with machine predictions of disruptions. In this context, the machine's predictions can be treated as input from an expert.

$Min\ Z_1 = \sum_{s} FS_s\, Y_s + \sum_{s,i} FBS_s^i\ BS_s^i + \sum_{m,i} FBM_m^i\ BM_m^i + \sum_{l} Pr'_l \left(\sum_{i,s,m,p} v_p\, BS_s^i\ X_{smp}^l + \sum_{m,r,p} v_p\ BM_m^i V_{mrp}^l \right) + \sum_{l} Pr'_l \left(\sum_{s,m,p} CS_{smp}\, X_{smp}^l + \sum_{m,r,p} CM_{mrp}\ V_{mrp}^l \right) + \sum_{r,p} h_{rp}\, Q_{rp}$		(1)
$Min\ Z_2 = \sum_{l} Pr'_l \left(\sum_{s,m,p} O_s^l (1 - SCS_s)\, X_{smp}^l + \sum_{m,r,p} A_m^l\, (1 - SCM_m) V_{mrp}^l \right)$		(2)
$Min\ Z_3 = \sum_{l} Pr'_l \left(\sum_{i,s,m,p} (CE_{sm} + CEB_i BS_s^i)\, X_{smp}^l + \sum_{i,m,r,p} (CE_{mr} + CEB_i BS_s^i)\, V_{mrp}^l \right)$		(3)
s.t.		
$\sum_{m.p} X_{smp}^l \leq \left(\alpha\, (1 - BS_s^0)(1 - O_s^l) + O_s^l\right) Y_s\, KS_s$	$\forall l, s$	(4)
$\sum_{r,p} V_{mrp}^l \leq \left(\beta\, (1 - A_m^l) + A_m^l\right) KM_m$	$\forall l, m$	(5)
$\sum_{r} V_{mrp}^l \leq \sum_{s} X_{smp}^l$	$\forall l, m, p$	(6)
$D_{rp} \leq Q'_{rp} + \sum_{m} V_{mrp}^l$	$\forall l, r, p$	(7)
$Q'_{rp} \leq (1 - U_r^l)\, Q_{rp}$	$\forall l, r, p$	(8)
$X_{smp}^l \leq M \times \lambda_{sm}^l$	$\forall l, s, m, p$	(9)
$V_{mrp}^l \leq M \times \mu_{mr}^l$	$\forall l, m, r, p$	(10)
$\sum_{i} BS_s^i = 1$	$\forall s$	(11)
$\sum_{i} BM_m^i = 1$	$\forall m,$	(12)
$BS_s^i \leq Y_s$	$\forall i, s$	(13)
$\sum_{p,l} X_{smp}^l \leq M \times N_{sm}$	$\forall s, m$	(14)
$BM_m^i \leq BS_s^{i-1} + M \times (1 - N_{sm})$	$\forall s, m, \mathrm{i}$	(15)
$BS_s^{(2)} + BS_s^{(3)} \leq 2 \times SCS_s$	$\forall s$	(16)
$BM_m^{(2)} + BM_m^{(3)} \leq 2 \times SCM_m$	$\forall m$	(17)
$\sum_{t} tE_l^t = \sum_{s} (1 - O_s^l) BS_s^{(3)} + \sum_{m} (1 - A_m^l) BM_m^{(3)}$	$\forall l$	(18)
$Pr'_l = \sum_{t} (\gamma)^t\ E_l^t\ Pr_l$	$\forall l \neq 1$	(19)
$Pr'_1 = 1 - \sum_{l \neq 1} Pr'_l$		(20)

The first objective function minimizes the total cost of the system. The first term represents the cost of selecting suppliers. The following two terms account for the fixed costs associated with integrating suppliers and manufacturing sites into the blockchain network. The operating cost of blockchain technology for both suppliers and manufacturers is included in the next summation. Transportation costs are the next component, and finally, the holding cost for emergency stock is considered the last item. Objective function Z_1 minimizes the total cost, while Z_2 minimizes the total distributed defective products across the system as the social objective. It is assumed that integrating the system with blockchain levels 2 and higher helps prevent the distribution of defective products. Finally, Z_3 is defined to minimize the total carbon emissions related to transportation and the use of various levels of blockchain.

Constraint (4) ensures that the supplier's capacity is sufficient in both normal and disrupted conditions. Similarly, constraint (5) accounts for the manufacturing capacity. Constraints (6) and (7) enforce the input-output balance for manufacturers and retailers, respectively. In constraint (7), emergency stock is used to meet demand during disrupted conditions when there is no available path for delivering the product to the retailer. However, the emergency stock can only be accessed by the retailer during these disrupted scenarios, as specified by constraint (8). A disrupted link cannot transfer products stated by constraints (9) and (10). Each facility can be equipped with only one level of blockchain technology, as defined by constraints (11) and (12). Level zero indicates that no blockchain technology is implemented at the facility. Blockchain enabling does not apply to non-selected suppliers, as guaranteed by constraint (13). Constraints (14) and (15) ensure that implementing a blockchain level at a manufacturing site requires a corresponding level of blockchain in all its suppliers. For facilities with blockchain levels higher than level 1, the manufacturing system can prevent the distribution of defective products, particularly in the case of quality disruptions at the facility.

To calculate the sustainability objective, the model requires the values for SCS_s and SCM_m, which are determined by constraints (16) and (17). Furthermore, the highest blockchain level helps facilities reduce the probability of quality disruptions compared to the normal probability, as defined by constraints (18), (19), and (20).

6.2 Optimization Model for Prioritizing Disruptions

As mentioned earlier, a high number of scenarios, in real-world problems, makes solving Model I more complex. Model II aims to identify the most significant disruptions in links or facilities that impact the performance of the manufacturing system after receiving a new estimation of disruption probabilities. This approach significantly reduces the number of scenarios. Upon receiving new data and running the model, the current state of the system must be considered. Therefore, the model takes into account the currently selected suppliers and the blockchain levels used by the facilities. It is important to note that the blockchain technology level can only be upgraded and downgrading the blockchain level is prohibited. As for the suppliers, the system has the flexibility to decide whether to make changes to the list of selected suppliers.

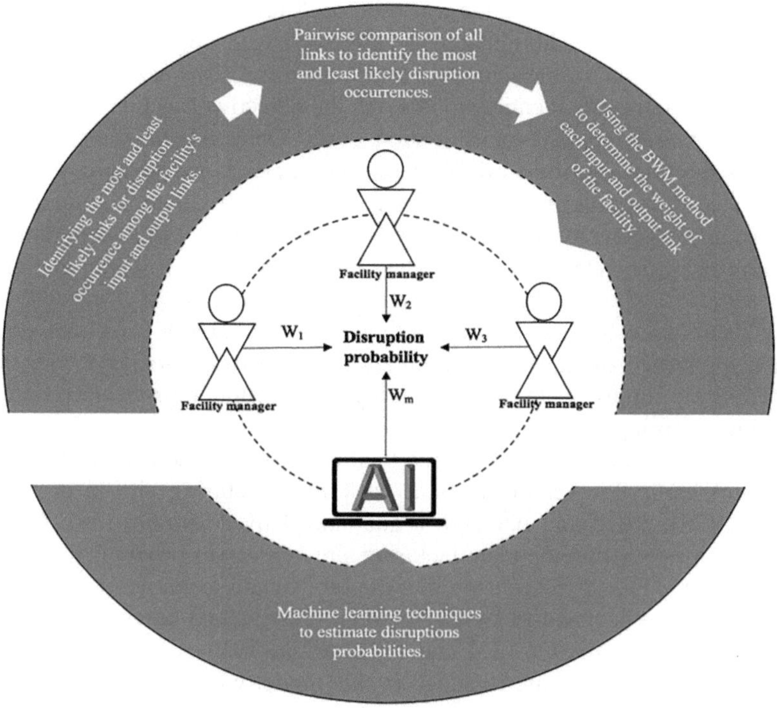

Fig. 4 Human–machine collaboration to estimate disruption probability considering their accuracy weights

Model II employs an estimation of the probability of disruptions in facilities and links by combining machine learning with human opinions. At each entity in the manufacturing supply chain, including suppliers and manufacturers, humans assess the likelihood of disruption occurring on the input and output links of the facility. As illustrated in Fig. 4, each facility manager provides their assessment of the relative likelihood of disruptions occurring within their facility, as well as the likelihood of disruptions along inbound and outbound links. Various Multi-Attribute Decision-Making (MADM) techniques can be applied for this purpose. In this context, we recommend using the Best-Worst Method (BWM), developed by Rezaei (2015). It is important to note that experts do not estimate the actual probability of disruptions at this stage; instead, they assess the relative likelihood of disruptions. On the other hand, machines use AI techniques to provide quantitative predictions that support decision-making. Objectives Z'_4 and Z'_5 are derived from human and machine outputs, respectively. The weights assigned to these objectives in the solution approach change over time, reflecting the previous prediction accuracy of both human and machine outputs.

In Modell II, the first term in the equation for Z'_1 refers to the cost of changing the selected suppliers. The additional costs associated with upgrading the blockchain level for suppliers and manufacturers are accounted for in the next two summations. The remaining terms are identical to those in the objective function Z_1 accounts for blockchain transaction and transportation costs. Objective function Z'_2, like Z_2, minimizes the total distributed defective products across the system except that it uses updated disruption probabilities for facilities. Furthermore, Z'_3 minimizes the total carbon emissions by considering updated information. Eq. (24) minimizes the maximum risk associated with the product flow at facilities. It is important to note that W represents the risk priority of the input or output links of the facility, or the facility itself, as estimated by the facility owner. Similarly, the final objective function, presented by eq. (25), minimizes the total risk associated with product flow across the supply chain, taking into account disruption probabilities predicted by machine learning.

Constraints (26) and (27) ensure the facility capacity. The balance between the output and input flow of products at each manufacturing site is provided by constraint (28). Satisfying demand is guaranteed by constraint (29). Constraint (30) determines whether the manufacturing site m receives products from supplier s. Constraints (31) and (32) ensure that the blockchain technology level can only be upgraded. The decision to switch between selected and non-selected suppliers is addressed by constraints (33) and (34). Additionally, constraints (11), (12), (13), (15), (16), and (17) are included in this model with the same definitions as before.

$Min\ Z_1' = \sum_s FS_s \sum_p Y_s' + \sum_s \left(\sum_i i \times FBS_s^i\ BS_s^i - \sum_i i \times FBS_s^i\ PBS_s^i \right) + \sum_{m,} \left(\sum_i i \times FBM_m^i\ BM_m^i - \sum_i i \times FBM_m^i\ PBM_m^i \right) + \left(\sum_{i,s,m,p} v_p\ BS_s^i\ X_{smp} + \sum_{i,m,r,p} v_p\ BM_m^i V_{mrp} \right) + \sum_{s,m,p} CS_{smp} X_{smp} + \sum_{m,r,p} CM_{mrp}\ V_{mrp}$		(21)
$Min\ Z_2' = \sum_{s,m,p} Pr_s\ (1 - SCS_s)\ X_{smp} + \sum_{m,r,p} Pr_m\ (1 - SCM_m)\ V_{mrp}$		(22)
$Min\ Z_3' = \left(\sum_{i,s,m,p} Pr_s\ (CE_{sm} + CEB_i BS_s^i)\ X_{smp} + \sum_{i,m,r,p} Pr_m\ (CE_{mr} + CEB_i BS_s^i)\ V_{mrp} \right)$		(23)
$Min\ Z_4' = \max\left(\max_s \left(\sum_{m,p} W_s\ X_{smp} + \sum_{m,p} W_{sm}\ X_{smp} \right), \max_m \left(\sum_{r,p} W_m\ V_{mrp} + \sum_{r,p} W_{mr}\ V_{mrp} \right) \right)$		(24)
$Min\ Z_5' = \sum_{s,m,p} Pr_{sm} X_{smp} + \sum_{m,r,p} Pr_{mr}\ V_{mrp} + \sum_{s,m,p} Pr_s \left(1 - BS_s^{(3)}\right) X_{smp} + \sum_{s,m,p} Pr_s'\ BS_s^{(3)}\ X_{smp} + \sum_{m,r,p} Pr_m \left(1 - BM_m^{(3)}\right) V_{mrp} + \sum_{m,r,p} Pr_m'\ BM_m^{(3)}\ V_{mrp}$		(25)
s.t.		
$\sum_{m,p} X_{smp} \le Y_s\ KS_s$	$\forall s$	(26)
$\sum_{r,p} V_{mrp} \le KM_m$	$\forall m$	(27)
$\sum_r V_{mrp} \le \sum_s X_{smp}$	$\forall m, p$	(28)
$D_{rp} \le \sum_m V_{mrp}$	$\forall r, p$	(29)
$\sum_p X_{smp} \le N_{sm}$	$\forall s, m$	(30)
$PBS_s^i \le BS_s^i$	$\forall s, i$	(31)
$PBM_m^i \le BM_m^i$	$\forall m, i$	(32)
$PY_s - Y_s \le Y_s'$	$\forall s$	(33)
$Y_s - PY_s \le Y_s'$	$\forall s$	(34)
Constraints (11), (12), (13), (15), (16) and (17)		

After solving Model II, facilities and links with the highest product of disruption likelihood and product flow will be selected to determine disruption scenarios. These scenarios will be used in Model III, which is described in the section.

6.3 *Optimization Model for Updating Decisions*

This model, Modell III, uses the most significant disruption scenarios, determined by Model II in the previous step, to update all decisions based on the newly received disruption probability data.

$\text{Min}Z_1'' = \sum_s FS_s\, Y_s' + \sum_s \left(\sum_i i \times FBS_s^i\, BS_s^i - \sum_i i \times FBS_s^i\, PBS_s^i \right) + \sum_{m,} \left(\sum_i i \times FBM_m^i\, BM_m^i - \sum_i i \times FBM_m^i\, PBM_m^i. \right) + \sum_l Pr_l' \left(\sum_{i,s,m,p} v_p\, BS_s^i\, X_{smp}^l + \sum_{i,m,r,p} v_p\, BM_m^i V_{mrp}^l) + \sum_l Pr_l' \left(\sum_{s,m,p} CS_{smp}\, X_{smp}^l + \sum_{m,r,p} CM_{mrp}\, V_{mrp}^l \right) + \sum_{r,p} h_{rp}\, Q_{rp} + \sum_{r,p} TC_{rp}\, DQ_{rp}$		(35)
$Min\ Z_2'' = \sum_l Pr_l' \left(\sum_{s,m,p} O_s^l (1 - SCS_s)\, X_{smp}^l + \sum_{m,r,p} A_m^l\, (1 - SCM_m) V_{mrp}^l \right)$		(36)
$Min\ Z_3'' = \sum_l Pr_l' \left(\sum_{i,s,m,p} (CE_{sm} + CEB_i BS_s^i)\, X_{smp}^l + \sum_{i,m,r,p} (CE_{mr} + CEB_i BS_s^i)\, V_{mrp}^l \right)$		(37)
s.t.		
Constraints (4) – (20)		
Constraints (31) – (34)		
$PQ_{rp} - Q_{rp} \le DQ_{rp}$	$\forall r, p$	(38)

This model optimizes the total cost of the manufacturing system and the total distribution of defective products across the system, considering disruption scenarios through objectives Z_1'' and Z_2''. The only new constraint in Model III, constraint (34) determines the amount of reduced level of emergency stock of product p at retailer r.

6.4 Solution Approach

To solve the multi-objective models presented in the previous sections, we apply the weighted Tchebycheff method to transform the conflicting objectives into a single-objective model. The augmented weighted Tchebycheff method for these multi-objective models is as follows:

$Min\ Z + \rho \left(\sum_{k=1}^{\lvert k \rvert} W_k \left(\frac{Z_k - Z_k^{min}}{Z_k^{max} - Z_k^{min}} \right) \right)$		(39)
$s.t.$		
$\psi_k \left(\frac{z_k^{max} - z_1}{z_k^{max} - z_k^{min}} \right) \le Z$	$\forall k$	(40)

Here the parameter ρ is a non-negative scalar, typically a small positive value. Additionally, ψ_k >0 with the condition that $\sum_k \psi_k = 1$ and Z_k represent the objective functions.

7 Implementing the Proposed Framework and Optimization Models

This section provides a numerical example of a pharmaceutical manufacturing supply chain, highlighting the risk of disruption in links and facilities, to demonstrate how the proposed framework can be applied in practice. Dealing with disruptions in the pharmaceutical industry is of great importance worldwide, and in Canada, considering disruptions is crucial for this sector. According to Health Canada, in 2023–2024, Canada experienced a significant increase in drug shortages, rising by approximately 15% compared to 2022–2023 (Health Canada 2024a). The top three reasons for drug shortages during this period were manufacturing issues, unexpected increases in demand, and shipping delays. The average duration of a drug shortage was 93.5 days, marking a 4.5-day decrease compared to the average length in 2022–2023. Two notable examples of high-impact shortages during this period were Ozempic and Prazosin. Ozempic faced a shortage from August 2023 to February 2024, driven by a global surge in demand linked to its use for weight loss. On the other hand, Prazosin has been in short supply since early 2023 due to disruptions in the supply of its active pharmaceutical ingredient.

To illustrate how the proposed framework functions in practice, this section presents a numerical example based on Canada's demand, supply, distance, and cost estimations for a pharmaceutical manufacturing supply chain. The supply chain comprises three suppliers, two manufacturing sites, and six retailers. For simplicity just one product, Prazosin, is considered to produce by the supply chain. The main conditions for which Prazosin is prescribed include high blood pressure, post-traumatic stress disorder, and benign prostatic hyperplasia.

Considering the average number of Canadians affected by these conditions, and acknowledging that these groups overlap to some extent, Table 3 presents the demand for Prazosin across different provinces. Furthermore, it is assumed that potential suppliers are located in Vancouver, Montreal, and Toronto while manufacturing sites are situated in Montreal and Toronto. Moreover, aggregated demand point are Vancouver, Calgary, Winnipeg, Toronto, Montreal, and Halifax. The values of other parameters and distance between cities are shown in Tables 4 and 5, respectively.

Health Canada aims to enhance Canada's resilience by strengthening collective capacity in three key areas:

(i). *Planning*: Developing systems and standard operating procedures that support preparedness and enable rapid response to shortages.
(ii). *Prevention*: Identifying potential shortage risks and taking proactive steps to prevent shortages before they occur.
(iii). *Response*: Detecting early warning signs of shortages, minimizing their impacts when they happen, and quickly restoring access to essential products (Health Canada, 2024b).

Table 3 Estimating demand across various markets

Aggregated demand point (Retailer)	Covered areas (provinces)	Demand: estimated tablet usage (million)
Vancouver	British Columbia	99.3
Calgary	Alberta	83.9
Winnipeg	Manitoba, Saskatchewan	49.6
Toronto	Ontario	280.5
Montreal	Quebec	166.4
Halifax	Nova Scotia, New Brunswick, Newfoundland and Labrador, Prince Edward Island	47.4

Table 4 The parameters' values in the optimization models

Parameter	Value	Parameter	Value
γ	0.8	CS_{smp}	
α	0.6	CM_{mrp}	
β	0.6	h_{rp}	0.2 × .51
v_p	2.351	TC_{rp}	0.02 × .51
$FBS_s^{(1)}$	420,000	KS_s	340,000,000
$FBS_s^{(2)}$	600,000	KM_m	500,000,000
$FBS_s^{(3)}$	740,000		
$FBM_m^{(1)}$	900,000		
$FBM_m^{(2)}$	1,200,000		
$FBM_m^{(3)}$	1,440,000		

Table 5 Distance between facilities (km)

	Vancouver	Calgary	Winnipeg	Toronto	Montreal	Halifax
Vancouver	0	971	2299	4388	4580	5830
Toronto	4388	3415	2062	0	543	1792
Montreal	4580	3612	2268	543	0	1250

This plan outlines eight actions designed to build resilience against health product shortages and protect Canadians from the most severe consequences of such shortages. These actions are as follows.

- Action 1: Publish a list of drugs that are critical and vulnerable to shortages.
- Action 2: Require shortage management plans and safety stocks for drugs that are critical and vulnerable to shortages.
- Action 3: Enhance regulatory responses and reporting requirements.
- Action 4: Improve reporting systems.
- Action 5: Strengthen data analytics and shortage management capacity.
- Action 6: Improve information sharing.

- Action 7: Enhance transparency and clear communication.
- Action 8: Modernize regulatory frameworks.

In line with this plan, we propose two resilient strategies for the studied pharmaceutical manufacturing supply chain. First, emergency stock is recommended as a proactive strategy to help the manufacturing system meet customer demand during disruptions. Second, we defined three levels of leveraging blockchain to improve transparency, information sharing, and the speed of response when disruptions occur. At the highest level of blockchain implementation, blockchain can also help reduce the occurrence of disruptions through better prediction. Thus, blockchain serves both as a proactive strategy (when utilizing the first two levels) and as a reactive strategy (at the highest level, for enhanced prediction and disruption prevention).

In the studied pharmaceutical supply chain (see Fig. 5), the probabilities of disruption occurrences fluctuate over time. As a result, the system must continuously update its decisions and respond to these changes. Updated data will be provided through both machine-driven inputs and human expertise.

In this supply chain, each facility manager, whether an ingredient supplier or manufacturer, plays the role of an expert in predicting disruptions both within their own facility and across the inbound and outbound links. On the other hand, a quantitative prediction is provided by machines using AI techniques to support decision-making. Optimization Model II, described in Sect. 6, integrates both expert and machine predictions through two objective functions aimed at minimizing the total flow of products at risk of disruption. When converting the multi-objective model to a single objective, the weights assigned to these functions are based on the prediction accuracy of human and machine disruption forecasts from previous predictions. Given that the prediction accuracy of both human and machine forecasts may change over time, these weights can be adjusted accordingly.

To gather the opinions of facility owners regarding disruptions, the BWM is used. To this end, each facility owner will be asked for comparing the likelihood of disruptions occurrence in their inbound and outbound links and the facility. For instance, for the manufacturer M1 who receives material from suppliers S1 and S3 and send the final product to retailers R1, R3, R4, the following questionnaire can be utilized for this purpose (the answers shown are solely for illustrative purposes).

Fig. 5 The structure of the studied pharmaceutical supply chain

Question 1. Select the most link or facility to be disrupted.						
	M1	S1-M1	S3-M1	M1-R1	M1-R3	M1-R4
Choose the facility or link with the most probability of being disrupted.	*					
Question 2. Select the least likely link or facility to be disrupted.						
	M1	M1-S1	M1-S3	M1-R1	M1-R3	M1-R4
Choose the facility or link with the least probability of being disrupted.					*	

Question 3. How much more likely is the occurrence of disruption in the facility (or links) compared to the least likely event, as determined in question 2?									
	1 (the same probability)	2	3	4	5	6	7	8	9 (far more likely)
M1									*
M1-S1							*		
M1-S3				*					
M1-R1				*					
M1-R3	*								
M1-R4						*			
Question 4. How much more likely is the occurrence of disruption in the facility (or links) compared to the most likely event, as determined in question 2?									
	1 (far less likely)	2	3	4	5	6	7	8	9 (the same probability)
M1									*
M1-S1						*			
M1-S3		*							
M1-R1		*							
M1-R3	*								
M1-R4					*				

After gathering expert opinions, the weight associated with the likelihood of disruption can be determined using the model proposed by Rezaei (2015). It is important to note that these weights do not represent exact probabilities of disruptions; rather, they reflect the relative comparison of the likelihood of disruptions occurring.

$Min\ \xi$		(41)
s.t.		
$\mid \frac{W_B}{W_j} - a_{Bj} \mid \leq \xi$	$\forall j$	
$\mid \frac{W_j}{W_W} - a_{jW} \mid \leq \xi$	$\forall j$	
$\sum_j W_j = 1$		
$W_j \geq 0$	$\forall j$	

Where W_j is a positive variable and j represents a set of facility and related links. Also, a_{Bj} represents the preference for the most probable disruption B over facility (or link) j, using a scale of 1 to 9 from Question (4). Similarly, a_{jW} indicates the preference for all links and the facility over the worst link (or facility), using a scale of 1 to 9 from Question (3). For the given sample questionnaire, the weights obtained after solving model (41) are as follows: M1 = 0.272, M1-S1 = 0.083, M1-S3 = 0.25, M1-R1 = 0.25, M1-R3 = 0.044, and M1-R4 = 0.1. It is important to note that these represent the relative likelihoods of disruption, not probabilities.

The proposed framework incorporates both human opinions and machine predictions in two distinct formats. In Model I, an integrated human–machine estimation of disruption probability is used to develop the base sustainable-resilient solution, while in Model III, it is applied to update decisions. In contrast, Model II utilizes human and machine predictions separately, each applied to different objective functions.

To integrate human and machine predictions in Models I and III, the following steps are required. Along with human estimations for relative likelihoods, we also have AI-generated probabilities for disruption events. For instance, consider the following disruption probabilities: Pr(M1) = 0.15, Pr(M1-S1) = 0.002, Pr (M1-S3) = 0.1, Pr(M1-R1) = 0.075, Pr(M1-R3) = 0.001, and Pr (M1-R4) = 0.003. These machine-generated probabilities can be integrated with human expertise to estimate disruption probabilities for use in Model I and III. To do this, the AI outputs are first converted to relative likelihoods, resulting in the following values: M1 = 0.45, M1-S1 = 0.01, M1-S3 = 0.3, M1-R1 = 0.23, M1-R3 = 0.003, and M1-R4 = 0.01. As the next step, the overall relative likelihood must be calculated using the geometric average, incorporating the weights for human and machine predictions based on their previous estimation accuracy. In this example, with accuracy rates of 90% for humans and 80% for machines, the resulting relative likelihoods are as follows: M1 = 0.36, M1-S1 = 0.03, M1-S3 = 0.29, M1-R1 = 0.26, M1-R3 = 0.02, and M1-R4 = 0.04. Finally, consider the disruption probability of the facility (or link) with the least fluctuation in its disruption probability, and calculate all related probabilities. For instance, assume that M1-R1 has a base disruption probability of 0.075. By using the relative likelihood, the corresponding probabilities are: Pr(M1) = 0.106, Pr(M1-S1) = 0.009, Pr (M1-S3) = 0.085, Pr(M1-R1) = 0.075, Pr(M1-R3) = 0.005, and Pr (M1-R4) = 0.012. In Model II, human and machine opinions are used separately within their respective objective functions. It is important to note that the coefficient ψ_k for both human and machine predictions is determined based on their previous prediction accuracy.

Using long-term predictions for disruption probabilities, Model I will be applied to determine decisions regarding resilient strategies and supply chain network design. Based on the integrated human–machine predictions, Vancouver, along with the links between Montreal and Toronto, and Toronto and Winnipeg, are identified as having the highest disruption probabilities and eight of the related most likely disruption scenarios are presented in Table 6.

Table 6 Disruption scenarios at the beginning of planning horizon

	Disrupted links	Disrupted facilities	Probability of occurrence
Scenario 1 (normal condition)	–	–	0.5975
Scenario 2	Montreal to Toronto	–	0.1
Scenario 3	Toronto to Winnipeg	–	0.15
Scenario 4	–	Vancouver	0.05
Scenario 5	Montreal to Toronto, and Toronto to Winnipeg	–	0.015
Scenario 6	Montreal to Toronto	Vancouver	0.005
Scenario 7	Toronto to Winnipeg	Vancouver	0.075
Scenario 8	Montreal to Toronto, and Toronto to Winnipeg	Vancouver	0.0075

Using long-term predictions for disruption probabilities, Model I will be applied to determine decisions regarding resilient strategies and supply chain network design. Based on the integrated human–machine predictions, Vancouver, along with the links between Montreal and Toronto, and Toronto and Winnipeg, are identified as having the highest disruption probabilities. Consequently, eight of the most likely disruption scenarios are presented in Table 6.

By solving Model I, the base sustainable-resilient solution can be obtained, as shown in Fig. 6b. Whenever an update for disruption probabilities is received, whether from machine predictions or human input, Model II will be run to identify the three most impactful disruptions at the facility or links. For randomly generated real-time data, solving Model II yields the results depicted in Fig. 6c. Finally, since the most likely disruptions have changed, Model III must be executed to update decisions related to resiliency and supply chain management, as shown in Fig. 6d.

8 Conclusions

In recent years, the frequency and severity of disruptions in manufacturing systems and supply chains have increased, making them more vulnerable to both man-made and natural events. The COVID-19 pandemic and various natural disasters have highlighted these vulnerabilities, emphasizing the urgent need for more resilient and sustainable manufacturing systems. Disruptions can severely affect the sustainability of supply chains by interrupting the flow of goods, resources, and information, leading to significant economic, environmental, and social impacts. While only focusing on sustainability makes supply chains more vulnerable to disruption, we should be aware of the environmental and social implications of resilient strategies. So, it is crucial to find a balance between resilience and sustainability.

Advanced technologies, introduced by Industry 4.0 and rapidly evolved further in Industry 5.0, like blockchain, IoT, and AI provide new tools that foster resilience and

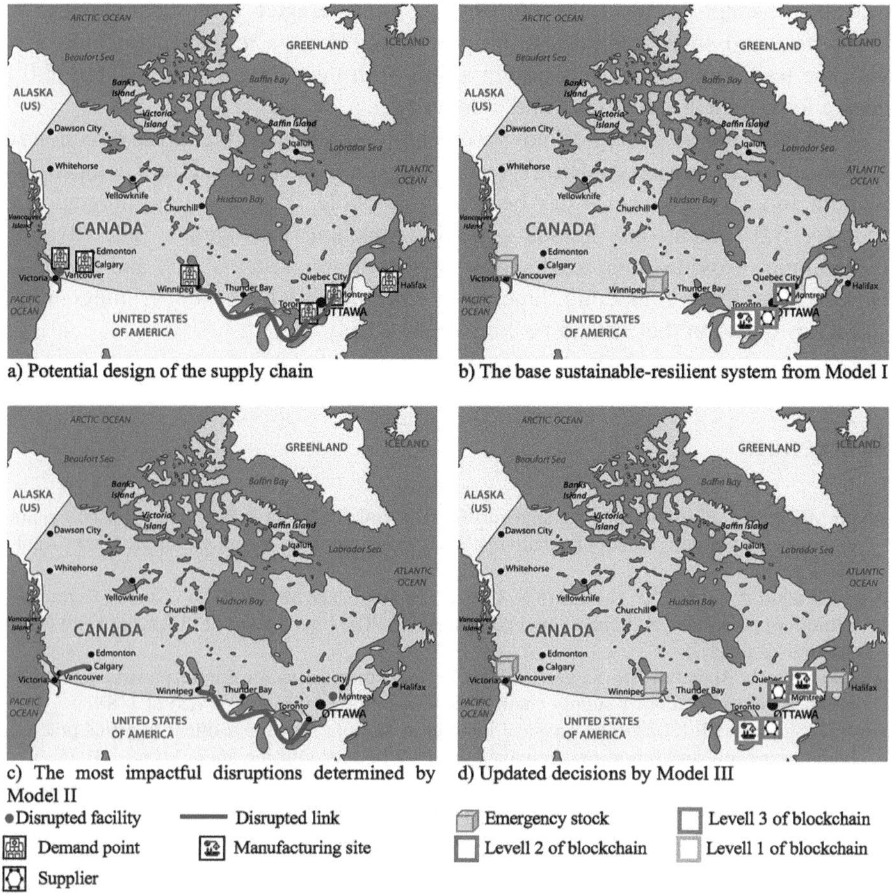

a) Potential design of the supply chain

b) The base sustainable-resilient system from Model I

c) The most impactful disruptions determined by Model II

d) Updated decisions by Model III

Fig. 6 Implementation of the proposed framework in the studied pharmaceutical supply chain

sustainability of supply chains and particularly provide real-time solutions for coping with disruptions. In the transition from Industry 4.0 to Industry 5.0, these technologies are upgrading to support resilience and sustainability strongly but the major revolution is defined around a human-centric approach which was overlooked in Industry 4.0. It is crucial to note that the goal of Industry 5.0 is not simply to keep humans in systems and define some superficial roles but rather it is a great opportunity to make synergic human–machine collaboration. So, defining some new roles for humans in collaboration with machines is needed. In addition, an AI-based system can serve as an expert collaborates with humans.

The framework developed in this chapter integrated three major dimensions of Industry 5.0 to support the resilience and sustainability of manufacturing supply chains. It converged human expertise and machine intelligence to facilitate the cooperation between them, incorporated real-time optimization decision-making models to adapt to disruption properly, and leveraged new technologies like blockchain and smart contracts to propose new effective resilient strategies along with the traditional ones. This design aligns with the principles of Industry 5.0 to ensure long-term sustainability and continuity.

With the rapid technological advancement in recent years, the future directions of businesses seem more complex. However, imagining success in future competitions without Industry 5.0 values has become impossible. The seamless integration of Industry 5.0 dimensions, human expertise, machine intelligence, and advanced technologies provides long-term success that supports sustainability and resilience. How to align these competing dimensions and make such a seamless integration is still a key question that should be considered deeply.

References

Abadi MQH, Sadeghi R, Hajian A, Shahvari O, Ghasemi A (2024) A blockchain-based dynamic energy pricing model for supply chain resiliency using machine learning. Supply Chain Anal 6: 100066

Abdul-Wadud A, Osei FAJ, Nurudeen S, Gawusu S, Abubakar M (2025) Blockchain technology: evolution, potentials, and operational challenges. In: The intersection of blockchain and energy trading. Elsevier, pp 47–74

Alptekinoğlu A, Bhandari AS, Sapra A (2024) Demand management using responsive pricing and product variety to counter supply chain disruptions. Eur J Oper Res 314(3):867–881

Aron C, Sgarbossa F (2023) The physical internet as an approach for resilient logistics practices: literature review and future research avenues. IFAC-PapersOnLine 56(2):11068–11075

Bai C, Zhu Q, Sarkis J (2024) Do blockchain capabilities help overcome supply and operational risks: insights from firm market returns during COVID-19. Omega 126:103049

Bakopoulos E, Sipsas K, Nikolakis N, Alexopoulos K (2024) A digital twin and data spaces framework towards resilient manufacturing value chains. IFAC-PapersOnLine 58(19):163–168

Brauner P, Ziefle M (2022) Beyond playful learning – serious games for the human-centric digital transformation of production and a design process model. Technol Soc 71:102140. https://doi.org/10.1016/j.techsoc.2022.102140

Burgos D, Ivanov D (2021) Food retail supply chain resilience and the COVID-19 pandemic: a digital twin-based impact analysis and improvement directions. Transp Res Part E Logist Transp Rev 152:102412

Carissimi MC, Creazza A, Borbon-Galvez Y, Colicchia C, Dallari F (2021) The sharing economy as an enabler of sustainable and resilient supply chains: a systematic literature review…. In: Summer School Francesco Turco proceedings, pp 1–7

Cheramin M, Saha AK, Cheng J, Paul SK, Jin H (2021) Resilient NdFeB magnet recycling under the impacts of COVID-19 pandemic: stochastic programming and Benders decomposition. Transp Res Part E Logist Transp Rev 155:102505

Cimino A, Longo F, Mirabelli G, Solina V (2024) A cyclic and holistic methodology to exploit the supply chain digital twin concept towards a more resilient and sustainable future. Clean Logist Supply Chain 11:100154

Corsini RR, Costa A, Fichera S, Framinan JM (2024) Digital twin model with machine learning and optimization for resilient production–distribution systems under disruptions. Comput Ind Eng 191:110145

Crane A, Matten D, Spence L (eds) (2014) Corporate social responsibility: readings and cases in a global context. Routledge

Dabija AM (2021) Principles of sustainability: History and evolution. In: *Alternative envelope components for* energy*-efficient buildings*. Springer International Publishing, Cham, pp 5–28
de Farias IV, dos Santos Alvim SL, de Simas D, Frazzon EM (2022) Visibility model for enhancing supply chains resilience. IFAC-PapersOnLine 55(10):2521–2525
Ekinci E, Sezer MD, Mangla SK, Kazancoglu Y (2024) Building sustainable resilient supply chain in retail sector under disruption. J Clean Prod 434:139980
Erol I, Ar IM, Peker I (2022) Scrutinizing blockchain applicability in sustainable supply chains through an integrated fuzzy multi-criteria decision making framework. Appl Soft Comput 116: 108331
Govindan K, Gholizadeh H (2021) Robust network design for sustainable-resilient reverse logistics network using big data: a case study of end-of-life vehicles. Transp Res Part E Logist Transp Rev 149:102279
Greis NP, Nogueira ML, Rohde W (2022) Towards learning-enabled digital twin with augmented reality for resilient production scheduling. IFAC-PapersOnLine 55(10):1912–1917
Health Canada (2024a) Drug shortages in Canada: Fiscal year 2023 to 2024 in review. Government of Canada. https://www.canada.ca/en/health-canada/services/drugs-health-products/drug-products/drug-shortages/2023-2024-review.html#oz
Health Canada (2024b) Building resilience: Health Canada's plan to address health product shortages, 2024 to 2028. Government of Canada. https://www.canada.ca/en/health-canada/services/drugs-health-products/drug-products/drug-shortages/plan-2024-2028.html
Hervani AA, Nandi S, Helms MM, Sarkis J (2022) A performance measurement framework for socially sustainable and resilient supply chains using environmental goods valuation methods. Sustain Prod Consum 30:31–52
Ipek K, Melisa OP, Kumar MS, Ajay K, Yigit K (2023) Using emerging technologies to improve the sustainability and resilience of supply chains in a fuzzy environment in the context of COVID-19. Ann Oper Res 322(1):217–240
Jafari-Nodoushan A, Sadrabadi MHD, Nili M, Makui A, Ghousi R (2024) Designing a sustainable disruption-oriented supply chain under joint pricing and resiliency considerations: a case study. Comput Chem Eng 180:108481
Khan MR, Khan I, Khan MR (2024a) Sustainable supply chain management in the post-COVID era. In: International encyclopedia of business management. Elsevier, pp 1–5
Khan MR, Masood F, Gul I, Khan MR (2024b) Blockchain technology in supply chain management: opportunities and challenges. In: Convergence of Industry 4.0 and supply chain sustainability, pp 275–295
Lotfi R, Khanbaba A, Alkhazaleh HA, Changizi M, Kadłubek M, Aghakhani S, SamarAli S (2024) A viable supplier selection with order allocation by considering robustness, risk-averse and blockchain technology. Comput Ind Eng 193:110319
Lupi F, Cimino MG, Berlec T, Galatolo FA, Corn M, Rožman N et al (2023) Blockchain-based shared additive manufacturing. Comput Ind Eng 183:109497
Masood F, Khan NR, Khan MR (2024) Blockchain-based supply chain management: A challenge yet an opportunity to gauge. In: Achieving secure and transparent supply chains with blockchain technology. IGI Global, pp 267–285
Mirzaee H, Samarghandi H, Willoughby K (2024) Comparing resilience strategies for a multistage green supply chain to mitigate disruptions: a two-stage stochastic optimization model. J Clean Prod 471:143165
Moadab A, Kordi G, Paydar MM, Divsalar A, Hajiaghaei-Keshteli M (2023) Designing a sustainable-resilient-responsive supply chain network considering uncertainty in the COVID-19 era. Expert Syst Appl 227:120334
Mohammed A, Zubairu N, Yazdani M, Diabat A, Li X (2023) Resilient supply chain network design without lagging sustainability responsibilities. Appl Soft Comput 140:110225
Moosavi J, Fathollahi-Fard AM, Dulebenets MA (2022) Supply chain disruption during the COVID-19 pandemic: recognizing potential disruption management strategies. Int J Disaster Risk Reduct 75:102983

Mostaghim N, Gholamian MR, Arabi M (2024) Designing a resilient-sustainable integrated broiler supply chain network using multiple sourcing and backup facility strategies dealing with uncertainties in a disruptive network: a real case of a chicken meat network. Comput Chem Eng 188:108772. https://doi.org/10.1016/j.compchemeng.2024.108772

Patidar A, Sharma M, Agrawal R, Sangwan KS (2023) Supply chain resilience and its key performance indicators: an evaluation under Industry 4.0 and sustainability perspective. Manag Environ Qual Int J 34(4):962–980

Rajabzadeh H, Rabiee M, Sarkis J (2024) Sourcing from risky reverse channels: insights on pricing and resilience strategies in sustainable supply chains. Int J Prod Econ 276:109373

Rame R, Purwanto P, Sudarno S (2024) Industry 5.0 and sustainability: an overview of emerging trends and challenges for a green future. Innov Green Dev 3(4):100173

Rezaei J (2015) Best-worst multi-criteria decision-making method. Omega 53:49–57

Robertson M (2017) A crash course in the science of sustainability. In: The community resilience reader: essential resources for an era of upheaval. Island Press/Center for Resource Economics, Washington, DC, pp 147–161

Safari L, Sadjadi SJ, Sobhani FM (2024) Resilient and sustainable supply chain design and planning under supply disruption risk using a multi-objective scenario-based robust optimization model. Environ Dev Sustain 26(11):27485–27527

Sathiya V, Nagalakshmi K, Jeevamalar J, Babu RA, Karthi R, Acevedo-Duque A et al (2023) Reshaping healthcare supply chain using chain-of-things technology and key lessons experienced from COVID-19 pandemic. Socio Econ Plan Sci 85:101510

Sazvar Z, Tafakkori K, Oladzad N, Nayeri S (2021) A capacity planning approach for sustainable-resilient supply chain network design under uncertainty: A case study of vaccine supply chain. Comput Ind Eng 159:107406

Sharifi, E., Amin, S. H., & Fang, L. (2024) Designing a sustainable resilient and responsive wheat supply chain under mixed uncertainty: a multi-objective approach. J Clean Prod 434:140076. https://doi.org/10.1016/j.jclepro.2023.140076

Singh G, Rajesh R, Misra SC, Singh S (2024) Analyzing the role of digital twins in developing a resilient sustainable manufacturing supply chain: A grey influence analysis (GINA) approach. Technol Forecast Soc Chang 209:123763

Srhir S, Jaegler A, Montoya-Torres JR (2023) Uncovering Industry 4.0 technology attributes in sustainable supply chain 4.0: a systematic literature review. Bus Strateg Environ 32(7): 4143–4166

Suryadi A, Rau H (2023) Considering region risks and mitigation strategies in the supplier selection process for improving supply chain resilience. Comput Ind Eng 181:109288

Torshizi E, Bozorgi-Amiri A, Sabouhi F (2024) Resilient and sustainable global COVID-19 vaccine supply chain design considering reverse logistics. Appl Soft Comput 151:111041

Tsao Y-C, Balo HT, Lee CKH (2024) Resilient and sustainable semiconductor supply chain network design under trade credit and uncertainty of supply and demand. Int J Prod Econ 274:109318. https://doi.org/10.1016/j.ijpe.2024.109318

Verma P, Srivastava R, Kumar S (2025) Blockchain technology: applications and challenges. Blockchain for IoT Systems, pp 1–12

Wang Y, Wang T, Cui W, Zhou G, Liu H (2024) Selecting resilient strategies for cost optimization in prefabricated building supply chains based on the non-dominated sorting genetic algorithm-II: facing diverse disruption scenarios. Sustainability (2071–1050) 16(14):6256

Xun X, Yuqian L, Vogel-Heuser B, Wang L (2021) Industry 4.0 and industry 5.0—inception, conception and perception. J Manuf Syst 61:530–535

Yang J, Liu Y, Morgan PL (2024) Human-machine interaction towards Industry 5.0: Human-centric smart manufacturing. Digit Eng 2:100013

Yavari M, Ajalli P (2021) Suppliers' coalition strategy for green-Resilient supply chain network design. J Ind Prod Eng 38(3):197–212

Yavari M, Bohreghi IM (2025) Developing a green-resilient power network and supply chain: integrating renewable and traditional energy sources in the face of disruptions. Appl Energy 377:124654. https://doi.org/10.1016/j.apenergy.2024.124654.

Yavari M, Zaker H (2019) An integrated two-layer network model for designing a resilient green-closed loop supply chain of perishable products under disruption. J Clean Prod 230:198–218

Yavari M, Zaker H (2020) Designing a resilient-green closed loop supply chain network for perishable products by considering disruption in both supply chain and power networks. Comput Chem Eng 134:106680

Yavari M, Enjavi H, Geraeli M (2020) Demand management to cope with routes disruptions in location-inventory-routing problem for perishable products. Res Transp Bus Manag 37:100552

Yavari M, Mihankhah S, Jozani SM (2024) Assessing cap-and-trade regulation's impact on dual-channel green supply chains under disruption. J Clean Prod 478:143836. https://doi.org/10.1016/j.jclepro.2024.143836

Developing Sustainable and Resilient Supply Chains Using Blockchain Technology

Gregory Barnes

Abstract With global supply chains becoming ever more complex, there is a need for a stable and resilient manner of exchanging data between partner organisations. Blockchain (BC) has been adopted in that regard by numerous organisations with a view to mitigating against a number of potential shocks, and to creating sustainable environmental value. This chapter reviews the current use of Blockchain (BC) technology in the development of supply chains with the aim of making them both resilient and sustainable. Following these principles, a framework is offered for the implementation of the technology in a supply chain that offers positive sustainability benefits. The chapter then discusses a range of applications for BC in a sustainable supply chain that may prove to be worthwhile for both practitioners and academics in the field.

1 Introduction

Global supply chains are the dominant organisational paradigm of modern business. Through dispersed manufacturing, international organisation, and cross border collaboration, the novel advantages of each country can be leveraged to produce real competitive advantage. As the supply chains develop and become more complex, the issue of resilience becomes more pressing. Similarly, the more resilient a supply chain is, the more effectively it will deal with environmental and social challenges. In the last 20 years, we have seen a series of global, systemic shocks that affect the fundamentals of supply chains. This includes the global financial crisis, multiple global conflicts, increasingly nationalistic procurement practices, and the Covid19 pandemic. These have been the result of challenges originating from geopolitical tensions, from health issues, and increasing impacts relating to environmental and climate considerations (Rogerson and Alves 2024). All these issues are becoming ever more prescient for supply chain professionals. This chapter addresses a range of

G. Barnes (✉)
Leeds Beckett University, Leeds, UK
e-mail: g.r.barnes@leedsbeckett.ac.uk

M. Venkatesh et al. (eds.), *Green Manufacturing Technologies in Industry 5.0*,
https://doi.org/10.1007/978-981-95-0363-6_6

issues by looking at them broadly under the auspices of 'sustainability' of the supply chain.

Blockchain Technology (BC) '…is a peer-to-peer network of information technology that keeps records of digital asset transactions using distributed ledgers that are free from control by intermediaries' (Hu and Ghadimi 2022 p42). In effect, this means that organisations can transfer a range of supply chain related information without external accreditations through a 'decentralised' ledger. This allows for significant advantages in complex supply chains that operate internationally, avoiding the problems of variable local/national accreditations, and standardising information between partners. The technology does have some major constraints however, which will be addressed alongside the potential solutions offered.

This chapter engages with the issues around BC and supply chains from a pragmatic perspective. Using evidence from academic and industry sources, a model is offered for consideration by organisations seeking to implement BC technology for sustainability purposes within their supply chain. First, a review of the recent perspectives on the use of BC in this context presented. Secondly, a framework for the adoption of the technology is described. This aspect is built around the view of Blockchain technology as a resource that may be utilised by the organisation in order to effect positive sustainability outcomes (Dubey et al. 2020) and ensure longevity of their supply chain outcomes. To conclude, some final thoughts are offered along with a view of the gaps in current research in the area that would benefit further exploration alongside validation of some of the academic arguments in this chapter.

2 Recent Perspectives on Blockchain Technology

Blockchain technology (BC) has been around for over a decade, since the original framing white paper was published by Nakamoto in 2008 (Alkhudary et al. 2024). Supply chain scholars have picked up on the potential significance of the technology, while practitioners have seen a potential operational benefit and acted accordingly. Several major companies have integrated the system into supply chain operations and are gaining benefits from the activity (more of which is discussed below in Sect. 5). The field is therefore mature enough to examine a range of perspectives on its application, and to critically evaluate the value offered by the system.

The academic work that frames the use of this technology in the context of supply chain risk has focused on two dimensions: 'Operational' risks and 'Disruption' risk (Tang 2005). These dimensions span the length of the supply chain and the way it may become less effective as circumstances impinge upon its operation. Operational risks are those aspects that are situated within the organisation and include production processes, compliance, manufacturing issues, material flow risks, and delays to market (Chelh and Ababou 2023; Wasim Layaq et al. 2019). Disruption risks are those external or business environment hazards that impact upon an organisation's activities. We have seen a number of disruptions in recent years from the Covid19 pandemic, red sea military conflict, to the war in Ukraine (Mircea et al. 2022). More recently, geopolitical trade disruption in North America may have additional impacts particularly for sustainability aspirations. Much of the work on the technology focuses on the improvement of operations within an existing network rather than more structural integration that speaks to broader aspects of supply management (Nandi et al. 2020). This is through the improvement of activities such as information sharing, coordination, and responsive action. As a result of the Operational and Disruption risks, many implementations of BC relate to mitigating hazards rather than building resilience. Four critical features of Blockchain best encapsulate the efficacy of the technology for supply chain management: Transparency, distribution of the ledger, immutability, and traceability.

Transparency across a supply chain is of crucial importance for effective management. BC provides several mechanisms that facilitate both transparency and traceability of activities through the chain. Firstly, Blockchain, depending on the implementation, allows for all the internal processes within the chain to be visible to the relevant partners and overcomes trust and confidence in the network. This is particularly important in the complex interrelations of supply that are now globally used (Kumar Singh et al. 2023). The principle of transparency allows for more accountability of the actions of supply chain partners. While this is important from a resilience perspective, the benefits for understanding efficiency of process and where value is added are also relevant. This aspect is particularly crucial when looking at productive and efficient supply chains that are sustainable, and not overly lean.

The distributed nature of the ledger means that no single point of record is at risk (Osato Itohan Oriekhoe et al. 2024). There are multiple records available and where traditional systems may be liable to corruption or lack of access to the system, the ledger may still be referenceable by an organisation. This redundancy allows for more confidence in the system and hopefully continued operation. All stakeholders have access to the system (according to the type of BC network) and may have real time visibility of the ledger entries from creation to distribution that should avoid the "challenge of information silos" (ibid. p151). As a result, a more holistic approach toward enterprises may be employed.

The immutability of blockchain means that once a transaction is recorded it may not be rescinded or retroactively amended either by error or malicious intent (Kshetri 2018). This is a function of the distribution of the ledger, and of the cryptographic nature of the data storage foundational to the technology. From a technology adoption perspective then, it meets many of the principles of secure data handling and management (Lohmer et al. 2020). A further dimension of security is offered through the consensus model (where implemented). Additions to the chain are required to be verified by a consensus of participants to ensure that there is validity to the entry (Fang and Zhang 2024). In principle, where using automation, this should mean that there are multiple redundancies regarding the addition to the chain (several IoT devices for example). In the case of agricultural products, or less developed manufacturing contexts, there may be opportunity for additions by multiple points early in the chain that are not accurate. Through the use of immutable data principles, BC may improve fraud and errors within supply chain information (Chelh and Ababou 2023). This improvement is aligned with a range of organisational strategic benefits, but from a purely sustainability perspective, the accuracy of the data coming from a supply chain is of critical importance to ensure a realistic response. The intention of these layers of overlapping validation is a level of credibility to the entries on the Blockchain and confidence between members.

Traceability is a major aspect of Blockchain Technology that is highly pertinent to the discussion of sustainability. In establishing a reliable and transparent trail of information relating to a supply chain, the BC provides a clear understanding of the digital footprint of activities, production, or manufactured goods (Beck et al. 2023). New data 'blocks' or tokens are produced to trace each transaction and should as a result facilitate agile responses to demand, and changes in operational capabilities (Wang et al. 2019). Key data is aggregated in the ledger, with ERM software such as SAP becoming increasingly capable of integrating these sources into their manufacturing workflows. With these aspects of BC in place, there should be a full traceability of products that in turn supports quality assurance processes in several industries as will be discussed later in this chapter.

Blockchain technology can support risk management in a number of ways that are both reactionary for more immediate issues, and in a more long-term manner that would support overall resilience of the supply chain. Supply chain resilience in this context is the ability for a supply chain to effectively recover from a shock or hazard event (Lohmer et al. 2020). Resilience encompasses the "capacity to anticipate, detect, and defend itself against risks" (p2) before then returning to a steady level of performance. Transparency allows for greater knowledge of the supply chain and therefore a capacity to identify when and where risk may emerge, and how to best recover from hazard shocks effectively. The use of Smart Contracts is also a boon for resilience in terms of how actions can be quickly taken by an organisation when mitigating for changing circumstances and managing risk profiles.

3 Blockchain and Sustainability

Blockchain has been examined from the perspective of improvements in sustainability in a number of ways. This is in part an effort to explore the unique facilities of the technology, although from an organisational perspective there is also an imperative to prove a business case for new innovative expenditure—including an additional factor of sustainability into the investment makes the case more effective. The enhancement of transparency throughout the supply chain, along with the traceability for the supplies makes a larger difference to the degree to which one may track the sustainability of a product or good (Fang and Zhang 2024). This is enhanced further by the potential for tracking the miles that an item has travelled in the supply chain. Such information may then be fed into process optimisation to reduce food miles for example (Etemadi et al. 2021). Tier 3 emissions are tracked in this manner as well by a number of organisations, and in particular those in the public sector (Mircea et al. 2022) and for carbon credits (Hua et al. 2020; Sipthorpe et al. 2022).

There are additional benefits in terms of the integration with other technologies. There are examples where IoT allows for triangulation of the sourcing information. QR and RFID may be used by consumers for further information about products and their sources in order to understand the journey of that product and become more informed in their choices. As a result, the full lifecycle of a product may be further understood by both consumers at the point of use, and waste handlers when disposing of a product (Wasim Layaq et al. 2019).

The green practices in supply chains may be improved through the process of deeper tracking and compliance monitoring. Environmental data becomes accurately traceable and reportable in a similar manner to carbon tracking, but more importantly, this allows for the tracing of any compounds or activities that may have further negative externalities (such as dangerous chemical compounds). Energy consumption is also a major issue within all supply chain contexts as it has become a major cost in the manufacturing process (Chelh and Ababou 2023 p2).

The level of inventory held by organisations may be more effectively monitored in a validated and secure manner. Having such information available to the supply chain allows then for an effectively automated kanban system to be operated across even complex networks (Lenarduzzi et al. 2018). This process has sustainability benefits in terms of effective resource allocation, lean manufacturing, and the reduction of waste across the system leading to overall resource efficiency.

Blockchain may also facilitate the implementation of circular economy principles such as 'reuse, recycling, and remanufacturing' (Upadhyay et al. 2021). These principles are inherently linked to many of the other aspects of sustainability and effective organisational performance, but the processes afforded specifically by BC make operationalising them even more effective.

The innovations discussed above have been recognised in the opportunities present in shared information networks such as the European Union 'data market'. That public policy agenda is seeking to develop improved sustainability outcomes through a Pan European data network built on Blockchain principles. This includes the development of aligned data set for training Artificial Intelligence (AI) (Konya and Nematzadeh 2024; O'Connor et al. 2024; Purtova and Van Maanen 2024; Sanjay et al. 2024). In doing so, the EU is embracing the competitive advantages of Industry 4/5.0 at a supranational level for positive social and environmental outcomes (Lynn H. Kaack et al. 2022; Tchetin 2022).

4 Adoption of Blockchain for Sustainability Outcomes

The following section looks at the implementation of the technology in an Industry 4.0/5.0 context from a critical perspective and examines how it matches the aspirations of improving sustainability and resilience in a supply chain. Importantly is the perspective that Blockchain can be used in a strategic manner once viewed as resource by the organisation (Yuthas 2021). The 'Resource Based View' (RBV) is one that focuses on the development of unique assets that the organisation can exploit. BC is in this perspective an asset that creates and supports informational assets that are unique and immutable to an organisation and are a source of competitive advantage. As a result, we can view the data held by the organisation as adding value to the management of risk at a foundational level.

This section critically views the factors around the implementation of Blockchain in a supply chain along the several lines: the types of risk that are to be addressed by the technology implementation, the type of chain to be used, the application of Smart Contracts, and the resource constraints facing an enterprise.

4.1 Types of Risk

In order to engage with the principles of the resource value, sustainability and resilience need to be seen in the context of the risks that may pertain to the organisation. It is worth looking beyond the definitions commonly offered that often reduce risks to a simple dichotomy between 'emerging' and 'principal' (Ashby 2019; Merna and Al-Thani 2008). Instead, it is worth viewing risk as being a factor of time, and proximity to risk becomes a factor that may be measured against the relevance of that risk to the organisation. This perspective sets 'temporal' (or immediate time-based risks) as being ones that are dealt with through flexibility and quick footed addressing of changes to the supply chain.

Structural risks are those associated with processes and the composition of organisations or institutions. They are within the purview of organisations to design and structure themselves. While subject to cognitive and institutional bias, they are generally within the realm of change that may be directly affected. Systemic risks however are those areas associated with the broader context within which organisations operate. As such they include the overall emphasis or understanding of certain implicit risks in how the business environment is ordered. An example of how the two compare would be the organisation of a specific bank, as opposed to the overall perception of risk within the subprime mortgage market leading to the financial crash (Fig. 1).

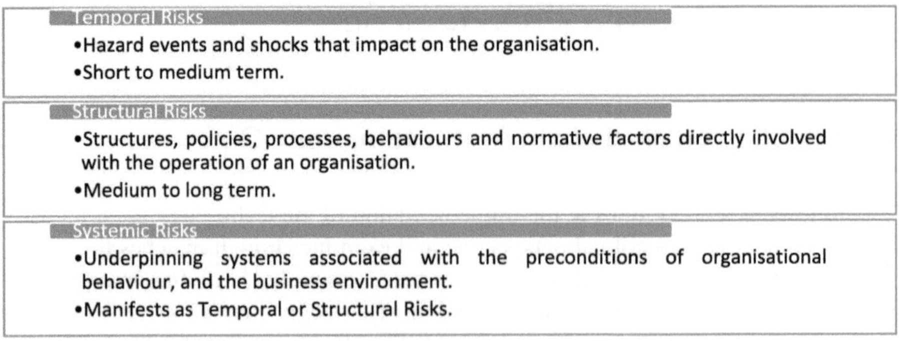

Fig. 1 Categorisations of risk (author's own)

With these points in mind, the issue of sustainability therefore needs to be considered from a Systemic perspective as it has a potentially paradigmatic impact on an organisation at both a Structural and Temporal manner. Systemic issues are difficult for an organisation to navigate; Organisations can prepare for there to be some systemic disruption, but their scope to minimise both the structural and temporal types of risk in the process is far greater. Risk mitigation through BC should be considered outside of the issue of immediate response and instead toward sustainable structures that support risk management holistically. This means moving beyond BC as a way to manage existing reactive issues and into the preconditions for organisational behaviour to be effectively managed.

4.2 *Types of Blockchain Implementation*

Much attention has been paid to the application of public blockchains, sometimes referred to as 'permissionless'. These are open and transparent ledgers that allow for public accessibility and understanding of their contents. While the transactions are public, the users may remain anonymous. This allows for the operation of cryptocurrencies and the like to exchange value tokens across the network and there to be an understanding of value. Semi-private and private blockchains exist as well that have a permission associated with their access. The level of permission may be totally restricted (a private ledger), semi-public (allowing for some elements of transparency), or consortium (associated with a particular network) (Alkhudary et al. 2024; SAP 2025). These varying types are important as the implementation of various supply chains lends itself to type of network supporting them.

While there are varieties of implementation, this also means that transparency more broadly needs to be considered. Having a totally open and porous approach to the chain would allow for all parties in a supply chain to have visibility. Where a hedging or multi source supply approach is utilised by an organisation, then there may be a disadvantage to sharing information. Similarly, where cost/price transparency may be inferred, this may also lead to an undesired level of information transfer to partners. The inverse is of course applicable, in that when dealing with certain types of contract relationship, the use of transparent 'on-cost' or 'open book' approaches may be desirable to understand the manner in which final price is constituted (Agndal and Nilsson 2010).

Data is at the heart of the process of addressing supply chain issues. Once these have been identified, then other Industry 5.0 technologies may be employed that rely on the core service being facilitated. As a decentralised ledger system, blockchain facilitates aspects of operations, but there are limitations to how this may be operationalised.

4.2.1 The Use of 'Smart Contracts'

Contracts as traditionally described are those agreements between two parties relating to their reciprocal responsibilities. Smart Contracts are used in this way, as a clear and effective replacement for the 'traditional' contracting process to improve business efficiency (Hu and Ghadimi 2022) but at a higher rate with significant benefits through automation. A second form however is the use of these contracts as a programming interface. These Smart Contracts are any 'Computer protocols which are developed and used to run decentralised applications on the blockchain' (Dietrich et al. 2020 p166). This may include elements of the 'traditional' contracting process, as well as the integration of technologies using the blockchain as an information pipeline. Smart Contracts may additionally be related to the tokenisation of physical processes (such as logistics or manufacturing) through Internet of Things (IoT) or RFID technologies (ibid. p168). As a result, the technology offers significant power for an organisation prepared to automate elements of their workflow in this manner.

The danger of 'smart' contracts is that where a 'hands off' approach to the qualification of information is taken (through public or semi-public chains) there may be a flaw introduced into the system. Similarly, where Smart Contracts are used to pivot in terms of supply options, there will be a need to ensure that companies are carefully vetted against agreed specifications. Therefore, while the system can *support* resilience, it does not *create* it, and managerial responses will still need to be considered.

The processing time for adding and accessing data onto the chain can be significant when dealing with larger chains. Each level of confirmation can add exponentially to the processing time. As a result, technologies have been integrated into the core of blockchains to facilitate longer documents into the relatively small ledger entry sizes. This allows for a search or verification of the entry without necessarily tracing the entire record that has been logged with it. Non Fungible Tokens (NFTs) are essentially built on the 'record and link' approach (Senkardes 2021; Wang et al. 2021).

Smart contracts in principle should not allow for this 'record and link' approach, but the size of the documents that may be included is an issue. The Ethereum specification necessarily limits documentation to 24 kb to allow for rapid dispersion and verification of data. Any larger size of document may lead to an increase in time to process additions to the chain, issues around verification, and scalability. Systems such as 'Solidity' have therefore been developed (Solidity 2025) that augment the Blockchain entries through a compressed programming language. This approach allows for a deeper level of integration than may otherwise be afforded by the 'record and link' which may be undermined through the corruption of addresses.

4.2.2 Resource Allocation and Constraints

The resources required to develop and implement a technology such as this are significant, especially when viewed across the whole enterprise. There are initial investments, but also ongoing variable costs that can increase as the system is scaled. While Blockchain technology contributes to the overall resource manifold that organisations operate within there is still a need to grow that resource base in order to operationalise the system. This can be in terms of the financial resources that are required, but also the ongoing skills and technical capability of organisations that need 'in demand' staff to manage them. The level of technical knowledge of supply chain managers needs to be significantly improved to retain relevance in the future.

The use of blockchain may be linked effectively to the information required for Enterprise Risk Management (ERM) principles (Barnes 2024). These principles are aligned with the dimensions of risk mentioned earlier in this section. Importantly, once we view the use of information within an organisation as a unique resource, and the blockchain as a repository of that resource, then the risk handling approach is of critical importance. ERM offers a way to manage the resources within the organisation and the broader network to secure the value inherent within it. The offset in terms of allocation of scarce organisational resources is however that managing an ERM necessarily suggests a larger organisation being involved. They are simply too unwieldy for most SMEs.

Working across blockchain and non-blockchain supply chains or other non-integrated markets (Ravindran et al. 2021 p494) is a problem. The reliance on the main informational and organisational paradigm being one revolving around a potentially bespoke system means that changing suppliers becomes difficult. Due to scale issues relating to ERM, we may unwittingly be 'locking out' potentially innovative alternative sources of supply, or sources of cost reduction and market comparative advantage. This is another balancing act that organisations need to tread to effectively weigh up the choice to employ BC in their supply chain. An important consideration when establishing an ERM or BC platform across several countries is the issue of investment. Companies and regions that are still developing a technology base, and may be less able to access financing could struggle to participate effectively (Tharatipyakul et al. 2022). Those costs are difficult for new suppliers to bear speculatively and may indeed prevent many SMEs from supplying to BC using clients. Additionally, the nature of closed BC systems (the type of system) may prevent one supplier from easily swapping between chains: Thus losing the scale efficiencies across several clients and supply opportunities.

Much has been stated in older articles and studies about the virtue of blockchain in resisting the impacts of cybercrime. This is now becoming more problematic, in part due to the ubiquity of the technology, the opportunity for bad actors to add entries (Sachan et al. 2023) but also the manner in which those items stored on the chain may in future be utilised. Malicious code is possible to hide on an Ether network or Solidity entry (Gimenez-Aguilar et al. 2023), while AI may be impacted by hidden prompts (IBM 2024). This means that more attention needs to be placed on how BC is implemented to prevent adverse effects (Xu 2016). All these factors are important to be considered when looking at the potential to use the technology for sustainability purposes, particularly when large data sets are likely to be drawn from BC derived networks.

When considered critically there are some benefits to the implementation of the technology in this context, but also some aspects that are a challenging fit. The key determinant factor will be the applicability of the technology to the context within which it is to be employed, as that will be a balancing factor against the resources required for its implementation. The next section of this chapter looks at some of the examples and applications of the technology to further explore its applicability.

5 Applications and Examples

While there have been examples used throughout the chapter, it is worth reviewing the application of the technology within three specific contexts, namely; Materials, Manufacturing, and Intellectual and Informational Processes. The latter of these three is more specifically aligned with the aspirations of Industry 5.0 contexts than some of the former where 4.0 technologies have taken predominance. The important factor to note is that both paradigms are facilitated by the technology and the benefits it has to offer.

Materials, Goods, and Logistics handling has been the focus of research in the field of blockchain integrated supply chains. Systematic literature reviews have focused on these as being the most obvious manifestation of the technology as there are the greatest number of potential scaled benefits.

Supporting food supply chains is possible using blockchain and there are many positive benefits. Tracing and tracking management allows for real-time monitoring of the levels of supply and leads to less wastage. This is becoming increasingly important for large retailers. For example, in the UK, the main 4 supermarkets are working to improve wastage levels from their supply chains in response to climate change legislation requirements (HMG 2023). Tesco is notable for having integrated their fish packaging waste stream using Oracle blockchain technology (ESM Magazine 2023) with the aspiration to reduce waste from this source by 50%.

There are good examples of where blockchain has allowed tracing of vegetables and plant matter in response to health concerns. Walmart used this technology to trace a salmonella outbreak in 2018 that was linked to a Romaine lettuce producer (Walmart 2018). This in turn provided a basis for the supermarket to improve overall food safety and traceability for sustainability outcomes. When combining this technology with IoT and RFID (Rajabzadeh and Fatorachian 2023), there are opportunities for improving transparency for consumers. Food and beverages can then have meaningful and accurate information provided as to their source, and their constituent components. This further allows information about the handling of food to be accurately tracked which is important when dealing with allergies and contaminants (BBC News 2021).

Environmental tracking is possible as BC allows for verification of farming practices and processes. This can result in the adoption of sustainable farming practices within the supply chain even in less developed economic contexts. This is due to the issue of environmental credits to supply chain partners using even relatively modest smartphone apps. Regenerative agricultural processes can be supported as the chain partners are able to evidence both the outcomes from supply and also the inputs into the farm (Upadhyay et al. 2021). These processes incentivise farmers that are working to promote soil health, battle climate change impacts, and improve local ecological systems. The blockchain allows for immutable evidence of such activities.

BC has been utilised to address the issues of Illegal, Unreported and Unregulated (IUU) fishing in several areas. By creating a transparent and immutable record related to each catch, the origins of each source may be traced by both processors and consumers. This system help enforce fishing rights regulations, labour regulations, and environmental regulations that impact the industry. Furthermore, the transparency holds producers and fishing companies to account publicly for their actions, and increases pressure when transgressions are identified. New Zealand (EllenR 2021) and Norway (Mathisen 2018) are specific examples where this has been applied for some time. The European Union has also developed a traceability system along BC lines to facilitate the Common Fisheries policy through the TRACE4EU project (Luna et al. 2024).

Coffee has been a major supply chain context within which blockchain has been explored and is also very well suited. The raw materials are subject to long supply distances, a concentration of countries from which they are available at scale, and well documented issues around labour rights. In addition, where 'fairtrade' and sustainability principles have been implemented, there are responsibilities placed on the end retailer to verify the sourcing in a transparent manner. Indeed, this is a cornerstone of that offer to the consumer. The data handling and processing applications of blockchain facilitate the information transfer effectively across the chain, which is crucial where rapidly changing global prices may additionally affect the overall chain, or where multiple sources are involved (Tharatipyakul et al. 2022). In addition, quick payments and reduction of legal contracts facilitated through Smart Contracts may be advantageous. This is particularly important in areas where political instability may result in unreliable government legislative frameworks (Ravindran et al. 2021). Ultimately, some trust is required among partners, but where the main systemic conflicts can be resolved, then interorganisational partnership is more likely to be effective.

Manufacturing applications of BC have included a range of activities and benefits for the supply chain. This may be viewed in two areas: the management of supplier-based risk, and the operational benefits that may be accrued from optimised manufacturing processes.

Enhancing supplier accountability through immutable records allows for effective interchange of orders, invoices, and payments through a chain. Even where traditional payment methods are utilised through banks, there are still benefits to the communication of these processes securely through chains. Construction contracting in particular benefits from this application as the use of open-book systems is widespread, and built into some major international standard forms of contract (Wannenwetsch et al. 2023). Vendor selection according to both pricing and sustainability indicators may be additionally possible because of an open approach to BC information sharing. This would radically alter the approach that companies use when making choices quickly regarding their supply. Emissions could be radically adjusted as a result.

Demand management is a key factor in the optimisation of manufacturing and operations. Effective demand management is also the best way to change the level of sustainability from a long supply chain for a purchasing organisation. BC allows for the communication through the whole chain of what levels of demand are required at any given time (Dubey et al. 2020) thus leading to production of what is needed at the time it is needed.

Traceability of the constituent components of manufactured goods has become an important aspect that addresses not only to stakeholder demands to ensure sources are environmentally accountable but also matching broader company aspirations. Clothing supply chains are particularly well researched. The issue is that attempting to trace networks that are increasingly complex and obfuscated for several reasons to do with the original sources. The reason that food *may* be more successful for application of BC is that the sourcing to create any single item of food should be more obvious and immediately traceable. A banana for example will only have one source, while a loaf of bread may have many dozens. When we look at clothing, the rise of production in countries like Vietnam is only possible through more invasive investigation of the make-up of the plant fibers that constitute the clothing. There have been investigations from 2022 onwards into such issues and calls for greater transparency of sources.

Intellectual and informational processes are likely those applications most aligned with the technological developments around Industry 5.0. As such they are more associated with the development and exploitation of information for sustainability ends, often utilising AI for the purpose (Kaack et al. 2022; Goralski and Tan 2020).

While healthcare has obvious physical and materials elements associated with it, the application of BC has been beneficial around information management. The security of patient data is an obvious imperative and there are several platforms built around this principle (Cyran 2018; Shen et al. 2019; Yaqoob et al. 2022). Underpinning each application is the same principle of information value, transparency, immutability, and security that BC is well suited toward.

In addition, the pharmaceutical industry is seeking to use the same principles to secure intellectual property (Akram et al. 2024; Bocek et al. 2017; Clark and Burstall 2018). For our discussion in this chapter, the importance of keeping the IP related to research and development intact makes the organisation more resilient, and sustainable. They can then retain the income from their R&D investment and reinvest to develop further. Of critical importance is also the ability to prevent counterfeiting, enforcing regulatory compliance in manufacturing, ensure tolerances around specification, and prevent wastage.

6 Areas for Future Research

While there are several dimensions for consideration in this chapter, there are more areas that may be explored in future research. The relative commodity suitability poses a further area for examination. While there are obvious applications in a variety of industries as outlined in this chapter, there are significant constraints for further applications.

There are further research opportunities to be explored in this field, but this is an area that professional applications are leading the charge regarding the possibilities of the technology. The integration of Blockchain with other technologies is one area for future development. Internet of Things (IoT) is a symbiotic technology that requires some of the benefits of the BC fundamentals. The European Union Data Market agenda in particular is developing the precursors for the use of BC derived data for processing through Artificial Intelligence (AI) (Beatrice Garske et al. 2021; Francisco and Linnér 2023; PromethEU 2023; Purtova and Van Maanen 2024).

There is a case to be made for the exploration of the potential return on investment for companies; in particular, SMEs for whom investments discussed herein would be difficult. Empirical studies into the potential Return on Investment (ROI) from implementation of BC would be beneficial to verify many of the points made in this chapter. For example, the improvement of climate change and scope 3 emission data leading to a more accurate purchasing of carbon credits within certain industries.

The issue of social sustainability has not been well explored in academic literature. There are many studies of the potential impacts of environmental sustainability (as have been discussed here) but the further development of social dimensions is limited (Chaudhuri et al. 2023). This is a shame as there are a number of risks, impacts, and mitigations that are linked inherently to 'green' sustainability that require further consideration to be more holistically relevant.

7 Conclusions

The aspiration in blockchain technology is the creation of a single 'source of truth' that a supply chain may refer to. This fosters and promotes trust among all the stakeholders associated directly or indirectly with the Supply Chain. By doing so, governance of an organisation can in turn be promoted and lead to improving organisational performance. The other aspects that benefit organisations include overall efficiency, productivity, transparency (for reporting on performance) and obviously organisational sustainability. As technology improves, there is a desire to continue to use this technology by addressing some of the constraints mentioned in this article. These include the scalability and interoperability challenges that are presently felt. Most of all is the challenge to gaining low technology contexts to buy into the technology platforms necessary to run such a challenging system. Areas where raw materials and emissions are generated are the last to be possessed of information technologies that are the precursor for Blockchains unless there is a corporate sponsor involved due to the potential level of investment required. An inverse issue is that when corporate systems are used, they may require a 'closed' protocol system, rather than a public platform which makes interoperability difficult.

There are many benefits for sustainability resulting from the use of BC in a supply chain. The role of BC as a technology that underpins data validity allows for better decision-making on the back of complete and accurate understanding of the impact of operations. Having a resilient and integrated data system allows for tracking impacts across the whole operation and therefore facilitates scope 3 emission reporting, and other environmental impacts.

The use of BC as a foundational technology for the operation of the whole enterprise moves resilience supporting activities from hazard response and reactivity toward a systemic approach. This is an important shift for any organisation seeking to address systemic risk areas such as climate change and environmentally responsible supply sourcing. The shift toward addressing systemic issues results in a more holistic management approach for organisations that matches with the aspirations of ISO31000:2018 and ISO14001 for Risk Management and for Environmental Management, respectively. Both frameworks encourage and support the accurate and timely management of relevant and transparent data. BC sits well within these frameworks (Barnes 2024) and gives an effective route to sustainable supply chain management.

References

Agndal H, Nilsson U (2010) Different open book accounting practices for different purchasing strategies. Manag Account Res 21(3):147–166

Akram W, Joshi R, Haider T, Sharma P, Jain V, Garud N, Singh N (2024) Blockchain technology: a potential tool for the management of Pharma supply chain. Res Soc Adm Pharm 20(6):156–164

Alkhudary R, Queiroz MM, Féniès P (2024) Mitigating the risk of specific supply chain disruptions through Blockchain technology. Supply Chain Forum Int J 25(1):1–11

Ashby S (2019) Risk management. ICSA Qualifying Programme. Chartered Governance Institute, London

Barnes G (2024) Risk managment frameworks and the implementation of Blockchain technology [online]. In: *Achieving secure and transparent supply chains with Blockchain technology*. IGI Global, p 15. https://www.igi-global.com/chapter/risk-management-frameworks-and-the-implementation-of-blockchain-technology/337349

Beck J, Birkel H, Spieske A, Gebhardt M (2023) Will the Blockchain solve the supply chain resilience challenges? Insights from a systematic literature review. Comput Ind Eng 185:109623

Bocek T, Rodrigues BB, Strasser T, Stiller B (2017) Blockchains everywhere – a use-case of Blockchains in the Pharma supply-chain [online]. In: *2017 IFIP/IEEE Symposium on Integrated*

Network and Service Management (IM), May 2017. Lisbon, Portugal. IEEE, pp 772–777. http://ieeexplore.ieee.org/document/7987376/ Accessed 13 Mar 2025
Chaudhuri A, Bhatia MS, Kayikci Y, Fernandes KJ, Fosso-Wamba S (2023) Improving social sustainability and reducing supply chain risks through Blockchain implementation: role of outcome and behavioural mechanisms. Ann Oper Res 327(1):401–433
Chelh S, Ababou M (2023) Improving green supply chain resilience through Blockchain innovation: a risk mitigation perspective. In: E3S Web of Conferences, vol 412. EDP Sciences, p 01056
Clark B, Burstall R (2018) Blockchain, IP and the pharma industry—how distributed ledger technologies can help secure the Pharma supply chain. J Intellect Prop Law Pract 13(7):531–533
Cyran MA (2018) Blockchain as a foundation for sharing healthcare data. *Blockchain in Healthcare Today* [Online], March. https://blockchainhealthcaretoday.com/index.php/journal/article/view/13. Accessed 13 Mar 2025
Dietrich F, Turgut A, Palm D, Louw L (2020) Smart contract-based Blockchain solution to reduce supply chain risks [Online]. In: Lalic B, Majstorovic V, Marjanovic U, Von Cieminski G, Romero D (eds) *Advances in production management systems. Towards smart and digital manufacturing*, vol 592. Springer International Publishing, Cham, pp 165–173. https://link.springer.com/10.1007/978-3-030-57997-5_20. Accessed 10 Mar 2025
Dubey R, Gunasekaran A, Bryde DJ, Dwivedi YK, Papadopoulos T (2020) Blockchain Technology for Enhancing Swift-Trust, collaboration and resilience within a humanitarian supply chain setting. Int J Prod Res 58(11):3381–3398
EllenR (2021) *Blockchain Supply Chain Traceability Project* [Online]. https://www.pmcsa.ac.nz/2021/02/21/blockchain-supply-chain-traceability-project/
ESM Magazine (2023) *Tesco uses discarded plastic collected from coastal areas for packaging* [Online]. ESM Magazine. https://www.esmmagazine.com/packaging-design/tesco-uses-discarded-plastic-collected-from-coastal-areas-for-packaging-231115 Accessed 13 Mar 2025
Etemadi N, Borbon-Galvez Y, Strozzi F, Etemadi T (2021) Supply chain disruption risk management with Blockchain: a dynamic literature review. Information 12(2):70
Fang L, Zhang X (2024) Optimizing resilience in agricultural supply chains under Blockchain. Acad J Sci Technol 12(2):226–229
Francisco M, Linnér B-O (2023) AI and the governance of sustainable development. An idea analysis of the European Union, the United Nations, and the world economic forum. Environ Sci Pol 150:103590
Garske B, Bau A, Ekardt F (2021) Digitalization and AI in European agriculture: a strategy for achieving climate and biodiversity targets? Sustainability 13(9):4652
Gimenez-Aguilar M, De Fuentes JM, Gonzalez-Manzano L (2023) Malicious uses of Blockchains by malware: from the analysis to smart-Zephyrus. Int J Inf Secur 22(5):1445–1480
Goralski MA, Tan TK (2020) Artificial intelligence and sustainable development. Int J Manag Educ 18(1):100330
HMG (2023) *The waste prevention programme for England: maximising resources, minimising waste* [Online]. GOV.UK. https://www.gov.uk/government/publications/waste-prevention-programme-for-england-maximising-resources-minimising-waste/the-waste-prevention-programme-for-england-maximising-resources-minimising-waste Accessed 13 Mar 2025
Hu Y, Ghadimi P (2022) A review of Blockchain technology application on supply chain risk management. IFAC-PapersOnLine 55(10):958–963
Hua W, Jiang J, Sun H, Wu J (2020) A Blockchain based peer-to-peer trading framework integrating energy and carbon markets. Appl Energy 279:115539
Kaack LH, Donti PL, Strubell E, Kamiya G, Creutzig F, Rolnick D (2022) Aligning artificial intelligence with climate change mitigation. Nature Climate Change
Konya A, Nematzadeh P (2024) Recent applications of AI to environmental disciplines: a review. Sci Total Environ 906:167705
Kshetri N (2018) Blockchain's roles in meeting key supply chain management objectives. Int J Inf Manag 39:80–89

Kumar Singh R, Mishra R, Gupta S, Mukherjee AA (2023) Blockchain applications for secured and resilient supply chains: a systematic literature review and future research agenda. Comput Ind Eng 175:108854

Lenarduzzi V, Lunesu MI, Marchesi M, Tonelli R (2018) Blockchain applications for agile methodologies [Online]. In: *Proceedings of the 19th international conference on agile software development: companion, May 21, 2018*. Porto Portugal. ACM, pp 1–3. https://dl.acm.org/doi/10.1145/3234152.3234155 Accessed 11 Mar 2025

Lohmer J, Bugert N, Lasch R (2020) Analysis of resilience strategies and ripple effect in Blockchain-coordinated supply chains: an agent-based simulation study. Int J Prod Econ 228:107882

Luna M, Fernandez-Vazquez S, Tereñes Castelao E, Arias Fernández Á (2024) A Blockchain-based approach to the challenges of EU'S environmental policy compliance in aquaculture: from traceability to fraud prevention. Mar Policy 159:105892

Mathisen M (2018) The application of Blockchain technology in Norwegian fish supply chains

Merna T, Al-Thani FF (2008) Corporate risk management. John Wiley & Sons Ltd, New Jersey

Mircea M, Stoica M, Ghilic-Micu B (2022) Analysis of the impact of Blockchain and internet of things (BIoT) on public procurement. IEEE Access 10:63353–63374

Nandi ML, Nandi S, Moya H, Kaynak H (2020) Blockchain technology-enabled supply chain systems and supply chain performance: a resource-based view. Supply Chain Manag Int J 25(6):841–862

O'Connor R, Bolton M, Saeri AK, Chan T, Pearson R (2024) Artificial intelligence and complex sustainability policy problems: translating promise into practice. Policy Design Pract 7(3):308–323

Oriekhoe OI, Ashiwaju BI, Ihemereze KC, Ikwue U, Udeh CA (2024) Blockchain technology in supply chain management: a comprehensive review. Int J Manag Entrep Res 6(1):150–166

Pret Allergy Death: Parents Welcome Natasha's Allergy Law (2021) *BBC News* [Online], 1 October. https://www.bbc.com/news/uk-58756597. Accessed 13 Mar 2025

PromethEU (2023) PromethEUs' joint publication on EU data strategy. The EU'S data strategy from a multifaceted perspective. Views from Southern Europe. PromethEU

Purtova N, Van Maanen G (2024) Data as an economic good, data as a commons, and data governance. Law Innov Technol 16(1):1–42

Rajabzadeh M, Fatorachian H (2023) Modelling factors influencing IoT adoption: with a focus on agricultural logistics operations. Smart Cities 6(6):3266–3296

Ravindran, U., Bhardwaj, P. and Vamsi, P. R. (2021) Blockchain design for securing supply chain management in coffee retailer network. International Journal of Scientific Research in Science and Technology, 7(4), pp. 492–502

Rogerson M, Alves K (2024) Climate change and modern slavery in public procurement. January [Online blog]. https://files.modernslaverypec.org/production/assets/downloads/MSPEC_Procurement_Research_Summary_2024-01-17-152646_sesm.pdf?dm=1736268038. Accessed 1 Feb 2025

Sachan RK, Agarwal R, Shukla SK (2023) Identifying malicious accounts in Blockchains using domain names and associated temporal properties. Blockchain Res Appl 4(3):100136

Sanjay T, Neha B, Ercan O (2024) What was, is and will be: a bibliometric analysis of European Green Deal (EGD). Int J Technol Manag Sustain Dev 23(1):85–104

SAP (2025) *Blockchain: the new technology for trust* [Online]. https://www.sap.com/products/artificial-intelligence/what-is-blockchain.html. Accessed 1 Jan 2025

Senkardes CG (2021) Blockchain technology and NFTs: a review in music industry. Pressacademia 8(3):154–163

Shen B, Guo J, Yang Y (2019) MedChain: efficient healthcare data sharing via Blockchain. Appl Sci 9(6):1207

Sipthorpe A, Brink S, Van Leeuwen T, Staffell I (2022) Blockchain solutions for carbon markets are nearing maturity. One Earth 5(7):779–791

Solidity (2025) *Solidity* [Online]. Solidity Documentation Hub. https://docs.soliditylang.org/en/latest/. Accessed 1 Jan 2025

Tang CS (2005) Perspectives in supply chain risk management: a review. *SSRN Electronic Journal* [Online]. http://www.ssrn.com/abstract=925274. Accessed 9 Mar 2025

Tchetin K (2022) European green deal. *Yearbook of the law department* [Online]. https://ojs.nbu.bg/index.php/YLD/article/download/648/538

Tharatipyakul A, Pongnumkul S, Riansumrit N, Kingchan S, Pongnumkul S (2022) Blockchain-based traceability system from the users' perspective: a case study of Thai coffee supply chain. IEEE Access 10:98783–98802

Upadhyay A, Mukhuty S, Kumar V, Kazancoglu Y (2021) Blockchain technology and the circular economy: implications for sustainability and social responsibility. J Clean Prod 293:126130

Walmart (2018) *In wake of Romaine E. Coli Scare, Walmart deploys Blockchain to track leafy greens* [Online]. https://corporate.walmart.com/news/2018/09/24/in-wake-of-romaine-e-coli-scare-walmart-deploys-blockchain-to-track-leafy-greens. Accessed 13 Mar 2025

Wang S, Ouyang L, Yuan Y, Ni X, Han X, Wang F-Y (2019) Blockchain-enabled smart contracts: architecture, applications, and future trends. IEEE Trans Syst Man Cybern Syst 49(11): 2266–2277

Wang Q, Li R, Wang Q, Chen S (2021) Non-fungible token (NFT): overview, evaluation, opportunities and challenges [Online]. arXiv. http://arxiv.org/abs/2105.07447. Accessed 12 Mar 2025

Wannenwetsch K, Ostermann I, Priel R, Gerschner F, Theissler A (2023) Blockchain for supply chain management: a literature review and open challenges. Procedia Comput Sci 225:1312–1321

Wasim Layaq M, Goudz A, Noche B, Atif M (2019) Blockchain technology as a risk mitigation tool in supply chain. Int J Transp Eng Technol 5(3):50

What Is a Prompt Injection Attack? | IBM (2024) [Online]. https://www.ibm.com/think/topics/prompt-injection. Accessed 12 Mar 2025

Xu JJ (2016) Are Blockchains immune to all malicious attacks? Financ Innov 2(1):25

Yaqoob I, Salah K, Jayaraman R, Al-Hammadi Y (2022) Blockchain for healthcare data management: opportunities, challenges, and future recommendations. Neural Comput & Applic 34(14): 11475–11490

Yuthas K (2021) Strategic value creation through Enterprise Blockchain. J Brit Blockchain Assoc 4(1):1–10

Blockchain Applications for Long-Term Sustainable Supply Chains

Tehreem Ansari and Syed Imran Zaman

Abstract This chapter critically analyzes how blockchain technology enhances accountability, transparency, and sustainability in supply chains, especially in green manufacturing and ethical sourcing. As businesses balance profitability with environmental responsibility, the need for secure and traceable supply chain solutions has intensified. The immutable nature of blockchain provides a viable method for documenting transactions within the supply chain, facilitating audits and thwarting tampering. This chapter focuses on utilizing blockchain to create long-term transparent and sustainable supply chains. It explores how blockchain can verify compliance with environmental and ethical standards throughout the production and sourcing processes. By granting stakeholders access to consistent data on sourcing and environmental impacts, blockchain enhances accountability among all involved parties. This increased transparency may foster consumer trust and encourage sustainable purchasing behaviors, enabling end-users to trace products back to their origins and evaluate their environmental effects. This chapter's primary goal is to outline a framework for integrating blockchain into sustainable supply chains and to analyze specific cases where blockchain has successfully advanced sustainability and ethical sourcing. By providing insights into current blockchain applications in green manufacturing, this chapter aims to highlight both the benefits and challenges, including data privacy concerns, scalability issues, and the need for unified industry standards. In contributing to academic discourse, this chapter emphasizes blockchain's potential to transform sustainable supply chain management and advance global corporate social responsibility discussions. It puts forth potential paths for future investigations aimed at improving blockchain's synergy with nascent technologies like the Internet of Things (IoT) and Artificial Intelligence (AI), fostering supply chains that are robust, clear, and environmentally sustainable.

T. Ansari (✉) · S. I. Zaman
Jinnah University for Women, Karachi, Pakistan
e-mail: tehreem_ansari@yahoo.com; imaranzaman@gmail.com

M. Venkatesh et al. (eds.), *Green Manufacturing Technologies in Industry 5.0*,
https://doi.org/10.1007/978-981-95-0363-6_7

1 Introduction

Industry 5.0 signifies a fundamental transformation from the technology-focused paradigm of Industry 4.0 toward a more human-centric, sustainable, and resilient industrial framework. This innovative methodology underscores the necessity of incorporating human-centric principles, sustainability, and resilience within industrial operations, with the objective of fostering a more inclusive and environmentally sustainable manufacturing landscape (Nazarejova et al. 2024). The emphasis on human-centricity entails the empowerment of workers through technological advancements and their active participation in decision-making processes, which is essential for the formulation of sustainable employment strategies (Bucci et al. 2024). In addition, Industry 5.0 harnesses advanced technologies like artificial intelligence (AI), the Internet of Things (IoT), and additive manufacturing to enhance production efficiency and promote sustainability (Rani et al. 2024).

Sustainable supply chains are essential for environmentally responsible manufacturing, as they promote resource efficiency and mitigate ecological consequences. Additive manufacturing, for example, presents a transformative opportunity for enhancing sustainable production frameworks by streamlining supply chains and facilitating product customization. The implementation of blockchain technology is integral to enhancing transparency and accountability within these supply chains. It establishes a secure, decentralized infrastructure for the tracking and verification of transactions, which is crucial for sustaining trust and integrity in industrial processes (Fraga-Lamas et al. 2024).

The capacity of blockchain technology to provide reliable and secure data is congruent with the objectives of Industry 5.0, thereby facilitating the advancement of human-centric, sustainable, and resilient applications (Fraga-Lamas et al. 2024). As the paradigm of Industry 5.0 progressively develops, it is imperative for manufacturing sectors to integrate these technologies and methodologies to augment their competitiveness and sustainability within the forthcoming industrial milieu (Ghobakhloo et al. 2024).

2 Sustainable Supply Chains and Blockchain Technology

Contemporary supply chains are progressively oriented toward sustainability objectives, propelled by the imperative to mitigate environmental degradation, manage resource limitations, and promote social equity. These objectives are congruent with the Sustainable Development Goals (SDGs) established by the United Nations, which underscore the necessity of minimizing carbon emissions, ensuring ethical sourcing practices, and fostering the adoption of circular economy principles. Nevertheless, the realization of these sustainability aims poses considerable obstacles, particularly in the domains of green manufacturing and ethical sourcing. Manufacturing enterprises encounter challenges such as the generation of waste, emissions of greenhouse gases, and the intricate nature of global supply chains, all of which hinder the effective implementation of sustainable practices (Jan et al. 2024).

Ethical sourcing is further complicated by challenges such as substandard working conditions and the absence of transparency within global supply chains (Okay et al. 2024). Blockchain technology presents a viable solution to these issues by improving transparency, traceability, and accountability throughout supply chains. Its decentralized and immutable characteristics facilitate the real-time dissemination of information, which can revolutionize traditional supply chain management into sustainable supply chain management, thereby fostering adherence to environmental Sustainable Development Goals (Jan et al. 2024; Park and Li 2021). Within the food supply chain, blockchain enhances effective sustainability communication by addressing constraints such as information overload and greenwashing, thereby ensuring accurate reporting of sustainability initiatives (Cao et al. 2023).

Although blockchain possesses considerable potential, its utilization in the realm of sustainability governance is accompanied by notable criticisms. Critics contend that it reinforces pre-existing audit cultures and managerial methodologies that have historically been inadequate in addressing labor and environmental exploitation, thereby potentially engendering a "veil of transparency" that obfuscates rather than ameliorates sustainability challenges. Nevertheless, the capability of blockchain to furnish a secure, transparent, and efficient framework for supply chain management presents substantial prospects for the promotion of sustainability objectives across diverse sectors, including food, healthcare, and manufacturing (Sahoo et al. 2022; Yontar 2023). Consequently, while blockchain technology provides considerable avenues for the enhancement of supply chain sustainability, its deployment necessitates meticulous oversight to ensure that it authentically contributes to ethical and sustainable practices.

3 Blockchain Fundamentals for Sustainable Manufacturing

Blockchain technology, distinguished by its foundational tenets of decentralization, immutability, and transparency, presents substantial transformative capabilities across diverse sectors, notably in manufacturing and supply chain management. The principle of decentralization guarantees that no singular entity possesses overarching control over the entirety of the network, thereby augmenting security and mitigating the likelihood of data breaches. Immutability ensures that once data is inscribed upon the blockchain, it remains unalterable, thus providing a dependable and tamper-resistant record of transactions (Khan et al. 2023). Transparency, an essential characteristic of blockchain, permits all stakeholders within the network to access identical data, thereby fostering an environment of trust and accountability (Yap et al. 2023; Mukherjee et al. 2022).

Within the manufacturing domain, the pertinence of blockchain is particularly accentuated in the establishment of ethical and transparent processes. By facilitating real-time, shareable, and transparent data on an immutable ledger, blockchain catalyzes the transition from conventional supply chain management to sustainable supply chain management practices, thus ensuring adherence to environmentally sustainable development goals (SDGs). This metamorphosis is vital for the minimization of waste and greenhouse gas emissions, aligning manufacturing operations with international environmental standards (Jan et al. 2024).

Moreover, blockchain-based certifications for green standards are surfacing as a formidable mechanism to verify and validate sustainable practices. These certifications exploit blockchain's transparency and immutability to furnish verifiable evidence of compliance with green standards, thereby bolstering consumer confidence and promoting sustainable practices (Correa Tavares et al. 2021). The incorporation of blockchain within supply chains not only enhances traceability and transparency but also bolsters the circular economy by safeguarding data privacy and fostering resilience (Mukherjee et al. 2022). Nevertheless, the implementation of blockchain in supply chains is not devoid of challenges, encompassing security vulnerabilities and the necessity for consensus protocols to ensure network efficacy (Agarwal et al. 2024; Sedlmeir et al. 2022).

Notwithstanding these challenges, the strategic deployment of blockchain can profoundly augment the ethical and transparent management of manufacturing processes, thereby contributing to a more sustainable and accountable industrial environment (Dutta et al. 2020).

4 Understanding Sustainability in Supply Chains

Sustainability, conceptualized through the framework of Environmental, Social, and Governance (ESG) considerations, constitutes a complex construct that interweaves diverse facets of corporate accountability and operational efficacy. The environmental component is primarily concerned with the mitigation of emissions as well as the stewardship of energy and resource utilization, whereas the social dimension prioritizes health, safety, and community involvement. Governance pertains to the frameworks and methodologies that guarantee accountability and transparency in organizational decision-making processes (Liu et al. 2024). The incorporation of ESG considerations into corporate practices is imperative for the realization of sustainable development, as it harmonizes corporate objectives with overarching societal and environmental ambitions (Whitelock 2019).

Traceability and ethical conduct within supply chain operations are indispensable for ensuring that sustainability pledges are fulfilled across the entire supply chain continuum. Corporations are increasingly subjected to scrutiny regarding the social and environmental ramifications of their suppliers, thereby necessitating a commitment to transparency and adherence to sustainability benchmarks. Innovative digital methodologies, including data-driven monitoring systems, are emerging to bolster supply chain transparency and traceability, although these technological advancements remain in a state of evolution and are frequently applied in pilot phases (Ebinger and Omondi 2020).

The integrity of data is of paramount significance in the quantification and enforcement of sustainability indicators, as precise and trustworthy data are crucial for the evaluation of ESG performance and the formulation of informed strategic choices. The convergence of financial and non-financial sustainability indicators within corporate reporting frameworks rectifies the drawbacks characteristic of standard sustainability assessments and aids in the synchronization of sustainability initiatives with financial disclosures (Seow 2024). Moreover, the integration of ESG considerations into financial decision-making frameworks has the potential to enhance the sustainability of financial systems, as evidenced by the increasing emphasis on ESG criteria within financial institutions across various OECD member states. In conclusion, the interplay among ESG integration, supply chain transparency, and data integrity is essential for the progression of sustainable business practices and the generation of shared value for all stakeholders involved (Sancha et al. 2023).

5 Enhancing Transparency in Green Manufacturing

Enhancing transparency within the realm of green manufacturing necessitates the meticulous tracking of carbon emissions and resource utilization, a process that can be proficiently administered through the application of blockchain technology. Blockchain serves as a formidable mechanism for the continuous oversight of emissions and waste management, ensuring the dissemination of credible data across cooperative manufacturing supply chains. This outcome is facilitated through a methodical framework comprising blockchain nodes, distributed ledgers, and smart contracts designed to standardize and authenticate data storage and inquiries, as evidenced by a case study that substantiates its efficacy in ensuring credible data dissemination within cross-enterprise supply chains.

Moreover, the incorporation of blockchain within sustainable fashion supply chains has indicated considerable potential in augmenting operational transparency and advancing environmental initiatives. By enabling fashion retailers to transparently communicate the environmental quality of their products via blockchain, it effectively addresses the credibility apprehensions of manufacturers and bolsters consumer trust (Guo et al. 2020). The implementation of blockchain in green manufacturing is further bolstered by its capacity to enhance product information, thereby fostering transparency and traceability throughout the materials supply chain. This methodology is congruent with Industry 4.0 technologies, such as augmented reality and cloud computing systems, which collectively provide comprehensive product lifecycle information specifically tailored for stakeholders (Barata and da Cunha 2021).

Illustrative instances of blockchain initiatives in green manufacturing encompass its utilization within sustainable fashion and materials supply chains, which underscore its potential to amplify transparency and bolster environmental sustainability endeavors. These initiatives exemplify the role of blockchain in promoting the disclosure of environmental initiatives and enhancing the traceability of sustainable practices, thereby contributing to the overarching objectives of green manufacturing (Guo et al. 2020; Barata and da Cunha 2021). In summary, blockchain technology emerges as a critical instrument in propelling transparency and accountability in green manufacturing, in alignment with the increasing demand for sustainable practices and adherence to regulatory compliance.

6 Blockchain Fundamentals for Sustainable Practices

Blockchain technology presents numerous salient characteristics that are profoundly pertinent to the advancement of sustainable practices, including immutability, transparency, and decentralization. These characteristics significantly improve the monitoring and validation of sustainable practices by guaranteeing that data inscribed on the blockchain is impervious to tampering and accessible to all relevant stakeholders, thereby cultivating trust and accountability. For example, within the domain of consumable goods, blockchain can be utilized to chronicle supply chain transactions and sustainability metrics in a transparent manner, thereby facilitating the computation of a product's sustainability index score through the implementation of smart contracts (Hasan et al. 2024).

In the agri-food sector, blockchain acts as a catalyst for the transition toward a circular economy by enhancing data utility, optimizing supply chain management, promoting eco-efficiency, and ensuring traceability, thereby addressing challenges such as information asymmetry and food safety issues (Pakseresht et al. 2022). Moreover, in the context of distributed generation, the decentralized nature of blockchain bolsters peer-to-peer energy trading and fortifies security, rendering it an invaluable instrument for the management of renewable energy sources alongside other elements of decentralized energy systems (Yap et al. 2023).

Prominent blockchain platforms such as Hyperledger, Ethereum, and VeChain play a crucial role in underpinning sustainable applications. Ethereum, for instance, is employed in water management frameworks to facilitate transparent interactions and enhance customer engagement through tradable incentives (Pahontu et al. 2023). Hyperledger is frequently utilized in industrial symbiosis to bolster informative flows and integration within circular supply chains (Ventura et al. 2024). VeChain, renowned for its supply chain solutions, has the capacity to augment the credibility of quality indicators on product labels, thereby empowering consumers to make informed decisions predicated on sustainability attributes (Contini et al. 2023). These platforms, with their robust frameworks and capabilities, are instrumental in propelling sustainable practices across diverse industries by harnessing blockchain's intrinsic features to ensure data integrity, transparency, and decentralized governance.

7 Blockchain's Role in Ethical Sourcing

Contemporary supply chain ethics in sourcing revolves around the approach to obtaining products that align with social conscience and ecological integrity, factoring in elements such as just labor conditions, environmental preservation, and transparent operations. This paradigm is increasingly vital as consumers and stakeholders necessitate heightened accountability and ethical benchmarks from corporate entities. The implementation of blockchain technology is instrumental in augmenting ethical sourcing by facilitating a decentralized, transparent, and immutable ledger that can authenticate the provenance of products, assure adherence to equitable labor practices, and verify the utilization of environmentally sustainable resources. As an illustration, the capability of blockchain to supply live, distributable data facilitates the conversion of typical supply chain management into sustainable supply chain management, thereby corresponding with the aims of sustainable development goals (SDGs) (Jan et al. 2024).

The attributes of the technology, including transparency and traceability, are essential for diminishing information asymmetry and bolstering trust among stakeholders, which is paramount for ethical sourcing (Najjar et al. 2023). In the realm of fashion, blockchain can be employed to attest to the environmental integrity of products, thereby affecting consumer confidence and market demand (Guo et al. 2020). Furthermore, the incorporation of blockchain into supply chains can yield enhanced sustainability performance by improving transparency, reliability, and efficiency, which are crucial for the attainment of social equity and environmental conservation (Park and Li 2021). Notwithstanding its potential, the implementation of blockchain within multi-tier supply networks may present complexities, necessitating meticulous management to surmount challenges such as integration and legal ambiguities (Najjar et al. 2023).

Nevertheless, the capacity of the technology to deliver comprehensive and frequently updated information positions it as a significant asset for suppliers aspiring to secure a competitive edge through the voluntary revelation of sustainability practices (Vinayavekhin et al. 2024). In summary, blockchain technology presents considerable promise in facilitating ethical sourcing by ensuring that supply chains are characterized by transparency, accountability, and alignment with sustainable practices.

8 Case Studies of Blockchain-Enabled Ethical Sourcing in the Apparel and Electronics Industries

The utilization of blockchain technology is progressively being adopted to augment ethical sourcing within the apparel and electronics sectors, with multiple corporations exemplifying this trend. Within the apparel sector, the Lenzing Group has executed blockchain-enabled traceability mechanisms to validate products and guarantee supply chain sustainability. This initiative effectively confronts the complexities associated with monitoring materials from the stage of raw fiber production through to garment fabrication, thereby underscoring the necessity for clearly defined traceability objectives and suitable scoping throughout supply chains (Ahmed and MacCarthy 2021). In the electronics domain, the attributes of blockchain regarding data immutability, transparency, and traceability are currently being investigated to bolster ecological sustainability practices. The technology's capacity to enhance governance frameworks and sustainability benchmarks is substantial, albeit its implementation is impeded by technological underdevelopment and trust-related concerns (Jimenez-Castillo et al. 2024). Furthermore, the incorporation of blockchain within sustainable supplier selection methodologies, as exemplified by a hybrid BWM-MULTIMOORA and bi-objective programming strategy, accentuates its function in improving procurement processes by ensuring transparency and sustainability in the evaluation of suppliers (Vaezi et al. 2024). These case studies elucidate the transformative potential inherent in blockchain to foster ethical sourcing by delivering a transparent, traceable, and accountable framework that empowers stakeholders while aligning with sustainability objectives across various industries (Lee 2021). Notwithstanding the obstacles, such as the necessity for regulatory support and the integration with complementary technologies like the Internet of Things (IoT), the continuous endeavors within these sectors reflect a burgeoning commitment to ethical sourcing facilitated by blockchain technology (Nath et al. 2022).

9 Improving Accountability Across the Supply Chain

Blockchain technology is increasingly acknowledged as a potent instrument for augmenting accountability and transparency throughout supply chains, especially in the verification of manufacturers' sustainability assertions. Its intrinsic characteristics of immutability and transparency render it an exemplary solution for tackling the issues of greenwashing and information saturation in sustainability communication, as evidenced within the context of the food supply chain (Cao et al. 2023). Blockchain enables real-time audits and verification of adherence to environmental standards by facilitating peer-to-peer verifiable carbon accounting, which guarantees confidentiality while upholding the transparency and verifiability of emission data (Heiss et al. 2023). This functionality is vital for precise carbon footprint evaluations and resonates with the overarching objectives of sustainable supply chain management (Park and Li 2021). Additionally, blockchain augments stakeholder visibility and consumer trust by offering a secure and tamper-resistant framework for tracking products from their origin to the end consumer, consequently enhancing traceability and accountability (Ellahi et al. 2023).

The technology's potential to converge with other Industry 4.0 technologies further magnifies its influence, facilitating dynamic evaluations and refined decision-making across supply chains (Okay et al. 2024; Ebinger and Omondi

2020). By establishing a united environment among stakeholders, blockchain effectively addresses legal disagreements and operational challenges while also bolstering trust and communication in business-to-business dealings (Basheer et al. 2024; Rejeb et al. 2021). As blockchain technology continues to advance, its utilization in supply chain management is projected to expand, providing a transformative methodology for realizing sustainability objectives and bolstering consumer confidence in sustainability claims (Sahoo et al. 2022).

10 Applications of Blockchain in Circular Supply Chains

The rise of blockchain technology represents a significant innovation in the advancement of circular supply chains (CSCs) throughout diverse industries by resolving core problems like transparency, traceability, and operational productivity. Within the sphere of lithium-ion battery supply chains, blockchain offers a robust infrastructure for safety management by facilitating the effective monitoring and validation of safety-related data, thereby alleviating safety concerns and promoting sustainable practices (Chen et al. 2024). In the broader context of circular economies, blockchain-facilitated supply chain finance methodologies have been recognized as crucial in surmounting the challenges posed by linear economic models, thereby aiding the transition to circular frameworks by endorsing the ten principles of 10R and augmenting financial strategies across intricate supply chains (Modgil et al. 2024).

Moreover, the significance of blockchain in industrial symbiosis is accentuated through its capacity to endorse closed-loop manufacturing systems, wherein waste generated by one industry is repurposed as the raw material for another, thereby fostering sustainability and minimizing industrial waste (Ventura et al. 2024). The technology also plays a fundamental role in enhancing customer engagement within circular economies by establishing a decentralized framework that connects blockchain developers with specialists in corporate social responsibility, thereby promoting digital innovation and sustainable practices (Cerchione 2024).

Notwithstanding its potential, the integration of blockchain within circular digital supply chains encounters obstacles such as data transparency, market competition, and the absence of standardization, which must be addressed to fully harness its advantages (Chaouni Benabdellah et al. 2023). In cultivating dynamic capabilities for circular supply chains, blockchain aids in the sensing, seizing, and reconfiguration of resources, which are imperative for adapting to the circular economy paradigm (Quayson et al. 2023). In the context of recycling chains, blockchain enhances operational performance by bolstering transparency, integration, and service transformation, despite the persistence of technical and organizational constraints (Xie et al. 2023). Within extensive industrial sectors, blockchain's decentralized ledger functionalities underpin sustainability and circularity by facilitating efficient source tracking and information management (Abbas and Myeong 2024).

In the agri-food sector, blockchain augments data utility, improves the efficacy of supply chain management, enhances eco-efficiency, and bolsters traceability, thereby facilitating the transition toward a circular economy (Pakseresht et al. 2022). Finally, in the realms of closed-loop supply chains and reverse omnichannel systems, blockchain guarantees transparency and security, thus enhancing operational capabilities and business performance through the implementation of smart contracts and incentive mechanisms (De Giovanni 2022). Collectively, these applications highlight the fundamental role of blockchain in propelling circular supply chains by addressing operational challenges and promoting sustainable practices across various sectors.

11 Challenges and Limitations of Blockchain for Sustainable Supply Chains

Blockchain technology encompasses both prospects and obstacles for the development of sustainable supply chains. While it contributes to enhanced clarity, traceability, and operational effectiveness, its assimilation into supply chain management (SCM) is limited by notable challenges. A predominant issue is the excessive energy consumption linked to blockchain networks, which contributes to carbon emissions and presents sustainability risks (Babaei et al. 2024).

Moreover, the elevated initial costs of implementation and the ongoing expenses associated with maintenance, in conjunction with interoperability challenges, establish barriers to its widespread utilization. Legal and regulatory impediments further exacerbate the situation, as the absence of standardization and the necessity for oversight obstruct seamless integration. In addition, the intricate nature of combining blockchain with pre-existing systems, alongside the limited awareness among stakeholders regarding its potential advantages, exacerbates the challenges faced (Sharabati and Jreisat 2024). Notwithstanding these obstacles, the capacity of blockchain to enhance supply chain transparency and diminish waste is considerable, as it has the potential to cultivate sustainable and ethical practices (Balcıoğlu et al. 2024). Nevertheless, the realization of these advantages necessitates the resolution of technical intricacies and scalability concerns, which require interdisciplinary cooperation and sustained innovation (Lin et al. 2024).

Furthermore, the ecological implications of blockchain, particularly its energy requirements, must be addressed to ensure alignment with sustainability objectives (Babaei et al. 2024). To alleviate these difficulties, hybrid systems that integrate on-chain and off-chain solutions have been suggested to decrease costs and improve scalability, thereby facilitating more sustainable supplier selection processes (Haffar and Özceylan 2024). In summary, while blockchain presents significant potential for the transformation of supply chains, addressing its inherent limitations is essential for maximizing its capabilities in fostering sustainability.

Blockchain technology introduces a range of technical, data-related, and regulatory obstacles when implemented in sustainable supply chains. Analyzing the technical aspects, blockchain struggles with issues of scalability, interoperability, and resource limitations, which obstruct its far-reaching integration in supply chains. The assimilation of blockchain with pre-existing systems and other emerging technological innovations constitutes a significant barrier, necessitating considerable infrastructure advancements and collaborative efforts across multiple sectors (Lin et al. 2024). Data-related impediments encompass the assurance of effective information dissemination, systematic data governance, and the preservation of transparency and auditability throughout the supply chain. These challenges are exacerbated by the imperative for compatibility with alternative technologies and the optimization of time metrics.

On the regulatory spectrum, the ambiguous legal status of blockchain technology represents a considerable obstacle, as current laws and regulations may not sufficiently address the distinctive attributes of blockchain, including decentralization and immutability (Yontar 2023). Moreover, the absence of standardized regulatory frameworks can obstruct coordination among stakeholders, industries, and regulatory bodies, which is vital for the successful deployment of blockchain technology (Sargent and Breese 2024). Confronting these challenges necessitates the formulation of strategic action plans, such as those proposed through methodologies like the Integrated Fuzzy TOPSIS–ISM approach, which can assist in alleviating supply chain impediments and fostering blockchain adoption (Al Amin et al. 2023).

Furthermore, the establishment of facilitating regulations and the promotion of collaboration among supply chain partners can aid in the development of essential infrastructure and support sustainable practices for the integration of blockchain (Dehshiri and Amiri 2024). In summary, surmounting these challenges is crucial for harnessing blockchain's potential to enhance sustainability, transparency, and efficiency within supply chains. While blockchain technology holds considerable promise for augmenting sustainability in supply chains, it encounters numerous adoption challenges, including human and organizational factors, environmental concerns, and issues related to governance and standardization.

The barriers to adoption are substantial, with financial limitations, technological underdevelopment, and security apprehensions serving as notable challenges. These impediments are intensified by a deficiency in knowledge and expertise, coupled with cultural disparities, which obstruct the effective application of blockchain in sustainable supply chains (Ali et al. 2024; Rejeb et al. 2021). Human and organizational dynamics, such as resistance to change and insufficient management support, further hinder the adoption of blockchain, as organizations grapple with integrating novel technologies into their pre-existing frameworks (Rejeb et al. 2021). Environmental considerations also emerge, particularly in relation to the energy-intensive characteristics of blockchain operations, which may conflict with sustainability objectives (Jiang et al. 2022). Governance and standardization concerns are pivotal, as the lack of definitive standards and regulatory frameworks can induce uncertainty and inhibit widespread adoption.

The construction sector, for instance, identifies transparency and security as principal barriers, underscoring the necessity for robust governance structures to guarantee compliance and traceability (Singh et al. 2023). Furthermore, while blockchain possesses the capacity to positively influence all 17 Sustainable Development Goals (SDGs), adverse effects, such as heightened energy consumption, can obstruct advancement toward these goals (Jiang et al. 2022). Addressing these challenges demand a thorough strategy that involves systematic planning, policy development, and the involvement of stakeholders to improve the effectiveness of blockchain within sustainable supply chains (Ali et al. 2024; Singh et al. 2023; Jiang et al. 2022).

However, disparate regions are tackling the legal and regulatory issues associated with blockchain adoption in sustainability initiatives employing diverse methodologies, which mirror their distinctive policy environments and priorities. The European Union (EU), United States (US), and Asia constitute pivotal actors within this domain, each employing unique strategies to incorporate blockchain technology into their sustainability frameworks. These initiatives are imperative as blockchain possesses the potential to augment transparency, efficiency, and accountability within sustainability endeavors.

11.1 European Union (EU)

The EU is proactively incorporating blockchain technology into its sustainability policies, emphasizing regulatory frameworks that are congruent with its environmental objectives. The European Securities and Markets Authority (ESMA) alongside other regulatory entities are instrumental in formulating policies that endorse blockchain's contribution to the attainment of net-zero emissions (Mulligan et al. 2024).

The EU's methodology encompasses investigating blockchain's capabilities in energy management and supply chain transparency, which are essential for fulfilling the Sustainable Development Goals (SDGs) (Jiang et al. 2022).

11.2 United States (US)

In the US, regulatory initiatives are characterized by fragmentation, with various states implementing disparate approaches. The Securities and Exchange Commission occupies a critical position in overseeing blockchain applications, especially within financial markets (Mulligan et al. 2024). The US is concurrently examining blockchain's potential to curtail energy consumption, emphasizing the promotion of sustainable blockchain networks that mitigate carbon emissions (Truby 2018; Rani et al. 2024).

11.3 Asia

Asian nations, particularly China, are leading the charge in blockchain research and its application in sustainable management. There exists substantial collaboration among academics and industries to harness blockchain for energy management and environmental sustainability (Du et al. 2022). Regulatory frameworks in Asia are progressing to facilitate the integration of blockchain into sustainability efforts, with particular attention to addressing scalability and power consumption challenges (Alshahrani et al. 2023).

While these regions are advancing in the integration of blockchain into sustainability initiatives, challenges persist. The considerable energy consumption associated with blockchain technologies, notably in networks such as Bitcoin, presents a significant obstacle. Initiatives aimed at developing more energy-efficient blockchain networks are underway, with projects like the Renewable Energy Certificate Mechanism and Green Digital Finance Alliance gaining momentum (Alzoubi and Mishra 2023). Nevertheless, the regulatory landscape remains in a state of evolution, necessitating the formulation of more comprehensive policies that harmonize innovation with environmental sustainability.

12 Future Trends and Innovations in Blockchain for Sustainable Manufacturing

The implementation of blockchain technology is pivotal in transforming sustainable manufacturing methodologies, propelled by its ability to enhance transparency, traceability, and accountability. As various sectors confront the complexities associated with fulfilling sustainability objectives, the significance of blockchain will markedly increase, driven by technological innovations and shifting market paradigms. This section elucidates forthcoming trajectories in the development of blockchain, the prospects of tokenized assets, and the advancement of international standards for sustainability within the framework of Industry 5.0.

The amalgamation of blockchain with nascent technologies such as Artificial Intelligence (AI), the Internet of Things (IoT), and edge computing is poised to revolutionize sustainable supply chains. These synergistic interactions facilitate instantaneous data acquisition, sophisticated analytics, and automated decision-making processes, thereby promoting enhanced efficiency and environmental stewardship. For instance, IoT devices integrated into manufacturing facilities can transmit data to blockchain networks, generating immutable records pertaining to emissions, energy utilization, and waste production. This continual monitoring guarantees adherence to sustainability criteria and reveals avenues for process enhancement.

Furthermore, advancements in the scalability of blockchain will remedy existing constraints associated with energy utilization and processing velocities. Innovations such as sharding, layer-two protocols, and proof-of-stake mechanisms are under development to mitigate the ecological footprint of blockchain networks while augmenting their capacity to process substantial transaction volumes. As these tech innovations unfold, blockchain will become more user-friendly for small and medium enterprises (SMEs), ultimately leading to a more equitable use and a wider-reaching effect in different fields.

The rise of decentralized finance (DeFi) platforms specifically designed for sustainable investments represents another encouraging development. These platforms harness blockchain to deliver transparent, secure, and decentralized financial services, thereby enabling investments in environmentally-friendly initiatives. For instance, manufacturers could utilize blockchain-based crowdfunding platforms to finance the implementation of renewable energy solutions, ensuring that all financial exchanges are verifiable and aligned with ecological goals.

13 Potential of Tokenized Assets for Green Investment and Carbon Credits

Tokenization, defined as the transformation of tangible assets into digital tokens within a blockchain framework, is set to fundamentally transform the domains of green investments and carbon credit markets. The development of tokenized assets, featuring green bonds, renewable energy credits, and carbon offsets, delivers a clear and efficient mechanism for the oversight and trade of environmental assets.

The inherent immutability of blockchain technology guarantees the security and tamper-resistance of tokenized assets, while its transparency affords all relevant stakeholders immediate access to transactional data. As an illustration, carbon credits—documents indicating the reduction of one metric ton of carbon dioxide emissions—may be digitized and afterward traded on blockchain systems. This advancement not only augments the liquidity of carbon markets but also ensures the prevention of double-counting and fraudulent claims regarding credits.

Moreover, the deployment of smart contracts can facilitate the automation of the issuance, trading, and retirement processes associated with carbon credits, thereby diminishing administrative expenses and enhancing market efficiency. Blockchain-enabled platforms additionally streamline cross-border transactions, thereby fostering global engagement in carbon markets and expediting advancements toward international climate objectives.

In addition to carbon credits, tokenized assets have the potential to bolster green investments by enabling fractional ownership of renewable energy initiatives. Investors may acquire tokens that represent ownership shares in solar farms, wind turbines, or other sustainable projects, thus democratizing access to opportunities for green investment. This methodology promotes wider participation in sustainability initiatives and channels capital toward projects that yield significant environmental advantages.

14 The Advancement of Blockchain Standards for Global Sustainability in Industry 5.0

As the adoption of blockchain technology accelerates, the imperative for standardized frameworks and protocols becomes markedly pronounced. With Industry 5.0 promoting human-centric values, sustainability, and resilience, establishing coherent blockchain standards will be vital to achieve interoperability and uniformity across multiple sectors.

Global standards will tackle critical challenges including data interoperability, regulatory compliance, and ethical considerations. For instance, synchronized data types and regulatory protocols will support the trouble-free integration of blockchain networks alongside other digital frameworks, including enterprise resource planning (ERP) tools and Internet of Things (IoT) devices. This interoperability is vital for the establishment of comprehensive end-to-end visibility across supply chains, thereby enabling stakeholders to meticulously monitor sustainability metrics.

Regulatory frameworks will assume a crucial role in directing the evolution of blockchain standards. It is imperative for political entities and international organizations to unite in shaping directives that harmonize innovation with governance, ensuring that blockchain solutions are consistent with sustainability goals. For example, such frameworks could mandate the employment of energy-efficient consensus mechanisms and enforce transparency in the disclosure of blockchain's environmental impact.

Ethical considerations will also play a significant role in shaping the development of blockchain standards. As the adoption of blockchain expands, it becomes critical to address issues pertaining to data privacy, labor practices, and environmental justice. Standards that prioritize these ethical principles will bolster stakeholder trust and promote the widespread adoption of blockchain technology for sustainable manufacturing.

The convergence of blockchain technology with the principles of Industry 5.0 will further stimulate the emergence of decentralized autonomous organizations (DAOs) dedicated to sustainability. These organizations operate on blockchain networks and utilize smart contracts to automate decision-making processes, thereby fostering collaborative initiatives aimed at achieving sustainability objectives. For instance, DAOs could orchestrate resource sharing among manufacturers, thereby minimizing waste and advocating for circular economy practices.

In conclusion, the prospective trajectory of blockchain technology within the realm of sustainable manufacturing is underscored by technological advancements, the widespread adoption of tokenized assets, and the formulation of international standards. By adopting these technological progresses, industrial areas can build supply chains characterized by heightened transparency, efficiency, and durability, thus contributing to a future that is sustainable and fair for all.

15 Conclusion

Blockchain technology has emerged as a transformative instrument for elevating transparency, accountability, and sustainability within supply chains. Its fundamental characteristics—immutability, decentralization, and transparency—facilitate unprecedented levels of traceability and trust among stakeholders, thereby aligning effectively with the objectives inherent to Industry 5.0.

Integrating blockchain with synergistic technologies like the Internet of Things (IoT) and Artificial Intelligence (AI) significantly increases its power to overcome essential obstacles in sustainable manufacturing. These collaborative connections allow for instantaneous tracking, forecasting insights, and self-operating resolutions, enabling manufacturers to enhance operations, cut down on excess, and meet green targets. The utilization of tokenized assets and smart contracts further augments the functionality of blockchain by streamlining the management and exchange of green investments and carbon credits, thereby incentivizing sustainable practices.

Nevertheless, the extensive adoption of blockchain is accompanied by significant challenges. Issues such as elevated energy consumption, constraints on scalability, and uncertainties in regulation must be effectively resolved to fully harness its potential. Strategies aimed at mitigating these obstacles encompass investments in energy-efficient blockchain solutions, the establishment of global standards, and the promotion of cross-sector collaborations. Moreover, educating stakeholders regarding the advantages of blockchain technology and addressing concerns related to data privacy and ethical practices are critical components for successful implementation.

Peering into what lies ahead, the significance of blockchain within green manufacturing is likely to increase as various industries steadily incorporate the concepts of Industry 5.0. The development of decentralized autonomous organizations (DAOs), the establishment of interoperable standards, and the rise of blockchain-driven platforms will stimulate innovation and advance a more sustainable industrial ecosystem. By leveraging the capabilities of blockchain technology, manufacturers can establish resilient supply chains that effectively balance economic growth with environmental stewardship.

In conclusion, blockchain technology constitutes a formidable enabler of sustainability within the manufacturing sector and supply chains. By addressing existing limitations and capitalizing on its transformative potential, industries can progress toward a greener and more equitable future. As we transition into an epoch characterized by human-centricity and environmental responsibility, blockchain will indisputably remain at the vanguard of initiatives aimed at achieving global sustainability objectives.

References

Abbas Z, Myeong S (2024) A comprehensive study of Blockchain technology and its role in promoting sustainability and circularity across large-scale industry. Sustainability 16(10):4232

Agarwal U, Rishiwal V, Yadav M, Aslhammari M, Yadav P, Singh O, Maurya V (2024) Exploring Blockchain and supply chain integration: state-of-the-art, security issues and emerging directions. IEEE Access 99:1–1

Ahmed WA, MacCarthy BL (2021) Blockchain-enabled supply chain traceability in the textile and apparel supply chain: a case study of the fiber producer, Lenzing. Sustainability 13(19):10496

Al Amin M, Nabil DH, Baldacci R, Rahman MH (2023) Exploring blockchain implementation challenges for sustainable supply chains: an integrated fuzzy TOPSIS–ISM approach. Sustainability 15(18):13891

Ali MR, Khan SA, Kayıkçı Y, Mubarik MS (2024) A three-phase framework for mapping barriers to blockchain adoption in sustainable supply chain. Ind Manag Data Syst 125(1):306–336. https://doi.org/10.1108/imds-03-2024-0257

Alshahrani H, Islam N, Syed D, Sulaiman A, Al Reshan MS, Rajab K et al (2023) Sustainability in blockchain: a systematic literature review on scalability and power consumption issues. Energies 16(3):1510

Alzoubi YI, Mishra A (2023) Green blockchain–a move towards sustainability. J Clean Prod 430: 139541

Babaei A, Tirkolaee EB, Ali SS (2024) Innovative supply chain network design with two-step authentication and environmentally-friendly blockchain technology. Ann Oper Res:1–31

Balcıoğlu YS, Çelik AA, Altındağ E (2024) Integrating Blockchain Technology in Supply Chain Management: a Bibliometric analysis of theme extraction via text mining. Sustainability 16(22): 10032

Barata J, da Cunha PR (2021) Augmented product information: crafting physical-digital transparency strategies in the materials supply chain. Int J Adv Manuf Technol 112(7):2109–2121

Basheer M, Elghaish F, Brooks T, Rahimian FP, Park C (2024) Blockchain-based decentralised material management system for construction projects. J Build Eng 82:108263

Bucci, I., Fani, V., & Bandinelli, R. (2024). Towards human-centric manufacturing: Exploring the role of human digital twins in industry 5.0. Sustainability, 17(1), 129

Cao S, Xu H, Bryceson KP (2023) Blockchain traceability for sustainability communication in food supply chains: an architectural framework, design pathway and considerations. Sustainability 15(18):13486

Cerchione R (2024) Design and evaluation of a blockchain-based system for increasing customer engagement in circular economy. Corp Soc Responsib Environ Manag 32(1):160–175

Chaouni Benabdellah A, Zekhnini K, Cherrafi A, Garza-Reyes JA, Kumar A, El Baz J (2023) Blockchain technology for viable circular digital supplychains: an integrated approach for evaluating the implementation barriers. BIJ 30(10):4397–4424

Chen Z, Sarkis J, Yildizbasi A (2024) Digital transformation for safer circular lithium-ion battery supply chains: a blockchain ecosystem-data perspective. Int J Prod Res 63(5):1–22

Contini C, Boncinelli F, Piracci G, Scozzafava G, Casini L (2023) Can blockchain technology strengthen consumer preferences for credence attributes? Agric Food Econ 11(1):27

Correa Tavares E, Meirelles FDS, Tavares EC, Cunha MA, Schunk LM (2021) Blockchain in the Amazon: creating public value and promoting sustainability. Inf Technol Dev 27(3):579–598

De Giovanni P (2022) Leveraging the circular economy with a closed-loop supply chain and a reverse omnichannel using blockchain technology and incentives. Int J Oper Prod Manag 42(7): 959–994

Dehshiri SJH, Amiri M (2024) Evaluation of blockchain implementation solutions in the sustainable supply chain: a novel hybrid decision approach based on Z-numbers. Expert Syst Appl 235: 121123

Du W, Ma X, Yuan H, Zhu Y (2022) Blockchain technology-based sustainable management research: the status quo and a general framework for future application. Environ Sci Pollut Res 29(39):58648–58663

Dutta P, Choi TM, Somani S, Butala R (2020) Blockchain technology in supply chain operations: applications, challenges and research opportunities. Transp Res Part E Logist Transp Rev 142: 102067

Ebinger F, Omondi B (2020) Leveraging digital approaches for transparency in sustainable supply chains: a conceptual paper. Sustainability 12(15):6129

Ellahi RM, Wood LC, Bekhit AEDA (2023) Blockchain-based frameworks for food traceability: a systematic review. Foods 12(16):3026

Fraga-Lamas, P., Fernandez-Carames, T. M., da Cruz, A. M. R., & Lopes, S. I. (2024). An overview of blockchain for Industry 5.0: towards human-centric, sustainable and resilient applications. IEEE Access, 12, 116162–116201.

Guo S, Sun X, Lam HK (2020) Applications of blockchain technology in sustainable fashion supply chains: operational transparency and environmental efforts. IEEE Trans Eng Manag 70(4): 1312–1328

Haffar S, Özceylan E (2024) Blockchain-based system for supplier selection in sustainable and leagile supply chains. IEEE Access

Hasan HR, Salah K, Jayaraman R, Omar M (2024) Blockchain-based sustainability index score for consumable products. IEEE Access

Heiss J, Oegel T, Shakeri M, Tai S (2023) Verifiable carbon accounting in supply chains. IEEE Trans Serv Comput 17(4):1861–1874

Jan A, Salameh AA, Rahman HU, Alasiri MM (2024) Can blockchain technologies enhance environmental sustainable development goals performance in manufacturing firms? Potential mediation of green supply chain management practices. Bus Strateg Environ 33(3):2004–2019

Jiang S, Jakobsen K, Bueie J, Li J, Haro PH (2022) A tertiary review on blockchain and sustainability with focus on sustainable development goals. IEEE Access 10:114975–115006

Jimenez-Castillo L, Sarkis J, Saberi S, Yao T (2024) Blockchain-based governance implications for ecologically sustainable supply chain management. J Enterp Inf Manag 37(1):76–99

Khan K, Hassan F, Koo J, Mohammed MA, Hasan Y, Muhammad D et al (2023) Blockchain-based applications and energy effective electric vehicle charging–a systematic literature review, challenges, comparative analysis and opportunities. Comput Electr Eng 112:108959

Lee HL (2021) Supply chains with a conscience. Prod Oper Manag 30(3):815–820

Lin WL, Yang Y, Leow NX, Lim WM, Khee PC, Yip N (2024) Trends in the dynamic evolution of blockchain and circular economy: a literature review and bibliometric analysis. Bus Strateg Environ 34(1):1062–1084

Liu M, Lu J, Zhao C, Luo J, Liu Q, Wang H et al (2024) Antecedents and consequences of environmental, social, and governance: a bibliometric analysis based on the web of science database. Corp Soc Responsib Environ Manag 32(1):984–1001

Modgil S, Gupta S, Zhang B, Bag S (2024) Blockchain-enabled supply chain finance strategies for a circular economy revolution. Bus Strateg Environ 33(7):7006–7020

Mukherjee AA, Singh RK, Mishra R, Bag S (2022) Application of blockchain technology for sustainability development in agricultural supply chain: justification framework. Oper Manag Res 15(1):46–61

Mulligan C, Morsfield S, Cheikosman E (2024) Blockchain for sustainability: a systematic literature review for policy impact. Telecommun Policy 48(2):102676

Najjar M, Alsurakji IH, El-Qanni A, Nour AI (2023) The role of blockchain technology in the integration of sustainability practices across multi-tier supply networks: implications and potential complexities. J Sustain Financ Invest 13(1):744–762

Nath SD, Khayer A, Majumder J, Barua S (2022) Factors affecting blockchain adoption in apparel supply chains: does sustainability-oriented supplier development play a moderating role? Ind Manag Data Syst 122(5):1183–1214

Nazarejova, J., Soltysova, Z., & Rudeichuk, T. (2024). Requirements and Barriers for Human-Centered SMEs. Sensors, 24(14), 4681.
Okay NC, Sencer A, Taskin N (2024) Quantitative indicators for environmental and social sustainability performance assessment of the supply chain. Environ Dev Sustain:1–21
Pahontu BI, Petcu A, Predescu A, Arsene DA, Mocanu M (2023) A Blockchain approach for migrating a cyber-physical water monitoring solution to a decentralized architecture. Water 15(16):2874
Pakseresht A, Ahmadi Kaliji S, Xhakollari V (2022) How blockchain facilitates the transition toward circular economy in the food chain? Sustainability 14(18):11754
Park A, Li H (2021) The effect of blockchain technology on supply chain sustainability performances. Sustainability 13(4):1726
Quayson M, Bai C, Sun L, Sarkis J (2023) Building blockchain-driven dynamic capabilities for developing circular supply chain: rethinking the role of sensing, seizing, and reconfiguring. Bus Strateg Environ 32(7):4821–4840
Rani P, Sharma P, Gupta I (2024) Toward a greener future: a survey on sustainable blockchain applications and impact. J Environ Manag 354:120273
Rejeb A, Keogh JG, Simske SJ, Stafford T, Treiblmaier H (2021) Potentials of blockchain technologies for supply chain collaboration: a conceptual framework. Int J Logist Manag 32(3):973–994
Sahoo S, Kumar A, Mishra R, Tripathi P (2022) Strengthening supply chain visibility with blockchain: a Prisma-based review. IEEE Trans Eng Manag 71:1787–1803
Sancha C, Gutierrez-Gutierrez L, Tamayo-Torres I, Gimenez Thomsen C (2023) From corporate governance to sustainability outcomes: the key role of operations management. Int J Oper Prod Manag 43(13):27–49
Sargent CS, Breese JL (2024) Blockchain barriers in supply chain: a literature review. J Comput Inf Syst 64(1):124–135
Sedlmeir J, Lautenschlager J, Fridgen G, Urbach N (2022) the transparency challenge of blockchain in organizations. Electron Mark 32(3):1779–1794
Seow RYC (2024) Determinants of environmental, social, and governance disclosure: a systematic literature review. Bus Strateg Environ 33(3):2314–2330
Sharabati AAA, Jreisat ER (2024) Blockchain technology implementation in supply chain management: a literature review. Sustainability 16(7):2823
Singh AK, Kumar VP, Irfan M, Mohandes SR, Awan U (2023) Revealing the barriers of blockchain technology for supply chain transparency and sustainability in the construction industry: an application of Pythagorean FAHP methods. Sustainability 15(13):10681
Truby J (2018) Decarbonizing bitcoin: law and policy choices for reducing the energy consumption of Blockchain technologies and digital currencies. Energy Res Soc Sci 44:399–410
Vaezi A, Rabbani E, Yazdian SA (2024) Blockchain-integrated sustainable supplier selection and order allocation: a hybrid BWM-MULTIMOORA and bi-objective programming approach. J Clean Prod 444:141216
Ventura V, La Monica M, Bortolini M, Cutaia L, Mora C (2024) Blockchain and industrial symbiosis: a preliminary two-step framework to green circular supply chains. Int J Environ Sci Technol 22(1):1–14
Vinayavekhin S, Banerjee A, Li F (2024) "Putting your money where your mouth is": an empirical study on buyers' preferences and willingness to pay for blockchain-enabled sustainable supply chain transparency. J Purch Supply Manag 30(2):100900
Whitelock VG (2019) Multidimensional environmental social governance sustainability framework: integration, using a purchasing, operations, and supply chain management context. Sustain Dev 27(5):923–931

Xie S, Gong Y, Kunc M, Wen Z, Brown S (2023) The application of blockchain technology in the recycling chain: a state-of-the-art literature review and conceptual framework. Int J Prod Res 61(24):8692–8718

Yap KY, Chin HH, Klemeš JJ (2023) Blockchain technology for distributed generation: a review of current development, challenges and future prospect. Renew Sust Energ Rev 175:113170

Yontar E (2023) The role of blockchain technology in the sustainability of supply chain management: Grey based dematel implementation. Clean Logist Supply Chain 8:100113

Analyzing Carbon Emissions and Sustainability Factors of Distribution Network

Sri Yogi Kottala and Atul Kumar Sahu

Abstract Research on sustainable supply chain management (SSCM) case-based approach in Indian context is emerging. The problem of climate change is increasingly being accepted as a major problem by policy makers around the world. The main objective of this chapter is analysis of distribution network of the case company considered in three dimensions, i.e. sustainability, logistics, and CO2 emissions. We has proposed a sustainable transport and emissions assessment framework to evaluate the carbon footprints of companies, focusing on the transformation processes of distribution, warehousing, and logistics. Additionally, in present study, 15 factors are evaluated from the insights of decision-makers from three gypsum plant situated in India. The findings of the study explicated that Plant 1 possesses strong characteristics in terms of "Less Packaging with green material." Additionally, Plant 2 possesses strong characteristics in terms of "Organizational Competence" with indeed "Availability of Water Supply" and Plant 3 possesses strong characteristics in terms of "Less Packaging with green material" with indeed "Indoor logistics design" capabilities.

S. Y. Kottala (✉)
Symbiosis Institute of Business Management Hyderabad Campus, Hyderabad, India

Symbiosis International University (Deemed University), Pune, India
e-mail: sriyogi.k@sibmhyd.edu.in

A. K. Sahu
Department of Industrial & Production Engineering, School of Studies in Engineering & Technology, Guru Ghasidas Vishwavidyalaya (A Central University), Bilaspur, Chhattisgarh, India
e-mail: atulkumar.sahu@ggu.ac.in

M. Venkatesh et al. (eds.), *Green Manufacturing Technologies in Industry 5.0*,
https://doi.org/10.1007/978-981-95-0363-6_8

1 Introduction

The logistics and transportation sector makes global trade possible, allowing the production of goods (manufacturers) to reach consumers. The problem of climate change is increasingly being accepted as a major problem by policy makers around the world (Datta 2008). However, they are among the worst offenders due to their reliance on fossil fuel combustion for energy, particularly with respect to carbon dioxide (CO_2) emissions. Given the much larger picture of global climate change, the environmental impact of this sector has become a looming issue. Logistic practices have also increasingly become more sustainable over time, and there is an increasing focus on developing distribution networks that are environmentally friendly while being operationally efficient (Sridharan et al. 2023). This is the process of sustainable supply chain management; that is, the coordination of supply and demand for the sustainable logistics and systems management of material, information, and human resources as well as energy and emissions to minimize carbon-intensive logistics and supply chain management capabilities in fuel-efficiency and low carbon technologies. What is driving the movement toward sustainability? There are increased and stringent regulations, along with a change in consumer and business behavior, as they respond to the increased threat of global warming.

A rapidly expanding economy and extensive road logistics network make India face challenges and opportunities related to sustainability and CO2 reduction and high dependency on road transport in its logistics sector (Sridharan et al. 2023). The distributed networks of gypsum—an elemental material in construction—sourced from three plants in different regions of India are systematically studied in this work. Because of its large volume and wide usage, gypsum is a suitable case for studying the carbon footprint generated by directly related logistics networks. The study gives insight into how gypsum plants are distributed over the country and its distribution networks; this would be a first step to understanding sustainability, logistics, and CO2. A case study-based methodology is employed in this research, which involves formulations by Ooishi (1996) for vehicle speed and type emissions. It uses an energy approach advised for transport companies to provide an integrated perspective (Sridharan et al. 2023). These results contribute to sustainability logistics research and deliver practical recommendations for lowering carbon emissions for the transport sector. These findings establish the basis for a comprehensive approach to understanding and reducing the carbon footprint of logistics networks, not only in the Indian context but also across nations and various operational scenarios and vehicle types (Sridharan et al. 2023).

2 Importance of Sustainability in Logistics

Due to the large portion of global CO_2 emissions, the logistic sector has a significant role in sustainability efforts to achieve long-term environmental, economic, and social goals. These include addressing sustainability issues, optimizing routes, adopting alternative fuels, and investing in fuel-efficient vehicles to mitigate greenhouse gas emissions, help combat climate change, and reduce the ecological footprint. The economic benefit is that it saves money by increasing fuel efficiency and being a more efficient operation, thus adding to the bottom line of the 3P (people, planet, and profit). On the other hand, sustainability translates into respect for regulations, reduces legal risks, and contributes to creating a responsible brand; sustainability also attracts green consumers and investors. It also fosters innovations and long-term resilience to make operations climate-resilient and resource-secure. Moreover, sustainable logistics is closely related to the United Nations' Sustainable Development Goals (SDGs), aiming to reinforce global sustainability and promote socio-economic development. Sustainable logistics is not merely about green practices; hence, the new research is related to improving sustainable logistics strategies, as sustainable logistics is essential for a balanced future where all parties experience benefits beyond the environmental responsibility (Tabanjeh et al. 2025).

This chapter consists of seven sections, each addressing a vital component of the study. Section 2 provides an overview of the carbon footprint of transport networks, emission sources, contributing factors, and issues related to sustainability. Section 3 defines the research problem with a focus on critical issues related to transportation emissions and the need for effective mitigation strategies. Section 4 provides a literature review summarizing current studies related to carbon emissions, sustainable mobility, and policy interventions. Section 5 discusses the methodology with a focus on the study design, data collection process, and analytic techniques used. Section 6 presents the results and analysis with a focus on critical findings related to emission trends and sustainability measures. Section 7 discusses the implications of the findings with a focus on their applicability to policy and industry practice. The chapter ends with Sect. 8, summarizing critical insights, contributions, and recommendations for future study and sustainable transport practice.

2.1 Overview of the Carbon Impact of Transport Networks

Modern transport networks are the lubricant of economies, allowing goods and people to travel worldwide. However, there are huge emitters of greenhouse gases, particularly carbon dioxide (CO_2). Transport networks have different carbon impacts, and all modes of transport have unique emissions profiles and problems.

Road Transport The most significant contributor to the transport sector when it comes to CO_2 emissions is road transport, which is based on the combustion of fossil fuels such as diesel or gasoline and is used heavily by the transport mode. These emissions are driven partly by freight transport by road, with the impacts of last-mile goods distribution by trucking being particularly large. Road transport fuel consumption is also a source of carbon emissions, and inefficiencies exacerbate this during traffic congestion and using less-than-optimal routes (Banister 2008).

Rail Transport Rail transport is typically more energy efficient and has less carbon emissions than road transport, especially for long-distance freight and passenger services. If the energy is from a renewable source, electric railways can significantly reduce emissions. The scale of rail network electrification and the energy mix of electricity generation are the critical factors for the carbon impact of rail transport (Hickman and Banister 2019).

Air Transportation Air transport, essential for long-distance and international travel, is one of the worst greenhouse gas pollutants of all modes of transportation. Aircraft powered by jet fuel are responsible for very high CO2 emissions per passenger kilometer. The growth of aviation has been so rapid that even though fuel efficiency has been improved, sustainable aviation fuels have been developed, and air traffic management is enhanced, efforts to curb the carbon impact of air transport have not been sufficient (Gössling et al. 2017).

Maritime Transport One of the most essential services for global trade, maritime transport is relatively low-energy intensive compared to air and road transport. However, the tonnage transported by ships contributes much to the CO2 pollution. Carbon emissions from ships: increased risk of failure by 32% affects oil prices in the shipping industry. The carbon footprint of most ships is essentially dependent on heavy fuel oil, a high-emission fuel. This is countered by work to find alternative fuels, such as LNG and hydrogen, along with tech-based initiatives to lower fuel consumption and emissions (Notteboom et al. 2019).

Intermodal Transport Intermodal transport, where multiple modes of transport are combined in one journey, is another possible route for reducing the carbon footprint of logistics networks. Intermodal solutions can reduce emissions by optimizing transport modes based on efficiency and carbon impact. Therefore, long-haul cannot be easily displaced by the other modalities as in the case in which modal transposition can reduce costs—through lower emissions (ibid.).

2.2 Research Problem

Distribution networks are a large contributor to carbon emissions, particularly in transport. Footprint Disclosure (FD) provides a detailed understanding of the carbon emissions associated with distribution networks, moving beyond generic assumptions and exposing the current lack of awareness necessary to support sustainable logistics behaviors. This study seeks to fill this gap by investigating the carbon footprint of logistics networks, with a particular emphasis on gypsum distribution from Indian plants. The research problem revolves around the most impactful factors in distribution networks' carbon footprint and how we can assess sustainable strategies to lower emissions. Thus, the study sheds light on the correlation between sustainability, logistics efficiency, and CO_2 emissions in gypsum distribution networks for an improved understanding and development of novel strategies for reducing CO_2 emissions from logistics operations in India (Ghadimi et al. 2020).

2.3 Research Objectives

- Analysis of distribution network considering three dimensions, i.e. sustainability, logistics, and CO_2 emissions using a case.
- Evaluation of critical sustainable factors from decision-maker's insights from three gypsum plants (selected as case) situated in India.

3 Literature Review

Sustainability in logistics in line with heightened attention to environmental performance and embracing sustainability practices in supply chain operations, the literature on sustainability in logistics has grown noticeably over the last few years. For example, Akyelken (2011) emphasizes the role of green logistics in mitigating carbon footprints through route planning and energy-efficient vehicle strategies, which have paved the way for further investigations. Dey et al. (2011) stress the importance of firms implementing sustainability in logistic processes, suggesting that the way forward should include more research in green technology and eco-friendly operations. Colicchia and Strozzi (2012) conducted one of the earliest systematic literature reviews (SLR) focused specifically on Supply Chain Risk Management (SCRM). Based on their literature review, Chen et al. (2014) also propose future research directions, including empirical research and education on sustainability. Lam (2012) highlights that integrating risk management with sustainability strategies enhances resilience in maritime logistics. Perhaps the most notable work on green supply chain management (GSCM) combining a methodology specifically for green commitment of logistics service providers (LSPs) with specific avenues through which LSPs can improve their environmental performance has been proposed by Sumbal et al. (2024). Ullah et al. (2024) reported that environmental sustainability targets improve supplier–buyer relationship performance, and the partners of this relationship take the responsibility of the owners. Arya et al. (2020) describe the impact on investment judgment of trade-offs between environmental and economic efficiency in the emergence of automation and energy-efficacy advancements. Martins et al. (2020) and Sanches et al. (2022) presented another recent study that addressed Brazil's logistics chain by focusing on sustainability principles and concepts, such as waste reduction and the use of alternative fuels. Pan (2019) elaborates on product-service systems in the context of the physical Internet as a gate to seize on business model innovation and optimize resource use. Fergani et al. (2020) together exhibit the economic and environmental benefits to which logistical innovations can aspire by accurately modeling cost and greenhouse gases averted in mobile modular production. These studies indicate and highlight that the field is alive, evolving, and needing ongoing innovation.

4 Carbon Emissions in Transport

The literature on carbon emissions in transport is broad and heterogeneous, as this is an essential field in which emissions need to be reduced, resulting in a rich set of potential topics for research and a wide range of methods to investigate impacts. Walks (2014), Carter and Drouaud (2022) describe how organized spatial planning can reduce CO_2 emissions, emphasizing that investments in infrastructure are fostering a less mobility-dependent and historically low-carbon society. Unlike existing studies, Aldegheishem (2024) explored the structural time series effect of oil consumption on Saudi Arabia's carbon emissions, stressing the importance of other energy and efficiency improvements (Ali et al. 2023; Walmsley et al. 2015).

Balancing supply solutions and demand reductions in a low-emission heat and power sector in a CO2-constrained world Finnish example based on detailed energy system analysis using the Carbon Emission Pinch Analysis (CEPA) method changes the perspective to investigating the possibility of New Zealand reaching 1990 emission levels by 2050, stressing the needs and strategies required to meet ambitious reduction targets. This study by Cheng et al. (2015) estimates the lifecycle carbon emissions of the transport service sector in China with a simplified LCA (life cycle assessment) model, providing comprehensive information for more sustainable transportation planning (Duan et al. 2015). The analyses by Cheng and Tang (2015) suggest that in urban China, significant carbon emissions savings can be achieved through transitions away from private car use for shopping to specific public transport modes, indicating the potential for suitable public transport quality measures to mitigate emissions. Zhu and Li (2017) studied the transportation carbon emissions and the economic growth in the Beijing–Tianjin–Hebei region based on the decoupling theory and structure decomposition model, and they designed the corresponding path of decoupling carbon emissions from economic growth. Li and Yu (2019) are concerned about a peak of CO2 emissions in China's urban passenger transport in 2030 by considering the effects of alternative fuels, fuel efficiency improvements, and public transportation promotion. Faheem et al. (2020) also examine the linkages of economic growth, transport infrastructure, fuel consumption, and carbon emissions in Pakistan to provide policy implications for sustainable infrastructure investments. Similarly, Rugani et al. (2021) examine the environmental efficiency of economic instruments, focusing on transport tax revenues to control and reduce emissions in the EU, emphasizing the effectiveness of fiscal policies. Finally, previous studies have emphasized the inclusion of the transport sector under a carbon emission trading scheme, where they discuss the merits and challenges of this market-based tool in managing emissions (Ali et al. 2023; Li et al. 2019). This highlights the necessity of an integrated approach combining infrastructure development, policy interventions, technological improvements, and market-based instruments for better control of carbon emissions for an overall reduction of carbon emissions in the transport sector, enabling a shift toward a sustainable low-carbon transport system.

4.1 *Case Studies in Sustainable Logistics*

Consequently, businesses are paying more and more attention to sustainable logistics—the balance between environmental goals and economic interests. Anderson et al. (1999) concluded that legislation concerning packaging waste in the UK had brought significant changes in logistics activities, as firms had to adapt transport operations to meet new environmental requirements. Reverse logistics (Dowlatshahi 2000) is a profitable but likely fully engrained long-term strategy to recycle and reduce waste with economic good through case study examples. De Giovane et al. (2015) explored reverse logistics in Colombian companies based on paper recycling and electronic waste reuse, stressing that functional solid return and recycling systems are necessary for applying these processes in emerging markets. Cruijssen (2020) proposed a structure for load factors, logistics spending, and a drop in the carbon footprint by expanding restrictions for horizontal cooperation for shipper cooperation. The importance of creating sustainable hubs within green transport corridors has been accentuated by Prause (2014), emphasizing the potential for logistics clusters to enable ecological transport alternatives (Liu et al. 2022); Ivanova (2019). Lean management principles and sustainability goals were integrated by Zhou et al. into a model for sustainability-oriented green improvement in maintenance logistics based on a demand-driven approach. Rosenberg et al. (2021) demonstrate that collaborative micro-depot networks in urban areas can significantly reduce congestion, optimize delivery operations, and lower carbon emissions through shared logistics infrastructure. Sun et al. (2022) emphasize that integrating IoT and AI into reverse logistics within Industry 4.0 enables greater operational efficiency and environmental sustainability by enhancing smart connectivity, traceability, and autonomous decision-making. These results underscore the importance of regulatory compliance, technological innovation, cooperation, and strategic direction in sustainable logistics. They provide essential reference points for companies to lower their environmental impact while improving their competitiveness.

5 Methodology

The case study technique is appropriate for this research because it is employed in real-life settings. The Asian Development Bank (2009) proposes a three-step approach for analyzing CO2 emissions: We analyze and monitor current transport activity and CO2 emissions, forecast future emissions based on existing levels, and evaluate plans to reduce both. Ellram (1996) proposes that case studies serve as a research methodology for explaining, exploring, or describing an interesting occurrence. Yin (2003) a case study is warranted when a researcher has the opportunity to observe and examine a phenomenon that was previously inaccessible to scientific research. The methodology for estimating CO2 emissions from transportation is customized, with many methodologies used. Consider the speed and type of vehicle. Ooishi (1996) formulated an activity-based strategy. An energy-based methodology, recommended as a calculation approach for use by transportation corporations. The Factors Performance Score (FPS) is calculated to determine the case company's coverage of sustainable factors in gypsum plant operations.

6 About the Company

ABCD Gypsum is a multi-national company and a global leader operating in 50 countries with 90 manufacturing plants. ABCD, a XYZ group company is the only manufacturer of gypsum-based products in the country. It has pioneered dry construction techniques in the country, as well as concept selling in Indian building industry. India Gypsum has three plants at Jind, Wada, and Sriperumbudur. The investigation of CO_2 emissions shall be presented with real data for a case company. The company operates 19 distribution centers across four regions in India, with the monthly demand for each location considered for calculation in square meters. The South region includes Chennai, Coimbatore, Cochin, Bangalore, and Hyderabad, while the East region consists of Kolkata, Bhubaneswar, Guwahati, and Patna. The West region covers Ahmedabad, Mumbai, Pune, and Goa, whereas the Central region comprises Nagpur, Indore, Delhi, Chandigarh, Jaipur, and Lucknow. A comprehensive analysis of CO_2 emissions will be conducted using real data from a case company.

7 Analytical Framework

An Activity Based Strategy by Ooishi (1996) considering speed and type of vehicle, proposed theoretical models considering the velocity and vehicle categories to evaluate transportation systems. The following is an improved summary with formulas that show the total CO2 emissions of each trip from the warehouses of the different origins to the distribution centers:

Speed Analysis: According to Ooishi (1996), the framework analyzes the speed characteristics of various vehicles (those that people generally ride) and whose transportation is required. This analysis helps in appreciating the rate at which goods or passengers are transported and the effectiveness of the transportation system. V : the average speed for a vehicle is given by:

$$V = \mathrm{D}/T\mathrm{V} \tag{1}$$

Where:

D is the distance traveled (km).

T is the travel time (hours).

Type of Vehicle Assessment.

The assessment framework also considers the types of vehicles used in transportation based on size, capacity, and fuel efficiency. This evaluation establishes vehicle type fit for certain transport activities and potential improvements.

7.1 Integration of Speed and Vehicle Type

In a framework introduced by Ooishi (1996), a combined analysis involving speed and vehicle type is used to evaluate the system-wise performance of transportation systems.

$$\textit{For Large truck fc} = \frac{539.0}{v} - 11.03v + 0.0758v^2 + 587.6 \tag{2}$$

$$\textit{For Small truck fc} = \frac{544.2}{v} - 1.194v + 0.0117v^2 + 81.2 \tag{3}$$

Bringing them together gives an overall view of the pros and cons of transport systems and identifies weak points and potential cooperation areas.

7.2 Performance Metrics

The framework may include services such as travel time, delivery efficiency, fuel consumption, and service level performance. Using these metrics, we can quantitatively evaluate their performance and compare different systems or scenarios.

CO2 Emissions Calculation: For calculating total CO2 emissions (*E*E) per trip from various warehouses to distribution centers, the following formula can be used:

$$E = F \times C \times DE \quad (4)$$

Where:

E is the total CO2 emissions (kg).

F is the fuel consumption rate (liters per km or gallons per mile).

C is the CO2 emission factor (kg CO2 per liter of fuel or kg CO2 per gallon of fuel).

D is the distance traveled (km or miles).

Example calculation for total CO2 emissions per trip.

Assuming:

Fuel consumption rate F for a small truck: 0.2 liters/km.

CO2 emission factor C for diesel: 2.68 kg CO2 per liter.

Distance D from Warehouse A to Distribution Center B: 100 km.

The total CO2 emissions *E*E can be calculated as:

E=0.2 liters/km × 2.68 kg CO2/liter×100 km E = 0.2liters/km × 2.68 kg CO2/liter×100 km *E*=53.6 kg CO2E = 53.6 kg CO2.

7.3 Energy-Based Approach for Transport Companies

For transport companies, an energy-oriented strategy requires detailed planning of energy consumption in all phases of the process of transportation. An edited version of the first analytical framework is attached with citations, as shown below.

$$\text{CO2 emissions} = \text{fuel consumption}^{*}\text{fuel emission conversion factor} \quad (5)$$

$$\text{Tonnes CO2 emissions} = \text{liters}^{*}\ \text{kg CO2 per liter fuel} \mid 1.000 \quad (6)$$

7.4 Energy Consumption Analysis

Started by conducting an extensive energy consumption analysis, tracking energy use at each stage of transportation, including vehicle operating and maintenance and auxiliary operations, such as loading and unloading. The activity-based approach, which is the recommended calculation method for chemical companies, is used to estimate CO_2 emissions. The total CO_2 emissions are calculated using the formula:

$$CO_2\text{emissions} = \text{tonnes} \times \text{kilometers} \times \text{g } CO_2\text{per tonne} - \text{km} \quad (7)$$

According to NTM (2010), the average distance emission factor for road transport is 62 g CO_2/km. The calculation is typically based on a combination of the weight of goods transported, distance traveled, and transport mode. The emission factors vary depending on the vehicle type used.

To determine how much CO_2 is emitted when transporting one tonne of goods for one-kilometer, specific vehicle-based data would be ideal. However, due to the unavailability of vehicle-specific emission data, general emission factors for various vehicle types—such as light lorries, heavy lorries, container ships, and bulk ships are considered (NTM 2010).

The total CO_2 emissions can be computed using the formula:

$$\text{Total } CO_2 \text{ emissions} = \sum(\text{Weight} \times \text{Distance} \times \text{Emission factor}) \tag{8}$$

This methodology ensures a systematic estimation of emissions, enabling better sustainability planning and carbon footprint analysis.

8 Identification of Key Factors

To identify the key factors influencing energy consumption—such as vehicle type, payload, route, driving behavior, and operational practices—this chapter examines the determinants of energy use as discussed by Janpoom et al. (2023).

8.1 Data Collection and Monitoring

Utilizing sensors, telematics systems, and monitoring tools to collect energy usage data from vehicles, infrastructure, and operational activities to monitor energy consumption in real time and compared to performance metrics (Brown and Lee 2019).

Performance Benchmarking A comparison of energy consumption to industry standards and best practices to rate performance and identify areas of inefficiency for tangible insight into opportunities for improvement (Gerrish et al. 2017).

Simulation and Modeling Leveraging simulation and modeling techniques to assess the effect of various parameters on energy consumption and discovering the best suitable settings and working strategies by scenario analysis (Wang et al. 2022).

Optimization Strategies Design strategies to reduce energy consumption according to the operational needs, for instance, route optimization, vehicle scheduling, and eco-driving training, to increase energy efficiency (García et al. 2019).

Implementation and Monitoring Implement energy optimization strategies and monitor the results through regular data collection and performance review to adapt to the feedback to realize energy savings (White et al. 2022).

Education and Training Implement education programs to educate drivers, operators, and workforce members to adopt energy-efficient behaviors, engaging the efforts of employees to conserve energy for continual improvement (Adams 2018, p).

Evaluation and Track Record Through ongoing, operational, and annual review, stakeholders are kept abreast of energy optimization efforts and potential, looking for both further opportunities and warning signs of the need to bring continuous improvement in the backdrop of reducing negative environmental impact (Miller 2021).

Using this energy approach and following the principle of operations correctly, transport forms can manage energy usage, save costs, and move toward sustainability.

9 Analysis and Results

Annual emissions from the transport sector road transport are essential sources of greenhouse gas emissions, impacting climate change (IPCC 2018). To understand this section in detail and to develop effective strategies to reduce carbon footprints annually, it is necessary to understand and analyze the annual emissions in this segment. Section 3 further explores the most recent data on emissions from all modes of transport, with a particular focus on road transport, and uses this to explore future projections under several scenarios. The analysis highlights how policy approaches like electrification, fuel efficiency, modal shift, and alternative fuel adoption can reduce emissions. This exercise attempts to combine aspects of the transport sector to provide a holistic perspective on its role in climate change and facilitate the identification of measures to nudge the system toward sustainability.

From Fig. 1, graph "Road Focused Scenario: Getting India's Transport Emissions to Zero" shows policy implications to attain transport emissions in India to zero by

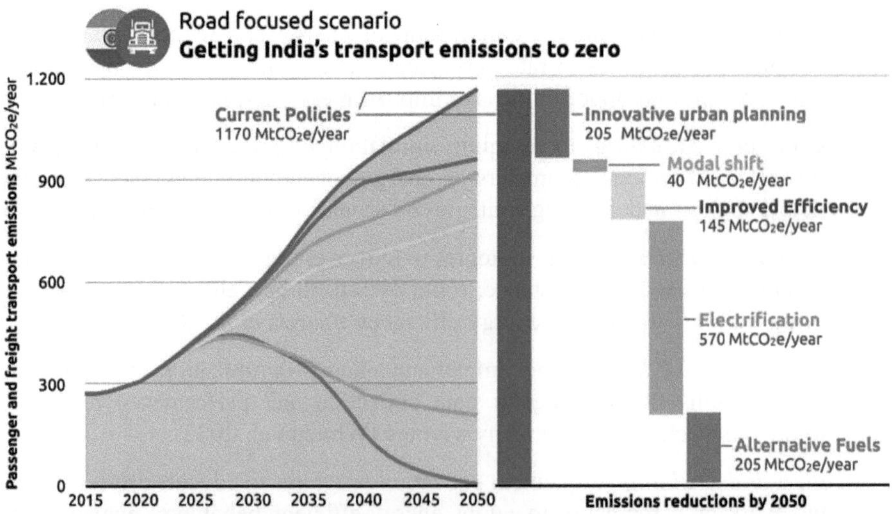

Fig. 1 Annual emissions from the transport sector until 2050 road-focused scenario (MtCO2e)

2050 to a level of 205 MtCO2e/year, which reduces 970 MtCO2e/year from the projected emissions under current policies. The most significant decrease comes from electrification, making our gasoline and diesel vehicles electric. This single measure could reduce emissions by 570 MtCO2e/year by 2050, showing the absolute necessity of electric vehicles to achieve real progress in this direction. Other policy moves could have a substantial effect in addition to electrification. Efficiency improvements, including increased fuel standards for gasoline and diesel vehicles, would result in a 145 MtCO2e/year reduction in emissions by 2050. At the same time, the UK is focusing on improving the fuel efficiency of existing internal combustion engine cars to reduce the carbon footprint of these vehicles.

In addition, a modal shift (shift to less energy-intensive modes of transport among bicycles, buses, and rail) would save 40 MtCO2e/year in 2050. Supporting these alternative means of transportation could reduce the dependence on private motor vehicles and decrease the total emissions. Now, there are alternative fuels, biofuels, and hydrogen, which by 2050 have the potential to reduce emissions of 205 MtCO2e/year as well. This use of cleaner fuels can be instrumental in providing a greener and more sustainable fuel for vehicles, thus lowering the overall environmental impact of the transportation sector. It may slash emissions even further when paired with innovative urban planning, though the chart does not say by how much. The first is to build cities conducive to walking, cycling, and public transportation rather than driving, which would translate into lower CO2 emissions. Sustainable transportation behaviors can be generated through urban planning strategies that encourage dense (compact), mixed-use development, and public transportation systems that are viable (capable of being duplicated or placed) and efficient. Note that the graph does not include the costs of implementing these policies. For example, electrification may require significant investment in infrastructure and vehicle technology, which offers fewer dispersion benefits. This graph, of course, does not take into account technological breakthroughs. Additionally, new technologies could help make these policy measures even more efficient, resulting in even more reductions in transportation emissions.

Next, the assumption is made that the Indian Transportation Sector will continue to rise. Finally, the graph shows that this would represent even more significant emissions cuts if the sector fails to grow as fast as anticipated. This demonstrates the need for ongoing monitoring and updating of policies that allow for increased growth and changes in transportation technology. Overall, the graphic drives home the point of why India needs a mosaic of solutions to curb transportation emissions. Increased electrification is the most impactful action, but efficiency improvements, modal shifts, alternative fuels, and more innovative urban forms are all required for an overall sustainable reduction in emissions.

9.1 Energy Demand by Fuel Type

Understanding how future transportation requirements can be met while simultaneously meeting sustainability goals relies critically on analyzing total energy demand by fuel type in the road-focused scenario. This scenario represents an

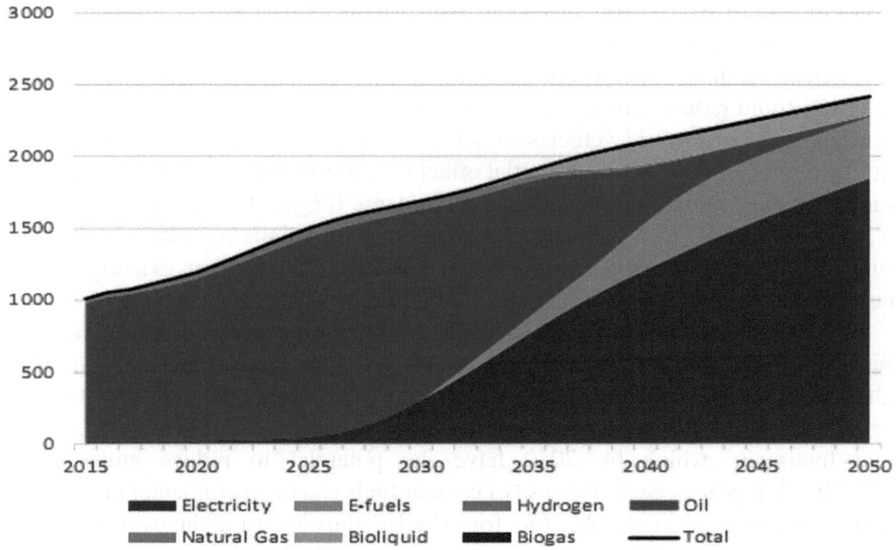

Fig. 2 Total energy demand by fuel type in the road-focused scenario

accelerated version of the changes in energy use, showing a transition from conventional to renewable energy sources. Such analysis helps to anticipate the future trends in demand and the implications for energy policy, capability, and the environment based on fuel type. The intro nicely leads to a more in-depth look at the various fuels that will matter in crafting a sustainable and efficient transportation system in future decades.

This line graph from Fig. 2 that shows our energy fuels from 2015 to 2050 tells a compelling story of the global energy transition. On the graph, the y-axis is the percent of total fuel consumed, and the x-axis lists the years from 2015 to 2050. A notable pattern that can be seen from the graph is the more significant consumption of higher energy sources. This category fits the Decarbonized Alternative Fuels and Bioenergy (DAB) sort of renewable electricity, biofuels, biogas, hydrogen, etc. The problems in Brazil are diverse in terms of energy, from the control of Direct Normal Irradiance (DNI) itself to the lack of far-sighted principles and policies to encourage the use of new energy by the average citizen. "Not me, who is a big power generator, " says the one! Another, anyone could be you or me! The challenge is to reverse this trend, always offering what is newer and better! Between 2001 and 2015, renewable sources achieved more than a share: 9% of total energy consumption. However, the graph predicts a much sharper increase, with renewables set to make up 81% of all energy consumed by 2050. This growth reflects the growing importance of sustainability and cleaner energy solutions as the world moves toward eliminating carbon footprints and the fight against climate change. On the other end of the spectrum are fossil fuels used to generate power, including oil, natural gas, and coal, where the graph shows dramatic drops in consumption. Historically, these traditional energy sources have been the most important, contributing 91% of 2015. However, the forecast expects a sharp drop to 19%—fossil fuels are expected to account for that percentage of overall energy demand by 2050. The fall highlights a broader trend toward cleaner sources of energy and away from carbon-heavy ones. The image generally represents change, which can be described as a change from fuel energy sources to renewable green energy. This change is part of a broader trend to promote energy sustainability, reduce global warming, and combat the climate crisis. Their move toward new, clean, and renewable technology and away from legacy fossil fuels reflects a commitment to a cleaner, more sustainable energy future.

9.2 Scenario Analysis for Large Trucks

This section is an in-depth analysis of scenario analysis for big trucks in terms of their CO2 emissions under various tailoring conditions. Since the transportation sector is a significant source of greenhouse gas emissions, understanding the factors influencing significant truck emissions is essential in targeting the strategies for their effective mitigation. Using data generated in cities like Delhi, Mumbai, and Chennai, the following study aims to analyze the CO2 emissions caused by speed variations and operational limits. We hope the analysis compares emissions at different speeds and cargo weights and reveals how significant truck emissions can be reduced to support more sustainable logistics practices.

The bar graph in Fig. 3 explains in detail the difference in CO2 emissions of large trucks—12-wheel type—when operating in three major cities in India: Delhi, Mumbai, and Chennai. These cities stand on the x-axis clusters, but as per the Truck operating per month, Delhi needs 671 trucks, Mumbai requires 1838 trucks, whereas Chennai needs 1377 trucks. The y-axis measures the CO2 emissions (mt). There are two sets of bars at each location, each representing different truck speeds. The blue bars show the CO2 emissions at 60 km/h. A higher weight limit (left 60/faster than 40) and the orange bars show the CO2 emissions at 40 km/h and the lower weight limit (right 40/slower than 60). This makes it possible to compare the emissions of different operating conditions. Figure 3 illustrates several important observations. Regarding the monthly CO2 emissions recorded in the three cities, Chennai has the lowest CO2 emissions in total, with 1377 trucks per month, to be specific. On the other hand, Delhi is more polluting than Chennai but is lagging behind Mumbai at 671 trucks per month. The total CO2 emissions are the highest in Mumbai, calculated with 1838 trucks per month (Note: there are many more trucks in Mumbai as well), indicating that the number of trucks also has a very high impact on overall emissions. The chart, again, visualizes the effect of the speed of emissions. Trucks, in contrast, have considerably higher CO2 output per truck at 60 km/h than at 40 km/h. That disparity is likely the result of higher-speed fuel burning rates, which should equate to higher CO2 emissions. Although the data clearly illustrates the trend of emissions from individual trucks, it does not specify which speed scenario leads to more CO2 emissions in total for each city. This would involve estimating the total

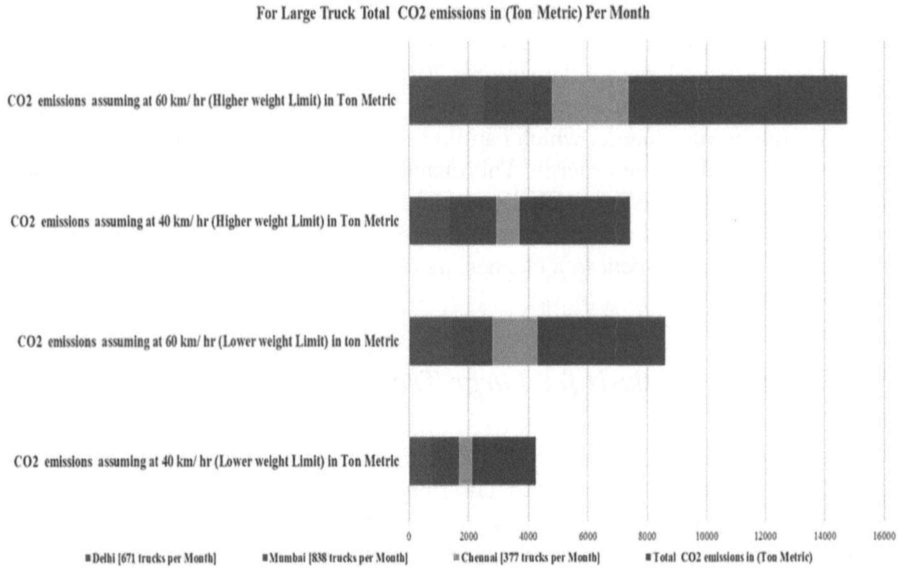

Fig. 3 Total CO2 emissions in (Ton Metric) per month

CO2 emissions from all trucks at all speeds and comparing the values. Moreover, the graph displays patterns for three areas in only one month. This snapshot in time and space may not be representative of CO2 emissions over the whole year or for the country as a whole. Even seemingly insignificant factors like seasonal changes, the mix of traffic for a city, trucking operational conditions, vehicle type, and routes Amazon's delivery trucks follow could have huge impacts on yearly emissions. Overall, this graph gives us some enlightening information regarding the effect of speed and truck number on CO2 in the observed cities. The results indicate that decreasing truck speed could reduce emissions. However, to get a complete picture of the emissions landscape and build a set of policies to reduce CO2 emissions from transport appropriately, you would need a more in-depth analysis, including annual data and a greater geographical sphere.

9.3 Scenario Analysis for Small Trucks

The following section includes a detailed scenario analysis of small trucks, including CO2 emissions under different operating conditions. Light trucks are a crucial element of several transport activities, and knowledge about the factors determining their emissions is necessary to create sustainable logistics offers. The study will examine several options, such as weight restrictions and changes in speed limits. This analysis aims to identify opportunities for mitigating the environmental effects of small truck operations by examining how these factors interplay with CO2

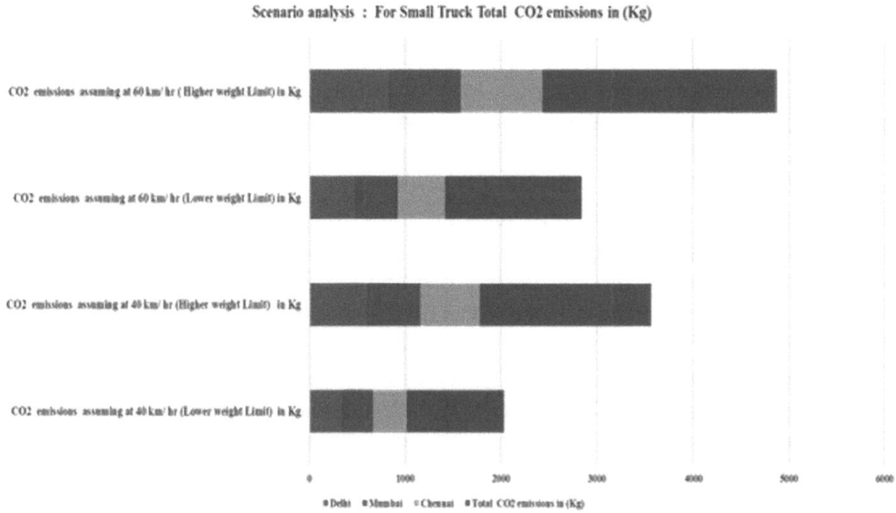

Fig. 4 Total CO2 emissions in (Kg)

emissions, particularly in urban transportation. In so doing, we hope to unpack findings that may assist policymakers and other decision-makers in developing policy measures to foster cleaner and more sustainable modes of transportation.

Figure 4 compares CO2 emissions for small trucks altered in at least one operational parameter (the driving condition, the development state), car2 (source). These can vary in weight and speed, but these are just designed scenarios. Another situation marked as "CO2 mg at 60 km over higher mass limit" showed emissions at 60 km/h above a higher weight limit. On the opposite end, potentially, measurable CO_2 emissions at 40 km/h with a lower weight limit (underweight lower limit) form another set of conditions. A graph with a vertical axis labeled "CO2 emissions (Kg)" describes the amount of CO2 emissions produced in kilograms. This axis is suitable for the comparison of emissions between the different scenarios. While the transcription text provided has a few typos here and there, the main point is conveyed: speed and weight of the car are correlated to emissions output. The broader point of the conclusion, then, is that small trucks generate more CO2 as they drive faster and heavier. The scenario is that more trucks with a higher weight limit traveling at 60 km/h would likely result in higher CO2 emissions than fewer trucks with a lower weight limit traveling at 40 km/h. The link is that speed, weight, and emissions are interrelated, and we must regulate these impacts of truck operations to save the environment. Policymakers and logistics planners need to be aware of these dynamics. Regulations that limit speeds and the laden weight of trucks could bring substantial economies in CO2 per ton of goods transported. This insight can inform the creation of more sustainable transportation strategies by achieving the right mix of efficient logistics and environmental stewardship. These analyses provide valuable information for compiling plans or policies, which are necessary to correct the environmental footprint of transport, in particular, more and more felt in an urban context since urbanization conditions most emissions.

10 Distribution Network

The modern world requires a certain level of efficiency, and distribution networks play a significant role in this efficiency. The problem is that these networks, despite being convenient, hide their ecological impact and the CO2 emissions they generate a secret. This contribution is directly proportional to the level of home deliveries it does each month. This note introduces the topic of distribution networks and the environmental consequences of their dynamics. It demonstrates the relationship between the number of vehicles on the road and the CO2 emissions produced as a result. This implies a trade-off between efficiency and sustainability and calls for more investigation to make these critical systems environmentally friendly.

The result from Fig. 5 gives a holistic treatment for the issue and aims at achieving two-thirds of CO2 reduction by 2050 across India's transportation sections. In the absence of new policies, the objective is to reduce an estimated 970 Mt. CO2e/year from the comparative level of emissions if possible, resulting in a target of 205 Mt. CO2e/year by 2050. These policies run the gamut, designed to reduce emissions throughout myriad forms of transportation. At the top of the list is electrification, referring to the shift from conventional gasoline and diesel to electric vehicles. Even with electrification alone, projections are that the measure could save up to 570 Mt. CO2e/year by 2050, suggesting it has a vital role in transport decarbonization. The graph also illustrates the potential for policy levers. Decarbonization due to improved efficiency (including those from fuel standards for conventional vehicles): 145 Mt. CO2e/year sequestered by 2050. Similarly, modal shift initiatives to promote energy-efficient modes of bikes, buses, and trains can deliver 40 Mt. CO2e/ year in emission reductions by 2050. The use of alternative fuels like biofuels and hydrogen is also forecast to save 205 Mt. CO2e/year by 2050, deepening the energy mix for transport. Though the graph serves as a complete list of these policy actions, it ultimately explains what innovative urban planning is doing now about cutting emissions. While not shown on the graph, land use planning

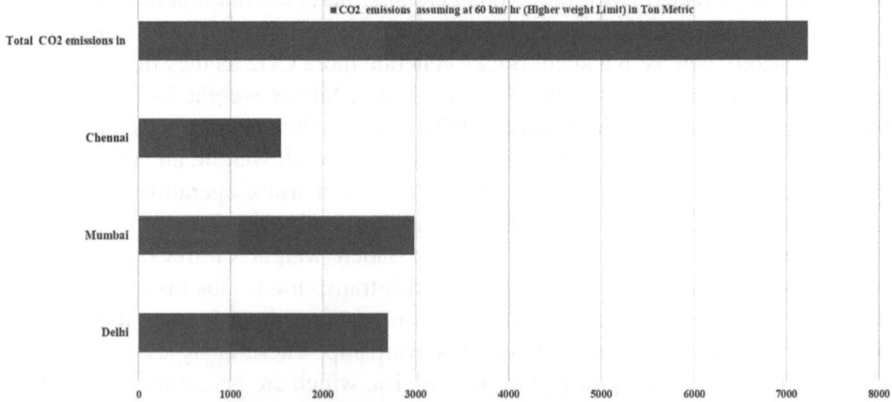

Fig. 5 CO2 emissions based on the volume of vehicles used per month

strategies that encourage walking, biking, and public transit may provide another means of mitigating emissions in more urban parts of a city. Interpreting the significance of these policy measures will depend on various factors. To the first point, the chart fails to directly capture the costs involved in implementation, which may differ very substantially from one measure to another. Moreover, the ability of these strategies to deliver emissions reductions may be further impacted by technological breakthroughs and unexpected disruptions. Secondly, this scenario depicts relentless growth in India's transport sector, but even better improvements might be possible if a different trajectory is taken. In closing, the chart highlights the multi-dimensional intervention necessary to dramatically reduce transport emissions in India. However, electrification is a significant strategy; other measures such as improved efficiency, modal shift, alternative fuels, and urban planning interventions are additional building blocks to tackle the overarching objective of reducing CO2 emissions and progress toward a sustainable transport future.

10.1 Trip-Based Emissions to Distribution Centers

CO_2 emissions per trip from three major source cities in India—Delhi, Mumbai, and Chennai—to multiple distribution centers. Comprehending the emissions from transportation activities is essential for evaluating logistic operations' impact on the environment and finding methods of consolidating the emissions produced. We investigate the emissions created per trip from these significant urban agglomerations to other distribution centers and make it possible to conclude the carbon footprint of inner-urban transport networks with this. This analysis lays the groundwork for driving sustainable transportation solutions that balance logistics' commercial and social considerations with an environmental focus. By scrutinizing trip-based emissions, we hope to guide stakeholders and policymakers in shaping environmentally friendly and efficient transportation behavior within cities.

The chart Fig. 6 compares the CO2 emissions generated by different types of vehicles used in distribution networks. Here, the vehicle types are identified on the x-axis (cars, vans, and Heavy Goods Vehicle (HGVs)), and emissions in CO2 in tons on the y-axis show the environmental cost of each category. After close inspection, distinct points are well visible in the graph which point out the nature of the emissions made by different types of vehicles. Trucks of various classifications are by far the highest CO2 emitters shown, followed by vans and then by cars. This hierarchy is probably because of the differing engine capabilities and load capacities; trucks are usually more extensive and powerful and carry more weight than vans and cars. Companies interested in reducing their distribution network environmental footprint should take to heart the critical implications of this research. Ways to cut CO2 emissions could involve better routing to reduce the distance a vehicle travels, using smaller, more fuel-efficient vehicles for some deliveries, and investigating alternative fuel sources that produce less greenhouse gas. The graph offers an enlightening view of how the different vehicle types compare regarding emissions.

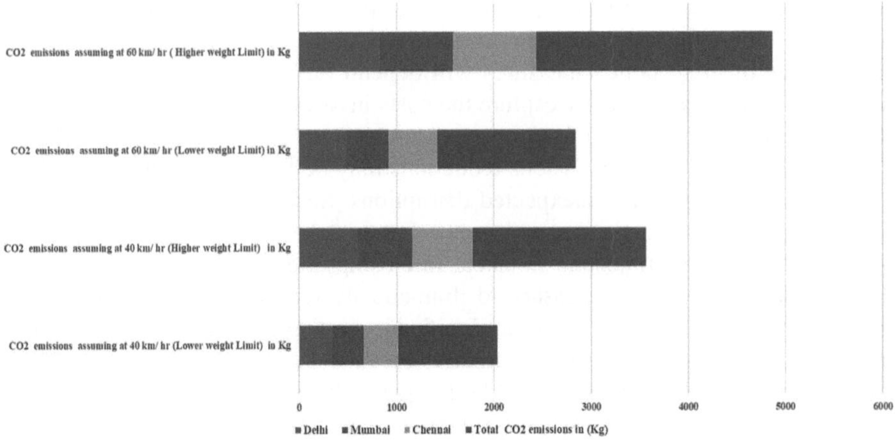

Fig. 6 CO2 emissions per trip from Delhi, Mumbai, and Chennai

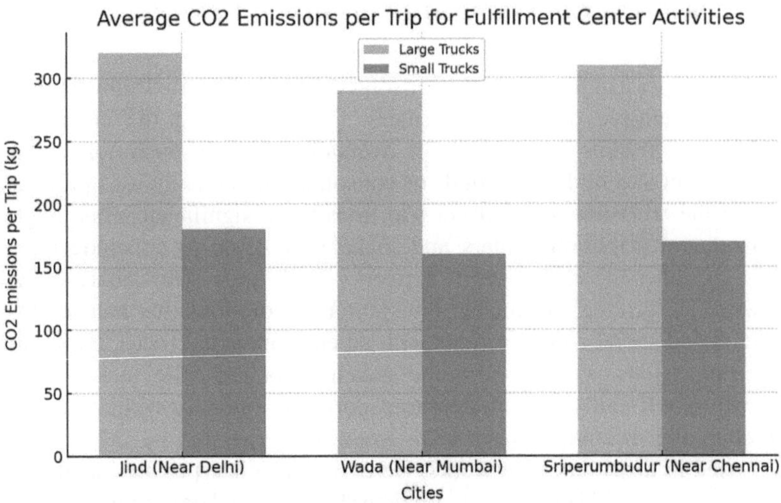

Fig. 7 Average CO2 emissions per trip for fulfillment center activities

However, it is essential to remember that the y-axis is without an actual emission quantity label. The numbers attached are the best available, but the exact emissions of each vehicle class without more precise data points are still unclear. However, the context this graph provides in terms of a relative comparison helps illustrate just how significant the choice of vehicle plays in the environmental footprint of distribution networks. Companies can harness these insights to take pre-emptive action to improve the sustainability of their logistics activities, thereby supporting broader conservation endeavors.

Fig. 7 provides the average CO2 emissions per trip involved in fulfillment center activities; the pictorial representation illustrates the average CO2 emissions per trip

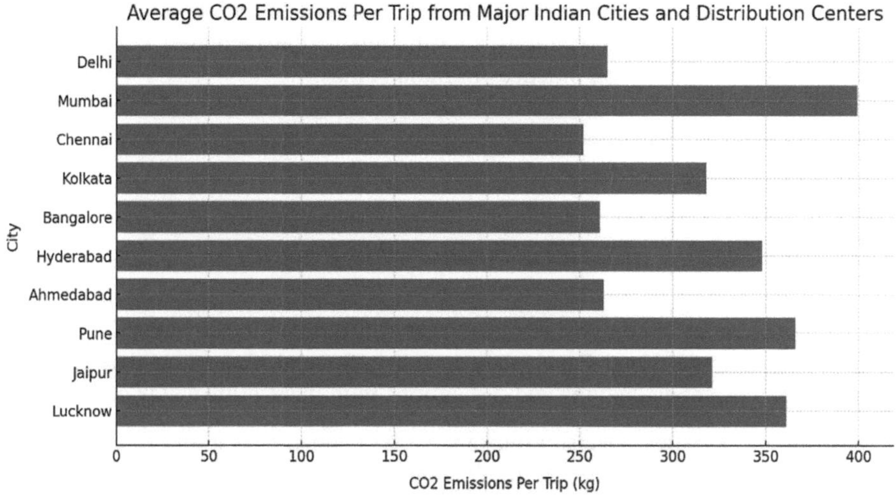

Fig. 8 Average emissions per trip from major Indian cities and distribution centers

for fulfillment center activities in three major Indian cities, comparing emissions from large and small trucks. The data indicates that large trucks consistently emit more CO2 than small trucks across all locations. Among the cities, Jind (near Delhi) records the highest emissions, suggesting factors like longer travel distances or higher fuel consumption. Wada (near Mumbai) has the lowest emissions, indicating potentially more efficient transportation routes or energy sources. Sriperumbudur (near Chennai) falls in between, with emissions slightly higher than Wada but lower than Jind. This comparison highlights the impact of logistics and location-based strategies on CO2 emissions, emphasizing the potential for optimization to reduce environmental impact.

Average CO2 emissions per trip (factors) from major Indian cities and distribution centers depicted in the graph of Fig. 8 shows the extent to which emissions levels differ from city to city, with Delhi posting the highest average emission per trip at 156 grams of CO2, followed by Mumbai at 139 grams of CO2 and Chennai at 121 grams of CO2. Some of these variations could be due to factors such as the level of congestion, trip distances, and fleet mix used in deliveries. For example, more congested cities may have higher emissions from idling longer, and cities with longer trip distances could have higher fuel use. However, differences in vehicle types used by the cities could also play a role, with some vehicles producing more CO2 than others. However, the exact cause for these disparities is challenging to determine, as the researchers do not have complete information on these metrics. The chart only shows the average emissions per trip and does not comprehensively represent the emission profile per distribution center and across all cities. Nevertheless, the findings reveal the scope for reducing CO2 emissions in urban areas using operational optimization of the distribution network. Future research and analysis that can elucidate the drivers responsible for this observed emission variation could help identify strategic interventions contributing to the increased environmental sustainability of distribution practices in Indian cities.

11 Discussion

11.1 Evaluation of Critical Sustainable Factors

Decision-makers at three gypsum plants share insights on the sustainability factors influencing the Board and Panel plant industry. Several interesting findings have come up in these consultations and discussions with these critical stakeholders:

Environmental Issues Decision-makers express concern over environmental sustainability and the environmental implications of the mining and processing natural gypsum. Naturally, however, the focus is on the production process, addressing environmental degradation, resource depletion, and carbon emissions, along with strategies for their mitigation.

Technological Innovations Decision-makers consider adapting to new technologies that help to improve sustainability performance. Production efficiency, waste reduction, and diminished environmental footprint are vital for market competitiveness. This dictates the necessity of investments in cutting-edge equipment, processes, and automation technologies.

Resource Optimization All decision-makers agree that it is critical to have resource optimization strategies. Such efforts focus on optimizing resource consumption, minimizing waste production, and adopting principles of circular economy to pave the way for more sustainable production.

Community Engagement Those making decisions underline the need to cultivate good relationships with the surrounding resident communities and other stakeholders. Securing a social license to operate requires recognizing the importance of building trust and gaining acceptance within local communities. A central component of this involves understanding and supporting initiatives focused on local job creation, social responsibility programs, and environmental stewardship—all of which are critical to achieving long-term sustainability.

Regulatory Compliance Decision-makers consider compliance with environmental regulations and industry standards a matter of high importance. Being proactive in compliance while actively participating in sustainability certification programs is perceived as important in maintaining operations and market integrity.

Value Chain Partnerships Decision-makers appreciate the role of collaboration and partnerships across the value chain to motivate sustainability performance improvements. By interacting with suppliers, customers, government agencies, and

Table 1 Factor Score Card (FSC) and related ranking for gypsum plants

Factors	Plant 1 (FPS)	Ranking
ARCF	6.133	6
FLAY	4.200	12
DFWP	4.933	10
ARRN	7.067	4
AWSU	3.400	14
GPUM	6.067	7
LPGM	7.867	1
COLN	5.333	8

industry associations, we share information, encourage innovation, and take united actions toward common sustainability goals.

Decision-Maker Key Takeaways The shared vision for sustainable progress and a pathway to it in a "difficult industry." Working together and using the best practices of Gypsum to Gypsum (GtoG) European initiative, gypsum plants hope to show how to reduce their environmental imprint and build more robust, more sustainable business practices.

From Table 1 (for abbreviation of factors, refer Appendix 1), Plant 1 rank influencing parameters analysis reveals strengths and areas of improvement in all the operational areas. Plant 1 takes the first position in "LPGM" and is high above other Plants regarding this dimension. This outcome often indicates that Plant 1 has been singled out for its best practice or operational excellence about "LPGM." The next one, "ARRN," can secure a notable fourth position, suggesting that factor management or activity execution regarding the same is carried out in a well-to-do manner. This means Plant 1 has a good way or mechanism for complying with the need for "ARRN." In addition, "ARCF" and "GPUM" also present good performances compared to others, placing Plant 1 among its competitors. Therefore, the "DFWP" and "COLN" values correspond to places where it might help for Plant 1 depressants to take the focus of further processing and optimization. While these factors score well, their ranking will direct management toward better engineering solutions to improve the performance of Plant 1. However, the fifth and sixth on the list—"FLAY" and "AWSU"—are low-hanging fruit regarding priorities and must be treated as such. This might be what Plant 1 needs to address if it aims to improve its operational efficiency and competitiveness within the industry landscape.

11.2 Comparison of Factors of Performance Scores (FPS)

Graphical representation of FPS scores for Plant 1, Plant 2, and Plant 3.

In Fig. 9, FPS performance analysis for Plant 1 reveals crucial disparities in operational efficiency across different functional areas. The top performing factors LPGM (7.867), ILDS (7.200), and ORCM (7.733) indicate strong efficiency in these

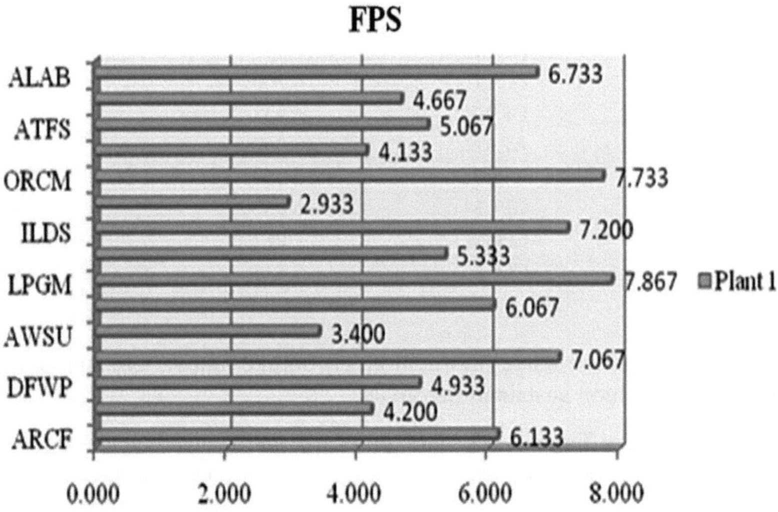

Fig. 9 FPS scores for Plant 1

areas. Other high-scoring factors, such as AWSU (7.067) and ALAB (6.733), suggest moderate to high effectiveness. However, certain areas underperform, with ILDS (2.933) and AWSU (3.400) showing critical inefficiencies that require immediate attention. DFWP (4.933) and ATFS (4.667) represent mid-tier performances, signaling potential for improvement.

When comparing Plant 1 and Plant 3, notable differences emerge:

LPGM (7.867 vs. 6.583)—Plant 1 outperforms Plant 3, demonstrating better operational efficiency.
ORCM (7.733 vs. 5.917)—Indicates stronger performance in Plant 1.
ILDS (7.200 vs. 6.500)—Suggests superior efficiency in Plant 1.
ALAB (6.733 vs. 5.000)—Shows Plant 1 is more efficient in this category.
AWSU (3.400 vs. 4.167)—Highlights a weaker score in Plant 1, necessitating improvement efforts.

The FPS scores reflect both strengths and inefficiencies in operational performance. High scores suggest well-optimized areas, while low scores reveal bottlenecks that require targeted interventions. Factors like congestion, trip length, and fleet composition impact performance, similar to CO_2 emission trends in Indian cities, where higher FPS scores align with higher operating intensity. Addressing low FPS areas, such as AWSU and ILDS in Plant 1, is essential for future improvements. Strategies like process optimization, better resource allocation, and workflow enhancements can drive efficiency gains. Further research should explore external influences on FPS performance, paving the way for sustainable operational improvements.

From Fig. 10, the Factors Performance Scores (FPS) for Plant 2 serve as a critical reference point for evaluating operational efficiency across various functional areas.

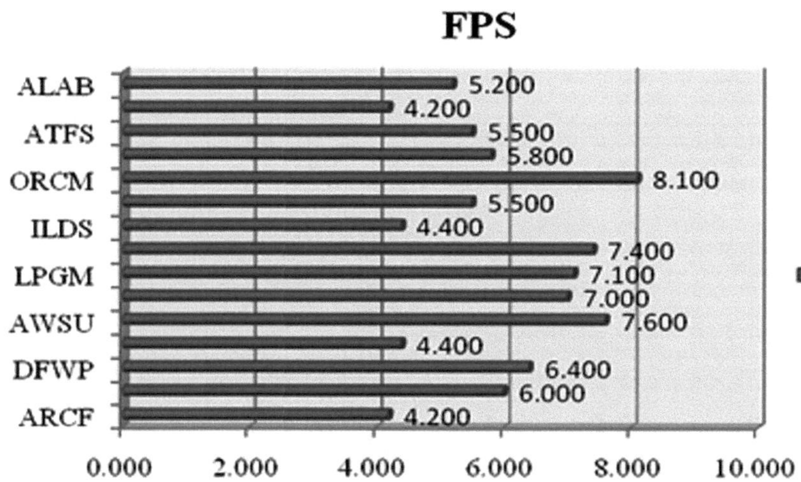

Fig. 10 FPS scores for Plant 2

The analysis reveals notable performance variations, with ORCM (8.100), LPGM (7.400), and AWSU (7.600) ranking as the highest-scoring factors, highlighting strong efficiency in these domains. Other well-performing factors, such as ILDS (7.100) and DFWP (6.400), further reinforce Plant 2's excellence in specific areas.

Key Takeaways from Plant 2's FPS scores are the top performers:

ORCM (8.100)—The highest-scoring factor, indicating peak operational efficiency.
AWSU (7.600), LPGM (7.400), and ILDS (7.100)—Demonstrate solid performance and efficiency.
Moderate performers:
DFWP (6.400) and ARCF (6.000)—Reflect stable operational effectiveness.
Lower performers:
ALAB (5.200) and ORCM (5.500)—Indicate areas requiring improvement.
ATFS (4.200) and ARCF (4.200)—The lowest FPS values, signaling the need for strategic enhancements.

Plant 2's superior performance in ORCM (8.100), AWSU (7.600), and DFWP (6.400) establishes it as the most efficient plant in these areas. However, the lower scores in ATFS (4.200) and ALAB (5.200) suggest the need for targeted optimization efforts. The comparative analysis indicates that variations in operational conditions, fleet efficiency, and emissions control strategies play a significant role in FPS differences across the plants.

Factors performance scores from Fig. 11 for Plant 3 reveal stark differences across categories, reflecting the differences between Indian cities' CO_2 emissions per trip. The LPGM (6.583) and ILDS (6.500) have the most favorable scores, with the best performances within these sectors, followed by ATFS (6.167) and ORCM (5.917), which also perform well. Moderate performances such as ALAB (5.000) and ARCF (5.917) reflect good efficiency, while DFWP (4.250) and AWSU (4.167)

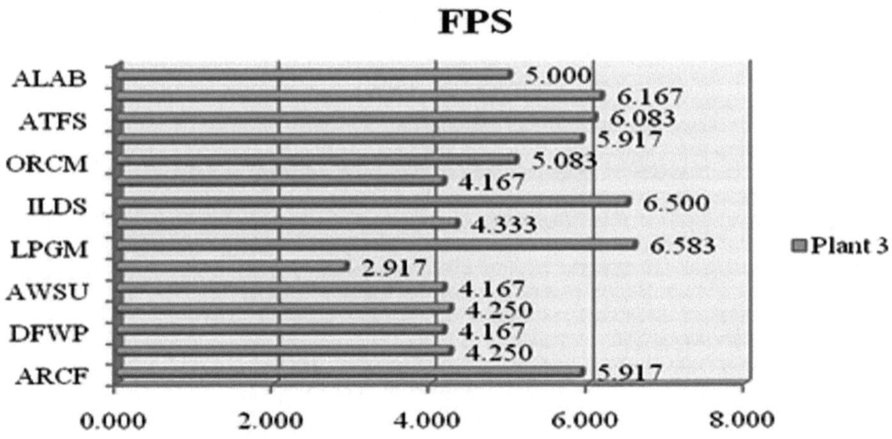

Fig. 11 FPS scores for Plant 3

Table 2 Comparative analysis: Plant 1 vs. Plant 2 vs. Plant 3

Factor	Plant 1 FPS	Plant 2 FPS	Plant 3 FPS	Observation
ALAB	6.733	5.2	5	Plant 1 leads, with Plant 2 slightly outperforming Plant 3.
ATFS	4.667	4.2	6.167	Plant 3 has the highest score.
ORCM	7.733	8.1	5.917	Plant 2 leads in efficiency.
ILDS	7.2	7.1	6.5	Plant 1 slightly outperforms Plant 2.
LPGM	7.867	7.4	6.583	Plant 1 ranks the highest.
AWSU	3.4	7.6	4.167	Plant 2 stands out significantly.
DFWP	4.933	6.4	4.25	Plant 2 emerges as the top performer.
ARCF	6.133	6	5.917	Minimal variation across plants.

reflect average performances. The worst score, LPGM (2.917), reflects inefficiencies requiring attention. These differences reflect CO_2 emission patterns in Indian cities, where Delhi (156 g CO_2 per trip) is linked with the best FPS scores, reflecting high operating intensity, while Mumbai (139 g CO_2) and Chennai (121 g CO_2) are linked with middle-range FPS scores, reflecting balanced performance. Just as operations are shaped by circumstances such as congestion, trip lengths, and fleet composition, FPS performance is shaped by operating circumstances. Improvement would increase performance by addressing inefficiencies, particularly in LPGM (Table 2), just as improving urban logistics streamlines emissions. Subsequent studies must address the identification of performance constraints and the strengthening of weak areas for overall improvement in efficiency.

Impact of technology on emissions which is one of the keys to sustainability is understanding how technology affects emissions (a crucial part of climate change and reducing our greenhouse gas production). The "Impact of Technology on Emissions" scenario tests the impact of changing market penetration of various

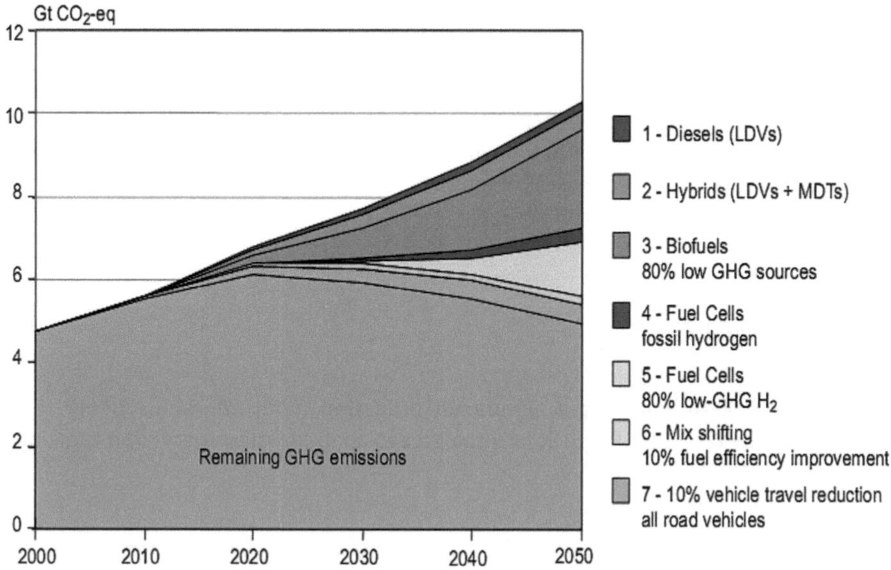

Fig. 12 Scenario postulating market penetration of all technologies. *From: Reproduced from Amano* et al. *Ingraham/CC BY 4.0* Fig. 12

technologies on emissions. While this tech-enabled future is an assumption, it provides critical insights into how technological progress could significantly reduce emissions in multiple sectors (and even several industrial clusters). It provides a complete examination of how modern technologies can influence the trajectory of emissions that supports making better-informed decisions and enacting specific policies to promote sustainable development in a community. This analysis showcases the monumental impact of technology in leading us to reduce emissions and develop a cleaner, more eco-friendly future.

From Fig. 12, this is a simplified graph of the required reductions in cumulative greenhouse gas (GHG) emissions from different sources tomorrow to stay on target from a baseline situation in 2050. It uses a bar graph whose x-axis is emission sources, and its y-axis is the emission metric in a gigaton (Gt) of CO2 equivalent. The graph discloses a few significant findings: Firstly, Emission Sources: It identifies the emissions that would be bisected across six primary sources being diesel light duty vehicles (LDVs), hybrid light and medium duty vehicles (LDVs & MDVs), biofuels, fuel cells with fossil hydrogen, fuel cells with low-GHG hydrogen, and mix shifting denoting a possible change in the ratio of the energy mix for electricity generation. Emitted Baseline Scenario—In practice, the emissions are relative and played against a presumed counterfactual, which is what emissions were expected to be under a baseline scenario without policy and technological change.

11.3 Large Reductions

Most strikingly, the graph shows the potential for massive reductions in greenhouse gas emissions by 2050 relative to Business As usual (BAU) Diesel LDVs, for example, is expected to virtually vanish by 2050 in what would represent a tremendous accomplishment of emissions abatement. Below is a closer look at the expected emissions reductions from each source.

11.4 Diesel LDVs

This category shows the most significant reduction, with emissions likely to be nearly eliminated by 2050 as official strategies to abandon diesel-powered cars come into effect.

11.5 Hybrid LDVs & MDVs

While not directly quantified by the chart, hybrids are expected to lead to a substantial decline in emissions, illustrating the significant role hybrids plays in controlling emissions from both light and mid-range vehicles. Biofuels: Biofuels can help lower emissions, but they have less impact than some other sources; having them in the mix reinforces the opportunity that renewable energy can play in decarbonizing transportation. Fuel Cells (Fossil Hydrogen): Showing modest emissions reductions relative to the BAU, fuel cells with fossil hydrogen are making some progress in emission abatement but are not as promising a possibility as other alternatives. Fuel Cells (Low-GHG Hydrogen): Fuel cells utilizing low-GHG hydrogen offer the most significant emissions reductions across all fuel cell options highlighting the need to move toward cleaner forms of energy. While these insights are undoubtedly important, there are a few significant caveats that need to be acknowledged: The figure does not specify a time frame for the baseline scenario, which can change how we think about the different emission-reduction trajectories; the graph does not account for potential future technological advancements that could help reduce emissions further. The extent to which the policies and technologies required to achieve these reductions are feasible or economically viable in reality is an open question, just like the implementation of these innovations would rely on the influence of socio-economic factors. The red-to-blue mapping scheme in the graph illustrates the potential paths for reducing emissions in the transport sector based on electrification, improved vehicle efficiency, and using more low-GHG alternative fuels—a sobering reminder of where we need to and can go. However, solving the complex problems connected to emission mitigation demands a more

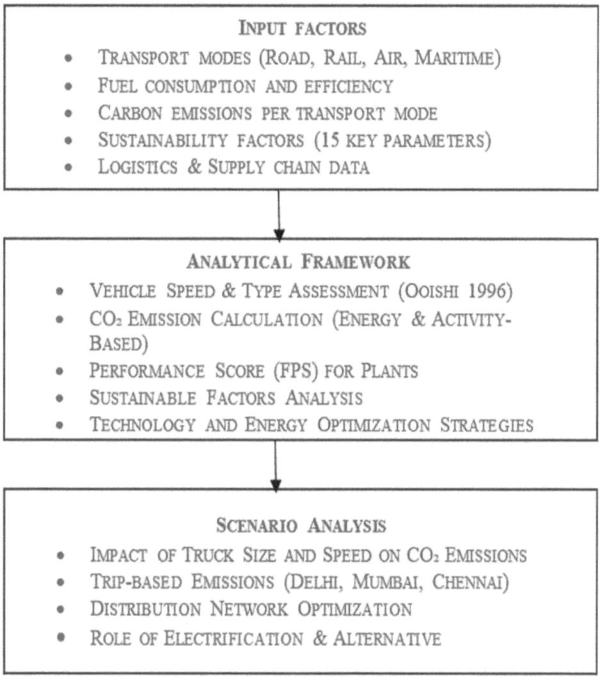

Fig. 13 Sustainable transport & emissions assessment framework

holistic treatment built on policies, technical innovations, and broader systemic changes.

From Fig. 13, this structured framework is designed to evaluate transportation sustainability and carbon emissions through three key sections. The input factors cover data on transport modes (road, rail, air, maritime), fuel consumption, carbon emissions, sustainability factors (15 major parameters), and logistics or supply chain details. The analytical framework focuses on vehicle speed and type assessment, CO_2 emissions calculation (both energy and activity-based), plant performance score (FPS) evaluation, sustainability parameter analysis, and strategies for optimizing technology and energy use. Lastly, the scenario analysis explores the impact of truck size and speed on CO_2 emissions, trip-based emissions across cities like Delhi, Mumbai, and Chennai, distribution network optimization, and the role of electrification and alternative energy in sustainable transportation. Together, this framework provides a systematic approach to assessing and enhancing sustainable logistics and emissions management.

11.6 Future Research Directions

To enhance operational efficiency further, future research should focus on identifying and addressing constraints in low-performing areas such as ATFS and ARCF, where FPS scores indicate potential bottlenecks. A deeper analysis of these categories can help uncover specific operational inefficiencies and inform targeted improvement strategies. Additionally, optimizing high-efficiency factors like ORCM and LPGM could yield further gains by refining best practices and implementing process enhancements that drive sustained performance improvements. Another critical research avenue involves examining FPS in conjunction with external variables such as CO_2 emissions, trip intensity, and fleet utilization, which could provide valuable insights into the broader operational impact on efficiency. By integrating these external factors into FPS assessments, organizations can develop more precise, data-driven strategies to enhance overall plant performance and sustainability.

12 Conclusion

Using multiple methods to calculate carbon emissions for the specified case data lowers weight restriction and speed resulting in lower emissions. Scenario study shows that vehicle speed strongly affects emissions. Carbon emissions for little (LDV) and large (HDV) cars in transportation networks are strongly correlated with weight limits. Both activity-based and energy-based approaches show that HDV emits more CO2 than MDV. Following scenario analysis, the case company should adopt multi-modal transportation depending on vehicle type. Companies should use City Logistics' freight consolidation hubs to reduce CO2 as innovative solutions. Arvianto et al. (2021) suggest innovative solutions such as parking, urban consolidation centers, fuel or charging systems, loading/unloading zones, logistics facility location, spatial network evaluation, infrastructure supply, land use considerations, and collection/delivery point systems.

The transport industry has 1600–2550 Mt. CO2 of mitigation potential by 2030 at a carbon price below 100 US$/t CO2. This partial assessment is based on biofuel use in transport and light-duty vehicle efficiency improvements. Lowering vehicle loads through the use of lightweight materials can improve fuel efficiency by 4–8% for a 10% weight reduction, depending on vehicle size and engine downsizing. Switching to high-strength steels (HSS), replacing steel with lighter materials like Al, Mg, and plastics, and improving design and forming methods can reduce vehicle weight. Biofuels can be used alone or in blends with other motor fuels. The development of biofuel technology aims to minimize greenhouse gas (GHG) emissions and reduce the transport sector's massive oil imports. Ethanol and biodiesel are global transport biofuels. Plant 1 has good "Less Packaging with green material" and "Organizational Competence" traits. Plant 1 has low "Top management Commitment and Engagement" under its perimeter. Plant 3 has strong "Less Packaging with green material" and "Indoor logistics design" skills. Under the circle of Plant 3, "Green products and utility of recycled materials" is weak. The study identified fifteen parameters for sustainable expansion that should be considered when choosing a gypsum plant site. Plant 2 has good "Organizational Competence" and "Availability of Water Supply." Under the circle of Plant 2, "Availability of Power" is weak. The study provided insights to sustain value. Improvements in energy efficiency, switching to lower-carbon fuels, and material efficiency are key techniques for reducing carbon emissions in cement production.

Appendix 1

Critical sustainable factors	Abbreviation
Availability of railways or civil aviation facility	ARCF
Flexibility in layout	FLAY
Disposal facility for waste products	DFWP
Availability of raw materials resources nearby	ARRN
Availability of water supply	AWSU
Green products and utility of recycled materials	GPUM
Less packaging with green material	LPGM
Cost of land	COLN
Indoor logistics design	ILDS
Top management commitment and engagement	TMCE
Organizational competence	ORCM
Proximity of markets from plant	POMK
Availability of transport facilities	ATFS
Availability of power	APOW

References

Adams R (2018) Energy-efficient practices in transportation. Sustain Transp J 5(2):78–92

Akyelken, N.: Green logistics: improving the environmental sustainability of logistics (2011)

Aldegheishem A (2024) The relationship between urbanization, energy consumption, economic growth, and carbon dioxide emissions in middle eastern countries: Saudi Arabia, Jordan, and Egypt. Environ Res Commun 6(6):065011. https://doi.org/10.1088/2515-7620/ad5791

Ali A, Khan MR, Ghouri AM (2023) Carbon neutrality concept and progress. In: Recent developments in green finance, green growth and carbon neutrality. Elsevier, pp 85–108. https://doi.org/10.1016/B978-0-443-15936-7.00003-7

Anderson S, Browne M, Allen J (1999) Logistics implications of the UK packaging waste regulations. Int J Log Res Appl 2(2):129–145

Arvianto A, Sopha BM, Asih AMS (2021) City logistics challenges and innovative solutions in developed and developing economies: a systematic literature review. Int J Eng Bus Manag 13:1–15. https://doi.org/10.1177/18479790211039723

Arya P, Srivastava MK, Jaiswal MP (2020) Modelling environmental and economic sustainability of logistics. Asia Pac J Bus Adm 12(1):73–94

Asian Development Bank (2009) Vehicle emissions control: a policy guide for Asia, Manila

Bagchi P, Sahu SK, Kumar A, Tan KH (2022) Analysis of carbon productivity for firms in the manufacturing sector of India. Technol Forecast Soc Change 178:12160. https://doi.org/10.1016/j.techfore.2022.121600

Banister D (2008) The sustainable mobility paradigm. Transp Policy 15(2):73–80

Brown A, Lee C (2019) Monitoring energy consumption in transportation systems. Energy Eff Rev 12(4):210–225

Carter C, Drouaud F (2022) Territory, ecological transition and the changing governance of ports. Territ Politics Gov 12(3):374–394. https://doi.org/10.1080/21622671.2022.2038661

Chen C, Xu M, Zhang T (2014) Understanding emission reduction in supply chains: evidence from the logistics sector. Resour Conserv Recycl 92:1–9. https://doi.org/10.1016/j.resconrec.2014.08.010

Cheng L, Tang UW (2015) Evaluating transport mode shift policies in an ageing society: a carbon footprint analysis. Transp Res Part D 41:227–238. https://doi.org/10.1016/j.trd.2015.09.004

Cheng Y, Yang J, Zhang X, Zhang J (2015) Measuring the lifecycle carbon footprint of the transport service sector in China using a simplified LCA approach. J Clean Prod 95:299–307. https://doi.org/10.1016/j.jclepro.2015.02.060

Colicchia C, Strozzi F (2012) Supply chain risk management: a new methodology for a systematic literature review. Supply Chain Manag Int J 17(4):403–418. https://doi.org/10.1108/13598541211246558

Cruijssen F (2020) Cross-chain collaboration in logistics. In: Collaboration in logistics: connecting multiple chains. Springer, Cham, pp 25–44. https://doi.org/10.1007/978-3-030-57093-4_3

Datta A (2008) The incidence of fuel taxation in India. Discussion Papers, 08–05. Indian Statistical Institute, Delhi

De Giovane CE, Duarte FDP, De Souza LC (2015) Recycling logistics practices: a study on Colombian companies. In: Reverse logistics: theoretical and practical aspects. Springer, pp 123–134

Dey A, LaGuardia P, Srinivasan M (2011) Building sustainability in logistics operations: a research agenda. Manag Res Rev 34(11):1237–1259. https://doi.org/10.1108/01409171111178774

Dowlatshahi S (2000) Developing a theory of reverse logistics. Interfaces 30(3):143–155

Duan H, Hu M, Zhang Y, Wang J, Jiang W, Huang Q, Li J (2015) Quantification of carbon emissions of the transport service sector in China using streamlined life cycle assessment. J Clean Prod 95:109–116

Ellram LM (1996) The use of the case study method in logistics research. J Bus Logist 17(2):93–138

Faheem M, Amin M, Shahbaz M, Mahalik MK (2020) Transport infrastructure and CO_2 emissions in Pakistan: new evidence using Fourier ARDL approach. Environ Sci Pollut Res 27:34678–34691

Fergani C, El Bouzekri El Idrissi A, Marcotte S, Pellerin R (2020) Optimization of hyperconnected mobile modular production toward environmental and economic sustainability. Environ Sci Pollut Res 27(33):41850–41867. https://doi.org/10.1007/s11356-020-09966-9

García M, Benavente J, Coloma JF (2019) Driving sustainability: analyzing eco-driving efficiency across urban and interurban roads with electric and combustion vehicles. World Electr Veh J 16 (3):143

Gerrish T, Ruikar K, Cook M, Johnson M, Phillip M (2017) Using BIM capabilities to improve existing building energy modelling practices. Eng Constr Archit Manag 24(3):426–442. https://doi.org/10.1108/ECAM-11-2015-0181

Ghadimi P, Wang C, Lim MK, Heavey C (2020) Intelligent sustainable supplier selection using multi-agent systems: a scenario-based approach. Comput Ind Eng 139:105552. https://doi.org/10.1016/j.cie.2019.106170

Gössling S, Cohen S, Hares A (2017) The psychology of climate change travel behavior: an analysis of travel mode choice. J Transp Geogr 62:122–130. https://doi.org/10.1016/j.jtrangeo.2017.03.003

Hickman R, Banister D (2019) Transport, climate change and the city: achieving sustainable urban mobility, 2nd edn. Routledge, London. https://www.routledge.com/Transport-Climate-Change-and-the-City/Hickman-Banister/p/book/9781138780373

Intergovernmental Panel on Climate Change (IPCC) (2018) Global warming of 1.5°C: An IPCC special report. https://www.ipcc.ch/sr15/

Ivanova S (2019) Sustainable logistics in emerging economies: lessons from transport infrastructure reforms. In: Sustainable development and innovation in the energy sector. Springer

Janpoom K, Suttakul P, Achariyaviriya W, Katongtung T (2023) Investigating the influential factors in real-world energy consumption of battery electric vehicles. Energy Rep 9:1706–1718

Lam JSL (2012) Risk management in maritime logistics and supply chains. In: Maritime logistics: contemporary issues. Emerald Group Publishing Limited, pp 101–120. https://doi.org/10.1108/9781780523415-007

Li X, Yu B (2019) Peaking CO_2 emissions for China's urban passenger transport sector. Energy Policy 130:418–427. https://doi.org/10.1016/j.enpol.2019.04.024

Li Y, Du Q, Lu X, Wu J, Han X (2019) Relationship between the development and CO_2 emissions of transport sector in China. Transportation Research Part D: Transport and Environment 74:1–14. https://doi.org/10.1016/j.trd.2019.07.011

Liu S, He N, Cao X, Li G, Jian M (2022) Logistics cluster and its future development: a comprehensive research review. Transp Res Part E 158:102585

Martins VWB, Anhlon R, Rampasso IS, da Silva D, Melo ACS (2020) Sustainability in logistics systems and its impact on the level of services definition: An exploratory analysis using structural equation modeling. In: Thomé AMT et al (eds) Ind. Eng. Oper. Manag., Springer Proc. Math. Stat. 337. Springer, Cham. https://doi.org/10.1007/978-3-030-56920-4_11

Miller SA, Habert G, Myers RJ, Harvey JT (2021) Achieving net zero greenhouse gas emissions in the cement industry via value chain mitigation strategies. One Earth 4(10):1395–1409. https://doi.org/10.1016/j.oneear.2021.09.012

Notteboom T, Pallis T, Rodrigue JP (2019) Port economics, management and policy. Routledge, London. https://porteconomicsmanagement.org

NTM (Network for Transport Measures) (2010) Emission factors for freight transportation. Stockholm, Sweden. https://www.transportmeasures.org/en/

Ooishi R (1996) Planning methodology for new freight transport systems in urban area. PhD dissertation. Kyoto University, Kyoto

Pan S, Trentesaux D, Ballot E, Huang GQ (2019) Eco-intelligent supply chain decarbonisation: a review. J Clean Prod 239:118074. https://doi.org/10.1016/j.jclepro.2019.118074

Prause G (2014) Sustainable development of logistics clusters in green transport corridors. J Secur Sustain Issues 4:59–68. https://doi.org/10.9770/jssi.2014.4.1(5)

Prause G (2019) Smart contracts for cooperative governance. Presented at the International Co-operative Alliance Committee on Co-operative Research Conference, Berlin, Germany

Rosenberg LN et al (2021) Introducing the shared micro-depot network for last-mile logistics. Sustainability 13(4):2067. https://doi.org/10.3390/su13042067

Rugani B, Benetto E, Zamagni A, Brandão M (2021) Environmental and economic performance of economic instruments for transport: a life cyclebased approach for the EU. Resources 10(8):81. https://doi.org/10.3390/resources10080081

Sanches JR et al (2022) Sustainable circular economy strategies: an analysis of Brazilian corporate sustainability reporting. Sustainability 14(10):5808. https://doi.org/10.3390/su14105808

Shahbaz M, Shahzad SJH, Mahalik MK, Hammoudeh S (2018) Does globalisation exacerbate CO_2 emissions in Pakistan? The role of agriculture, industry and fossil fuel consumption. Environ Pollut 234:421–429

Sridharan V, Nair R, Kumar M (2023) Carbon footprint analysis of distributed gypsum logistics in India: a systems approach to sustainable supply chain management. Sustain Transp Logist J 11 (2):133–155. https://www.sciencedirect.com/science/article/pii/S2352146523000156

Sumbal MS, Agha MH, Nisar A, Chan FTS (2024) Logistics performance system and their impact on economic corridors: a developing economy perspective. Ind Manag Data Syst 124(3): 1005–1025. https://doi.org/10.1108/IMDS-03-2023-0151

Sun X, Yu H, Solvang WD (2022) Towards the smart and sustainable transformation of reverse logistics 4.0: a conceptualization and research agenda. Environ Sci Pollut Res 29:31027–31042. https://doi.org/10.1007/s11356-022-22473-3

Tabanjeh STA, Pott C, Reining C (2025) The SuLo framework: a systematic literature review of drivers and barriers to sustainable logistics practices. Sustainability 17(12):5575. https://www.mdpi.com/2071-1050/17/12/5575

Ullah A, Zhang Q, Ahmad S, Sana S (2024) Leveraging Blockchain and big data-optimized supply chain for eco-innovative firm performance. Available at SRN: https://ssrn.com/abstract=5014917 or https://doi.org/10.2139/ssrn.5014917

Walks A (ed) (2014) The urban political economy and ecology of automobility: driving cities, driving inequality, driving politics, 1st edn. Routledge, London. https://doi.org/10.4324/9781315766188

Walmsley M, Walmsley T, Matthews L, Atkins M, Neale J, Kamp P (2015) Pinch analysis techniques for carbon emissions reduction in the New Zealand industrial process heat sector. Chem Eng Trans 45:1087–1092. https://doi.org/10.3303/CET1545182

Wang D, Chen Y, Zhu A, Guo Z (2022) Optimization of the Beijing economy-energy-emissions system 2021–2035: a scenario simulation analysis based on a system dynamics model. Sci Prog 105(3):1–20

White CR, Alton LA, Bywater CL, Lombardi EJ et al (2022) Metabolic scaling is the product of life-history optimization. Science 377(6606):834–839

Yin RK (2003) Case study research: design and methods, 3rd edn. Sage Publications, Thousand Oaks

Zhu X, Li R (2017) An analysis of decoupling and influencing factors of carbon emissions from the transportation sector in the Beijing-Tianjin-Hebei Area, China. Sustainability 9(5):722. https://doi.org/10.3390/su9050722.

Eco-Design Strategies: Tools and Challenges

B. R. Manoj Kumar, Naveed R. Khan, J. Soonu Aravindan, K. B. Theja, and Nikhil Gangadhar

Abstract Eco-design and life cycle assessment (LCA) are crucial in enhancing sustainable innovation across different industries. This chapter analyzes how eco-design principles—material efficiency, modularity, recyclability, and energy-efficient production—reduce the environmental impact along the whole product lifecycle. Using actual examples such as Borbonese and Tesla, this chapter demonstrates how companies are attaining sustainability without jeopardizing their performance or brand worth. It also addresses the application of tools like SimaPro and GaBi in assessing environmental effects and informing sound decision-making. ISO 14062 standards, circular economy, and digital technologies like 3D printing and smart systems are addressed. Based on examining challenges such as data shortages and cost pressures, the research proposes recommendations for overcoming these problems and indicates the future potential of eco-design to achieve regulatory compliance, better customer satisfaction, and profitable sustainability.

B. R. Manoj Kumar (✉)
Xavier Institute of Management and Entrepreneurship, Bangalore, India
e-mail: phd.brmk@gmail.com; manojbr@xime.org

N. R. Khan
UCSI University, Kuala Lumpur, Malaysia
e-mail: naveed.r.khan@gmail.com

J. Soonu Aravindan
Kumaraguru College of Liberal Arts and Science, Coimbatore, India
e-mail: soonuaravindan@gmail.com

K. B. Theja
Acharya Bangalore B-School, Bangalore, India
e-mail: thejakb22@gmail.com

N. Gangadhar
Amity Global Business School, Bangalore, India
e-mail: ngangadhar@blr.amity.edu

M. Venkatesh et al. (eds.), *Green Manufacturing Technologies in Industry 5.0*,
https://doi.org/10.1007/978-981-95-0363-6_9

1 Introduction

Global environmental challenges and increasing resource limitations have forced companies to adopt green practices along the entire value chains. One highly efficient way to achieve this is through eco-design, an extended design philosophy incorporating environmental considerations through all stages in the life cycle of a product. Rather than being regarded as an ancillary element, eco-design is, however, treated today as an essential strategy joining product innovation to environmental responsibility. Eco-design, or design for the environment (DfE), is aimed at resource efficiency, material sustainability, modularity, and recyclability. It involves designing products to meet user needs while minimizing environmental impacts—from raw material extraction through production, use, and end-of-life disposal. The model promotes environmental and financial sustainability and increases long-term competitive advantage for businesses. With increasing consumer consciousness and global initiatives toward sustainable development, companies are embracing Life Cycle Assessment (LCA) to evaluate the environmental impacts of their products. LCA offers a systematic, data-driven approach to quantify resource consumption, emissions, and waste associated with every stage of a product's life. When complemented with eco-design, LCA helps in identifying hotspots, improving design decisions, and validating sustainability claims. The United Nations Sustainable Development Goals (SDGs) highlight the need for such integrated practices, urging industries to innovate responsibly. Guidelines such as ISO 14062 provide structured guidance on integrating environmental considerations into product development. In such a case, companies that specifically adhere to green standards not only enhance their environmental image but also meet regulatory and consumer needs. As global markets converge toward sustainability, eco-design and LCA become integral tools for building resilient and future-proof businesses. This chapter delves into their principles, real-world uses, software tools, challenges, and innovation—providing an end-to-end guide for academics, practitioners, and policymakers to enable sustainable transformation in product design.

2 Eco-Design Strategies

It is through the integration of sustainability in all stages of the design process that the environmental footprint is reduced while consumer requirements are satisfied. The companies, by now, are capable of developing products that are not only green but also productive and cost-effective using the principles of eco-design. A key strategy is selecting eco-friendly materials. This will include application of bio-based resources, such as bamboo, hemp, biodegradable plastics, and recycled materials such as metals, glass, and post-consumer plastics. Life Cycle Assessment, among other tools helps to assess the environmental footprints due to the materials throughout the product lifetime and this prompts better and sustainable selection

(Nematchoua et al. 2022). Yet, the disassembly and repair also needs designing strategy for disassembly process. Companies provide modular products of interchanging constituents. It neither employs permanent fasteners. Consequently, consumers avoid some of those designs which disallow access of them toward the ability of repairing as well as recycling them. In such a manner, the life spans of products become elongated by contributing toward a circular economy as less waste happens, and as resources get preserved (Fukushige et al. 2025).

Energy efficiency and resource use efficiency are also two of the basic eco-design principles. Products could be designed for less energy use in their manufacture, and operation, and that in turn diminishes their total ecological burden. Techniques such as precision design and 3D printing reduce waste material and thereby promote sustainability. A lifecycle strategy structured this way ensures sustainability is considered at every stage of the product life cycle. By using materials that are sustainably sourced, designing for transportation optimized for use, and ensuring recycling or refurbishment at the end of its life cycle, this approach encourages a closed-loop system with little waste and maximum resource usage (Proctor 2006).

Eco-design practices have much to offer. It reduces environmental harm, cuts costs by utilizing resources effectively, and meets the growing consumer demand for sustainable products. Businesses also benefit from compliance with environmental laws and the promotion of innovation, which can improve their image and attractiveness in the marketplace (Nuryakin and Maryati 2020). In these contexts, integrating all these strategies can create environmentally friendly products, made to be viable and for a modern-day customer's requirements toward the attainment of a better, more sustainable future.

3 Tools and Methods for Implementation

Meaningful integration of eco-design and life cycle assessment (LCA) into product design will require instrumentation and methods for assessment of environmental costs of the company's products, as well as tools to help companies make better decisions for sustainability. Eco-design refers to the designing of the products that minimize environmental impacts (Wimmer et al. 2004). LCA is a term used to describe assessments of these effects throughout the whole lifecycle of the product, including extraction of raw materials and disposal at the end of its lifecycle. A range of tools and techniques allow designers to perceive effects, support sustainability, and make proper decisions over the course of the product's lifecycle (Nematchoua et al. 2022).

3.1 Software Tools for Eco-Design and LCA

There are numerous software solutions that exist today to aid in the implementation of eco-design and LCA into the product development process by organizations. These applications give precious information about the environmental impact of material usage, manufacturing processes, logistics, and the disposal cycle of a product.

SimaPro is one of the most popular computer systems used to perform the LCA. It is widely recognized to offer the ability of analyzing environmental contributions of products across its whole life cycle due to its large number of databases and strong modeling. As far as complex product system modeling goes, SimaPro can connect with and be used in the assessment of effects across an entire product life cycle. The exposure of full and detailed environmental impact databases, such as Ecoinvent and Agri-footprint, has greatly allowed researchers to analyze the potential in detailed scenario analysis and evaluate alternate designs, materials, and processes. This tool applies approaches of effects measurements such as CML, TRACI, and ReCiPe that examine different environmental impacts throughout the development point of products in a process. SimaPro is suitable for applications across industrial giants and researchers as well as consultants in achieving a more developed LCA analyses of complicated products.

GaBi is another extremely used LCA tool with an excellent and friendly interface having rich material as well as processing repositories. This kind of tool is commonly employed in automotive, electronics, and consumer goods industries. GaBi is made suitable for both beginners and experienced in the field of life cycle assessment (LCA) by providing easy navigation coupled to advanced functionalities. The tool enables the user to model the full product system including the transportation aspects, energy use, and waste management, providing a deep and comprehensive view of the environmental impact. Detailed reports generated by GaBi help the user understand which aspects of the product life cycle are having the most severe environmental impact and which aspects have to be modified, thus forming a vital tool for those sectors that require detailed modeling.

OpenLCA is an open-source software, free-of-charge for small enterprises and businesses requiring an inexpensive LCA solution. It is convenient, easy to use, free and that is why it is very attractive to small- to medium-sized enterprise (SMEs) or educational establishments. The flexible structure of OpenLCA enables the integration of several databases and LCA approaches thus making it applicable to a vast array of industries and products. It also provides support for a very wide range of established LCA databases, including Ecoinvent and ELCD. The customization is enabled by adjustable reports and visual graphics which can facilitate the process for users to present their results and therefore positioning OpenLCA as a suitable choice for small and medium enterprises, research programs, and teaching/academic activities.

Umberto deals with MFA and process-based LCA. Its greatest advantage lies in the analysis of material and energy flow of a product system that can be applied to assist companies identify resource consumption and waste production. The software has a graphical interface through which it enables the users to design complex systems and to find opportunities for enhancing environmental conditions. More importantly, it allows for interoperability with other applications, such as Excel and ERP software, enhancing the efficiency of data management and reporting. Umberto has proven particularly relevant to sectors that want to improve resource utilization, cut waste, and become efficient especially in manufacturing, construction, and packaging.

EcoAudit has been developed by the U.S. Environmental Protection Agency, serves as a user-friendly tool that helps organizations understand the impacts of their environmental activities such as emissions, energy, and water usage, among other resources. It is configured in a way that is usable for businesses with little experience in LCA and especially beneficial for small to medium enterprises. EcoAudit is focused on emissions and resource consumption, allowing users to analyze different scenarios to identify the most environmentally beneficial actions. It also helps organizations to follow environmental statutes as well as to follow the legalities that allow them to follow the law. EcoAudit is an ideal tool for companies at the start of their sustainability journey and those wanting a simple way to look at their environmental footprint (Proctor 2006).

3.2 Methods for Implementing Eco-Design and LCA

Apart from the availability of software tools, there exist several approaches by which eco-design and LCA can be easily integrated into a product development process. These strategies assist designers to make environmentally sensitive decisions and avoid environmental footprint of an entire product.

Cradle-to-Cradle Design (C2C) Products will be designed in such a way that all materials are reutilized or fully recycled after its life cycle has ended. The approach is toward reduction of waste and usage of renewable energy so that all the materials employed in the product can be part of a circular chain where no resource is lost and everything is repurposed (McDonough and Braungart 2002).

Design for Environment is that process of putting environmental factors in the product design process. The designers are encouraged to select environment friendly materials that increases energy efficiency and reduces waste during manufacturing. DFE aims to decrease the net environmental impact of products by accounting for ecological considerations at the early stages of designing (Chan 2011).

Eco-Design Thinking integrates ecological considerations with innovative product design. It involves the use of multidisciplinary cooperation and innovative thinking in order to create products that are not only functional but also environmentally sustainable. The focus of eco-design thinking is on the necessity to integrate environmental, social, and economic considerations to create sustainable business solutions to protect the environment (Proctor 2006).

Eco-Labeling and Certification programs like ISO 14001 and Energy Star provide a platform for firms to make their position as sustainable evident. Such certifications arm the consumer with information for making more informed choices and nudge firms toward the application of eco-design principles. Acquisition of these certificates ensures that the firm has followed all applicable environmental regulations and made significant strides to reduce its ecological footprint.

3.3 Guidelines for Incorporating Eco-Design

For companies seeking to reduce their environmental footprint and make their products more environmentally sustainable, the implementation of eco-design within the product development process is critical. There are many global standards and frameworks that provide useful recommendations for companies on how to embed environmental aspects into the design process (Wimmer et al. 2004). Adhering to these principles enables corporations to produce products that are less environmentally damaging, more socially responsible, and financially feasible, and at the same time enables them to gain market advantage.

3.4 ISO 14062: Environmental Management—Integration of Environmental Aspects into Product Design and Development

Another international benchmark critical to the integration of eco-design is ISO 14062, whose resultant product focuses on the need to integrate environmental considerations into all phases of design and development. ISO 14062 is a critical standard in the adoption of environmental aspects in the design and development of products (Quella and Schmidt 2003). It offers an established process that enables organizations to manage sustainability in the lifecycle of a product. By exploiting the concept of lifecycle thinking, ISO 14062 requires businesses to evaluate the environmental responsibility of their products all the way from raw material extraction through to production, use, and end of life. This holistic view is important, on the one hand, to recognize where businesses can curtail the use of resources, waste, and energy leading to more sustainable end products (Quella and Schmidt 2003).

The standards find a significant presence of Eco-design measures in the design process, particularly in the use of sustainable materials, improvements in energy efficiency, and reduced production waste. Adhering to the principles will help businesses develop products that are not just in line with the environmental-minded consumers but also efficiency of operations. ISO 14062 also adds that it is necessary to engage stakeholders and that various stakeholders, including suppliers and customers, need to be incorporated in the product design. This engagement will help associate the sustainability objectives of the product with broader environment and social criteria (Quella and Schmidt 2003).

Transparency and documentation are the key components of ISO 14062. Based on the assumption that companies need to have complete track of environmental assessment and design choices, and thus toward transparency of their sustainability endeavors, this guideline supports such points. This transparency can, further, enable ongoing optimization, allowing the companies to follow the ecological footprint of their product even after the launch of those products, and seek to further minimize it (Quella and Schmidt 2003). According to ISO 14062, companies can live up to the global environmental demands at the same time while being in anticipation of future legal challenges and market pressure for products with a more sustainable approach.

Standard enables early recognition of environmental risks in the design of a product. If these risks are taken care of at the beginning, then firms avoid expensive mistakes and lawsuits (Subramanian et al. 2023). Generally, ISO 14062 recommends sustainable supply streams wherein companies ought to act in partnership with their suppliers who embrace environmentally sustainable practices as long as possible toward making the whole product cycle as environmentally friendly as possible (Quella and Schmidt 2003). The economic benefits are also included in embracing ISO 14062. Sustainable product design tends to have cost-saving advantages due to the improvement of the consumption of resources and selling less waste (Chan 2011). Therefore, a company that offers sustainable products will be in a position to become competitive, hence improving the market share which it will hold. Ultimately, ISO 14062 is easily integrated with other management systems like ISO 9001, on quality management, and ISO 14001, which is on environmental management. It ensures that a unified approach to sustainability will be integrated in the business operation (Proctor 2006).

In summary, ISO 14062 gives a holistic approach to the integration of environmental considerations in product design. The guidelines enable companies to minimize their environmental footprint, increase the sustainability of their products, and enhance innovation while simultaneously gaining economic benefits and establishing customer wants for eco-friendly products (Quella and Schmidt 2003).

4 Challenges in Eco-Design and LCA

While eco-design and Life Cycle Assessment (LCA) offer significant advantages for reducing environmental impact, there are several challenges that can prevent their widespread adoption. These challenges include technical, economic, and market-related barriers that businesses must address to fully integrate these practices into their product development processes (Pati 2023).

4.1 Technical Challenges

Among the major technical challenges is the lack of standard methods and tools for the implementation of LCA. In cases where standardized methods for evaluating the environmental impacts of products and organizations, especially very young organizations, do not exist, eco-design can be very disturbing. Also, LCA requires correct and complete data to draw meaningful conclusions (Wimmer et al. 2004). Due to differences in the quality of available data, the assessments become confounded. A fair number of companies are experiencing difficulty in getting valid LCI data which is fundamental input to meaningful assessments. Erroneous or incomplete data may lead to wrong inferences and, consequently damage the efficacy of the eco-design process. On top of that, LCA comes with the intricate environmental modeling which may well pull into this area more specialized expertise beyond the reach of some organizations in general, more so small ones. The involvement of skilled people in this end thus poses a further challenge for firms looking forward to embracing the culture (Pati 2023).

4.2 Economic Challenges

Preliminary costs paid by implementing eco-design and performing LCA can be a significant obstacle, especially for small- and medium-sized enterprises (SMEs). To the small and medium enterprises, it would appear highly expensive to incur the costs for training personnel and purchasing the software required, the information gathering process, and acquiring advice from the experts when such cost cannot quickly be justified with a return on investment. In addition, it is often unclear whether there are financial incentives, for example, subsidies, to offset the costs of implementing eco-design, which reduces the attractiveness of it for a company (Pati 2023). Many companies may refuse to invest in sustainability without a clear financial incentive, seeing it as an unnecessary cost and not as a long-term investment. This challenge is more acute in those industries where cost competitiveness is of prime importance, and where the direct cost-benefits of eco-design cannot be immediately seen.

4.3 Market Challenges

One of the key market-based issues in green to implement eco-design is that consumers do not, usually, take sustainability into account in making a purchasing decision. When customers focus on factors like price and convenience, businesses are likely to pay less attention to environmental impacts and try to meet these immediate demands (Wimmer et al. 2004). Thus, companies are likely to hold back on eco-design investments because they feel that the market does not recognize such efforts. This is especially true when sustainable products lack strong demand or when consumers are unaware of the environmental benefits of their purchasing decisions (Proctor 2006). The evolution of consumer trends can indeed make it hard for companies to justify the cost associated with eco-design, particularly when competitors have not followed the same track.

4.4 Overcoming Resistance to Change

The challenges will be overcome if businesses have a comprehensive strategy. Training and development are crucial to the development of internal eco-design and life cycle assessment expertise. If the workforce is given the ability to comprehend the value of such practices, the organizations sustainability program will become more successful. An alliance with external experts can allow access to state-of-the-art data and therefore overcome the problem of access to it. Moreover, sustainability culture within the company is of significant importance (Pati 2023). The leadership itself has an active obligation to start this way as the participation of corporate chief executives is the foundation of the organization's commitment. Employees at all levels can be isolated in order to implement ownership and enthusiasm for implementing eco-design (Proctor 2006).

Economically, companies have to show the long-term value proposition of sustainability—less resource usage, better energy usage, and fewer wastes generated—and, more importantly, the potential for huge and sustainable savings. Presenting how it can enhance brand perception, customer loyalty, and competitive differentiation can help move away from short-term cost-to-ICM value considerations (Pati 2023). Besides, with governments all over the world introducing ever stricter environmental regulations, companies that embrace eco-design techniques can remain ahead of possible legal requirements. Sustainable practices can be an added competitive advantage for companies in line with the growing demand for ecologic products (Faud-Luke 2009).

Ultimately, the fight against resistance to change is that companies should see eco-design as an innovation space, not as an expense. By emphasizing the sustainability benefits that do not fade over time, organizations may gain support from management and labor force. With knowledge and environmental consciousness of consumers ever on the increase, companies that present themselves as leaders in sustainability are well placed to thrive within a market whose awareness of environmental implications is continuously growing (Sayal and Pant 2023).

5 Future Trends and Innovations in Eco-Design

The eco-design field is constantly changing as a result of technological gains in materials, technologies, and changing regulations. With sustainability integrated within business activities, companies establish new ways to reduce the environmental footprint and maximize the functionality of their product and consumer preference. The following section presents some of the significant trends and innovations that are shaping eco-design into the future.

5.1 Emerging Materials

Perhaps, one of the most important trends in eco-design is the increased use of bio-based and recycled materials that reduce reliance on fossil fuels and decrease negative impacts on the environment. Bioplastics, derived from renewable sources such as corn starch, sugarcane, or algae, are gaining increasing popularity as alternatives to conventional plastics based on petroleum. These possess the same properties as regular plastics but are biodegradable or recyclable more easily (Foschi et al. 2020). Companies are increasingly using recycled materials, including plastic waste recovered from the ocean, in their products. Adidas has made shoes from reclaimed ocean plastic as part of supporting the circular economy (Keivanpour 2023). With the newfound interest in product sustainability by consumers, this is probably one sector that more sectors will embrace the use of these materials to meet not only the needs of the consumers but also the regulatory pressure.

5.2 Innovative Technologies

Technological innovation associated with digital tools and manufacturing methods is changing eco-design. One of them is additive manufacturing, also known as 3D printing, which has the ability to utilize materials very precisely, minimizing waste and excess raw material use. This approach also enables companies to produce intricate, customized products that are impossible for traditional manufacturing techniques to be able to accomplish. Moreover, advancements in digital tools, such as Life Cycle Assessment (LCA) software, enable firms to assess their product's environmental impacts and take decisions on design for sustainability. These technologies improve the access and efficiency of the eco-design process for businesses to produce items with a smaller ecological footprint (Nematchoua et al. 2022).

5.3 *Evolving Regulatory Landscape*

The adoption of stricter environmental regulations by governments worldwide has pushed businesses to engage in sustainable practices. The application of eco-design principles ensures compliance with regulations while offering a competitive advantage for attracting environmentally conscious consumers. Sometimes, sustainable practices will result in cost savings, for example, lower energy consumption and lower waste management costs. Governments are offering incentives such as tax breaks and grants to encourage sustainable business practices, thus increasing the uptake of eco-design (Proctor 2006).

5.4 *Circular Economy and Product-as-a-Service Model*

One of the most important changes in eco-design is a shift toward a circular economy, which seeks to minimize waste and promote reuse, repair, and recycling. The circular economy is a new model that encourages products to be developed for longevity, ease of disassembly, and reuse, rather than the traditional "take-make-dispose" model (Keivanpour 2023). The product-as-a-service model, where products are leased or rented instead of sold, is gaining popularity. Companies such as Patagonia and Fairphone are using this model to enable customers to rent or repair products, thereby reducing the demand for new materials and decreasing environmental impacts.

5.5 *Smart and Connected Products*

Another driver is the rise of smart devices and the Internet of Things (IoT). Smart devices are equipped with sensors and digital connectivity, which enables them to track their energy consumption, maintenance needs, and general functionality. Such devices can lead to better resource management and extended product lifespans (Chekima and Gbadamosi 2016). For example, Nest smart thermostats optimize the energy consumption of homes by adjusting settings automatically. Smart devices also alert users of maintenance needs, thus extending product lifespans and reducing the need for replacement.

5.6 *Sustainable Manufacturing and Automation*

Advancements in sustainable manufacturing, including robotics and AI-driven processes, are helping businesses reduce waste, energy consumption, and raw material use. Technologies like 3D printing enable highly customized products with minimal waste, unlike traditional manufacturing methods, which often create excess material waste. Green manufacturing practices, such as using renewable energy sources and recycling production waste, are also becoming more common, supporting the goals of a circular economy.

5.7 Eco-Labeling and Transparency in Supply Chains

Consumers are demanding increasing transparency in products and understand the environmental and societal effects of those products. Increasing widespread eco-labels, such as Energy Star, Fair Trade certifications, helps consumers buy the right sustainable products. Another way that blockchain and digital tracking technologies enhance supply chain transparency is in making sure the environmental effects of products can be traced from the raw-material sourcing to final disposal.

5.8 Biodegradable and Compostable Packaging

As plastic waste becomes a growing issue, biodegradable and compostable packaging is emerging as an eco-design solution. Materials like seaweed, mushroom roots, and cornstarch are being developed as alternatives to traditional plastic packaging, which breaks down naturally in the environment. Companies like Loop are pioneering reusable packaging systems that minimize waste generation.

5.9 Green Chemistry and Eco-Friendly Chemical Substitutes

Green chemistry focuses on designing products and processes that minimize or eliminate the use of hazardous chemicals. This research is leading to safer chemical alternatives that maintain product quality while reducing environmental harm. Companies are increasingly adopting these eco-friendly substitutes to reduce pollutants and improve product sustainability (Saxe et al. 2022).

5.10 Artificial Intelligence (AI) and Machine Learning in Eco-Design

AI and machine learning are playing an increasingly important role in eco-design. These technologies can optimize product designs for sustainability, predict environmental impacts, and help create products with fewer materials and less energy usage. AI can also be used to streamline manufacturing processes and assess the effectiveness of eco-design solutions, making production more efficient and environmentally friendly.

5.11 Product Transparency and Consumer Engagement

Consumers are demanding more transparency about the sustainability of products. Companies are responding by providing detailed product information, such as digital product passports that outline the environmental impact of materials, carbon footprint, and end-of-life options. This transparency helps build trust and loyalty, as well as encourages more sustainable purchasing decisions.

5.12 Zero-Waste Design

Zero-waste design aims at eliminating all wastes from the whole lifecycle of a product. It is about creating products and packaging that can be reused, composted, or returned to the environment without leaving any waste behind. Some companies are researching on zero-waste clothing production by cutting fabrics to minimize scrap, while others are focusing on packaging systems that produce no waste.

5.13 Upcycling in Eco-Design

Upcycling is the process of changing waste materials into higher-quality products that support the circular economy. While traditional recycling results in low-quality products, upcycling turns old materials into new, valuable products. Several businesses are embracing upcycling practices, such as making new furniture from reclaimed wood or clothing from fabric scraps, which minimizes waste and the need for new materials (Cinar 2005).

5.14 Sustainable Business Models: B Corporations and Sustainability Reporting

Sustained business models, such as B Corporations, are likely to become more widespread. These businesses fulfill stringent social and environmental performance standards, aligning profit with ecological and social objectives. Additionally, sustainability reporting is on the rise. Companies disclose their environmental and social effects, prompting companies to implement eco-design into their strategies in line with investor, regulator, and consumer expectations (Wimmer et al. 2004).

5.15 Biodegradable Electronics

As reliance on electronics increases, the environmental impact of e-waste is becoming a greater concern. Biodegradable electronics, designed to break down more easily when no longer needed, represent a developing trend in eco-design. Companies are exploring biodegradable materials for electronic components and packaging, aiming to reduce the environmental footprint of e-waste.

5.16 Localized and Decentralized Production

Localized and decentralized manufacturing are getting attention as a sustainable approach to reduce emissions associated with transportation. 3D printing technologies can facilitate on-demand, small-scale production, reducing the need for large factories and long supply chains. Localized manufacturing also benefits communities by creating jobs and reducing reliance on global supply chains.

5.17 Ethical and Socially Responsible Eco-Design

Apart from environmental considerations, increasing attention is being paid toward ethical and socially responsible eco-design. Social impacts of products are being targeted by companies, including fair labor practices, away from human rights violations. Since a lot of consumer demand is now in terms of socially responsible products, Fair Trade certifications, and more, businesses are increasingly paying attention to equitable wages and gender equality initiatives.

6 Conclusion

Eco-design and life cycle assessment (LCA) play a crucial role in driving sustainability within the manufacturing industry. Despite technical and economic barriers, the long-term benefits of integrating sustainable practices—ranging from reducing environmental impact to enhancing competitiveness—far outweigh the challenges. The future of eco-design looks promising, given advancements in materials, technology, and legislation. If adopted, these trends will enable companies to maintain their competitive edge in an increasingly green-conscious world. Eco-design not only provides a competitive advantage but also helps optimize profits and build strong goodwill among customers. Organizations must implement green practices and integrate sustainability to create value for all stakeholders. Additionally, policymakers are developing frameworks to support and guide organizations in embedding sustainability within manufacturing firms. This includes providing financial aid, training, and technological support to encourage the adoption of sustainable practices in the manufacturing sector.

Case Study: Sustainable Practices at Borbonese

Introduction to Borbonese

Founded back in 1910, this is the oldest of the Italian luxury brands and earned iconic status owing to its specialization with respect to the manufacturing of handbags, jewels, and ornaments. Rich with heritage, Borbonese emerged quite swiftly with regard to making an identification and being popularly used among Italian high class, high-fashion customers. The O. P. (Occhio di Pernice or Partridge Eye) pattern is Borbonese's hallmark, being one of the core elements of its collections since the 1970s. The discreet pattern is developed from a unique process of shaping lamb skin, which remains the essence of the brand identity in luxury and craftsmanship.

Innovative Designs and Artistry

Combining the highest quality of materials and know-how in craftsmanship, Borbonese designs pieces in which, at the same time, there is a thought for small refinement. The O. P. texture has developed over the years by the brand in applications to different colors and materials. There are some productions as an example like the Luna Bag, which have become icons of the art of expression. The Luna Bag is inspired by the Futurist artist, Giacomo Balla. His manifesto shows a possible replacement of static garments with dynamic and current one.

Sustainability Through Eco-Design and Recycling

In keeping abreast of the ever-increasing sustainability trend within the fashion community, Borbonese has moved toward an environmentally conscious production process. Because fashion is seasonal, the company is also overwhelmed with leftovers such as leather, as well as unsold items of old collections. Instead of discarding these leftovers, Borbonese has embraced the Eco-Design approach, which involves recycling leftover materials to give them a second life.

For this purpose, the company listed all of its stock material and quickly collected eco-friendly bags and accessories as outerwear. These were stitched from scrap leather, metal pieces, and shoulder straps, respectively. This way, it provided an avenue to reduce waste and also create new and different designs. Using this so-called recycling-based approach, it can be seen that luxury fashion could indeed be sustainable, contrary to the stereotype that high-value items could not be environmentally friendly (Kishita et al. 2021).

Life Cycle Assessment (LCA) of the Savile Bag

The Savile Bag in suede is another interesting product from Borbonese, boasting a striking partridge eye print. A life cycle assessment (LCA) was conducted on the product to determine its environmental footprint at every production and distribution stage. The LCA was conducted using open-source software called openLCA and it was based on internal documents of Borbonese provided by Borbonese's own company, employee interviews, and data commingled from data sources in the industry. The main issues studied were energy consumption in the production process, waste generation, emissions during transportation, and disposal processes.

It emerged that the transport stage has the highest environmental impact, which is mainly due to the CO2 emissions arising from the product distribution at the final stage. Although the screening did not include the production of raw material, an equivalent paper Innovation for a Made Green IN Europe (I. M. A. G. I. N. E.) drew attention to the fact that processing one of the most polluting productions of the fashion industry, that is leather manufacturing: tanneries are notorious for their significant amounts of emissions and waste and the corresponding environmental load in the manufacturing of luxury leather goods.

Conclusion

Borbonese's commitment to sustainability in the form of eco-design, recyclability, and consideration of the environmental impact of its products demonstrates the brand's attempt to reconcile luxury with responsibility. Utilizing waste and a complete waste management evaluation as a tool, Borbonese leads the way for other luxury groups in adopting a sustainable approach in the management of its business operations.

Discussion Questions

1. How does Borbonese integrate sustainable practices into its production while maintaining its luxury status?
2. What are the challenges luxury brands face when implementing recycling strategies in their product lines?
3. How do life cycle assessments (LCAs) help in understanding the environmental impacts of fashion products?
4. In what ways can other luxury brands use Borbonese's approach as a model for their own sustainability initiatives?
5. How can the fashion industry address the environmental concerns associated with leather production and treatment?

Case Study: Eco-Design in the Automobile Industry—Tesla, Inc.

Introduction to Tesla, Inc.

Tesla Inc. was founded in 2003. It is at the forefront of the electric vehicle (EV) industry, working toward a fast transition to renewable energy globally. Through its electric cars, Tesla Inc. is in an effort to tackle urgent environmental concerns such as air pollution and climate change. The company is known to produce high-performance electric vehicles, such as the Model S, Model 3, Model X, and Model Y. Tesla's innovative business model integrates sustainability into the entire operation—from product design to manufacturing—making it a role model of eco-friendly designs in the automotive industry (Maradin et al. 2022).

Tesla's Commitment to Eco- Design

Ecodesign is a design approach through which products are designed to minimize environmental impact over their entire life cycle. At the core of Tesla's strategy is the adoption of the eco-friendly design concept in its vehicle design setting, such as decrease of energy consumption, employment of ecological materials, and decrease of waste (Maradin et al. 2022).

Designing for Sustainability

Tesla designs its vehicles with consideration toward the environment. Each design model features reduced energy by lightweight pieces and parts low in energy content. The general framework of cars manufactured by Tesla is mostly produced in aluminum content. Besides that, this lightweight component ensures significant performance improvement for both the vehicle's movement and increase the range without depleting fuel. Consequently, considering its lightweight attributes, the issue of increased fuel efficiency and ultimate carbon footprint release must be important in terms of improvements. In Tesla, energy-conserving functions are provided, according to regenerative braking that scavenges the energy from kinetic energy of the vehicle being driven on road and converting it into electrical energy thus increasing the driving range of the car. Moreover, Tesla designs its vehicles with durability in mind, ensuring that components last longer, which reduces the frequency of repairs or replacements and contributes to a lower overall environmental impact (Maradin et al. 2022).

Sustainable Manufacturing Practices

Tesla has made important efforts to lessen the environmental effect of its production methods, conforming to the ideals of eco-design. A central aspect of Tesla's production strategy is the creation of its Gigafactories. These large-scale production plants are engineered for maximum energy efficiency and operate using renewable energy sources, such as solar panels. For instance, the Gigafactory located in Nevada utilizes solar energy for part of its power needs, aiding in minimizing the factory's dependence on non-renewable energy sources. Tesla's environmentally friendly design also applies to its production methods. The firm has put resources into minimizing water use, waste creation, and energy expenditure in the manufacturing process. Tesla's dedication to reducing environmental impact is clear in the materials it selects—whenever feasible, the company incorporates recycled and recyclable materials in its vehicle manufacturing, decreasing the demand for new raw materials and supporting a circular economy (Maradin et al. 2022).

Battery Design and Recycling

A key element of eco-design in electric vehicles is battery technology. Tesla's batteries, used in its electric vehicles, are created with a focus on sustainability. The organization has progressed in enhancing the recyclability of its batteries, seeking to minimize the ecological effects of battery manufacturing and disposal. Tesla's modular battery configuration permits the replacement of single cells instead of the whole battery pack, minimizing waste and prolonging the product's lifespan. To tackle the issue of battery disposal, Tesla has been at the forefront of creating battery recycling solutions. Tesla aims to reclaim essential materials from old batteries, including lithium, nickel, and cobalt, which can be repurposed in creating new batteries. This approach not only diminishes the need for raw materials but also stops battery waste from piling up in landfills. Aside from recycling, Tesla has investigated the idea of "second-life" uses for batteries. This involves repurposing used EV batteries for energy storage solutions, such as the Powerwall, Powerpack, and Megapack. These products use repurposed EV batteries to store energy from renewable sources, such as solar power, providing a sustainable solution for energy storage while reducing the need for new batteries (Maradin et al. 2022).

Tesla's Eco-Design Impact on the Supply Chain

That commitment to sustainable design at Tesla extends out of manufacturing of cars and batteries, etc. In the context of raw material sourcing, the company requires its suppliers to work closely with it so as to ensure that procured materials are obtained sustainably. Tesla stresses that responsibly acquired materials (e.g., lithium, cobalt, nickel) should be assured to respect both environmental and human rights standards. Tesla has also made contributions toward sustainability in its logistics and transportation options. The company structures its supply chain in order to reduce the carbon emissions that are generated in the production, shipping, and delivery of raw materials and end products. Through its ongoing Gigafactory creation within areas that are closer to larger populations, Tesla reduces the mileage covered by vehicles and therefore reduces the associated transportation emissions with regard to its vehicles (Maradin et al. 2022).

Challenges in Implementing Eco-Design

Although Tesla has contributed to the green design, there are also still challenges ahead to reach a really sustainable automotive industry. Although the general production of lithium-ion battery is less environment damaging than the production of the gasoline vehicle, it is still broadly carbon-intensive, primarily because of the mining and processing of species such as lithium, cobalt, and nickel.

Tesla is currently addressing these challenges by embracing advanced technologies, like solid-state batteries, which are expected to be more energy-efficient and have a lesser environmental footprint. Particularly, Tesla is also actively developing novel means for recycling and reuse of materials for the sustainable prolongation of its vehicle and batteries.

Tesla's Future Sustainability Goals

Looking to the future, Tesla has very ambitious sustainability goals. The company is working toward an expansion of the output of electric vehicles, battery storage technologies, and solar products—all using renewable energy. Key ambitions for Tesla include the capabilities of its battery recycling initiatives to create new second-life uses for used batteries, and to spur eco-design inventions that reduce waste and optimize resources (Maradin et al. 2022).

One of the good examples in respect to eco-friendly design is Tesla; therefore, all other automotive manufacturing companies could take quite a good lesson from its works. The firm proved that it can work on creating luxurious cars at the same time, as on highly developed technologies, decreasing ecological footprint at the same time.

Conclusion

Tesla, Inc. is also pioneering the automotive industry, through the application of eco-design principles, or designing within environmental parameters with the objective to reduce the human footprint. The company continues to set the limits of eco-friendly automotive production through the manufacture of energy-efficient vehicles using eco-friendly materials and through the utilization of renewable energy-based manufacturing processes. The company emphasizes design for batteries, recycling, and ethical sourcing; in this sense, it reduces its environmental footprint, contributing to the goal of sustainable solutions in the automobile industry.

Discussion Questions

1. What are the key environmental benefits of Tesla's use of lightweight materials and energy-efficient features in its vehicles?
2. How does Tesla's modular battery design and recycling efforts contribute to its sustainability goals?
3. What are the challenges Tesla faces in ensuring the sustainability of its battery production and recycling efforts?
4. How can other automobile manufacturers adopt Tesla's eco-design practices to reduce their environmental footprint?
5. What role do eco-design and battery recycling play in Tesla's broader sustainability strategy, and how can these practices evolve in the future?

References

Chan HK (2011) Green process and product design in practice. Procedia Soc Behav Sci 25:398–402. https://doi.org/10.1016/j.sbspro.2012.02.050

Chekima B, Gbadamosi A (2016) Consumer values and green products consumption in Malaysia: a structural equation modelling approach. In: *Handbook of research on consumerism and buying behavior in developing nations*. IGI Global

Çınar, H. (2005). Eco-design and furniture: Environmental impacts of wood-based panels, surface and edge finishes. For Prod J, 55(11), 27–33.

Faud-Luke A (2009) Eco-design: the sourcebook. Thames & Hudson

Foschi E, Zanni S, Bonoli A (2020) Combining eco-design and LCA as decision-making process to prevent plastics in packaging application. Sustainability (Switzerland) 12(22):9738. https://doi.org/10.3390/su12229738

Fukushige S et al (2025) EcoDesign for circular value creation. Springer Nature Singapore

Keivanpour S (2023) Circular economy in engineering design and production: concepts, methods, and applications. Synthesis lectures on sustainable development. Springer, Cham. https://doi.org/10.1007/978-3-031-44652-8

Kishita Y, Matsumoto M, Inoue M, Fukushige S (eds) (2021) EcoDesign and sustainability II: social perspectives and sustainability assessment. Sustainable production, life cycle engineering and management. Springer, Singapore. https://doi.org/10.1007/978-981-15-6775-9

Maradin D, Malnar A, Kaštelan A (2022) Sustainable and clean energy: the case of tesla company. J Econ Financ Manag Stud 05(12):3531–3542. https://doi.org/10.47191/jefms/v5-i12-10

McDonough W, Braungart M (2002) Cradle to cradle: remaking the way we make things. North Point Press

Nematchoua MK, Orosa JA, Reiter S (2022) Life cycle assessment of waste products of a green-neighbourhood. Waste Biomass Valori 13(8):3639–3651. https://doi.org/10.1007/s12649-022-01749-2

Nuryakin N, Maryati T (2020) Green product competitiveness and green product success. Why and how does mediating affect green innovation performance? Entrepreneurship and sustainability Issues 7(4):3061. https://doi.org/10.9770/jesi.2020.7.4(33)

Pati SN (2023) Life cycle assessment: future challenges. Taylor & Francis

Proctor R (2006) The sustainable design book. Laurence King Publishing

Quella F, Schmidt WP (2003) Integrating environmental aspects into product design and development the new ISO TR 14062. Int J Life Cycle Assess 8:113–114. https://doi.org/10.1007/BF02978438

Saxe JK, Hoffman L, Labib R (2022) Method to incorporate green chemistry principles in early-stage product design for sustainability: case studies with personal care products. Green Chem 24(12):4969–4980. https://doi.org/10.1039/d2gc00842d

Sayal A, Pant M (2023) Examining the moderating effect of green product knowledge on green product advertising and green product purchase intention: a study using SmartPLS SEM approach. Int J Asian Bus Inf Manag 13(1):1–16. https://doi.org/10.4018/IJABIM.316175

Subramanian V, Peijnenburg WJGM, Vijver MG, Blanco CF, Cucurachi S, Guin JB (2023) Chemosphere approaches to implement safe by design in early product design through combining risk assessment and life cycle assessment. Chemosphere 311:137080

Wimmer W, Züst R, Lee K-M (2004) ECODESIGN implementation: a systematic guidance on integrating environmental considerations into product development, Alliance for global sustainability bookseries. Springer, Dordrecht. https://doi.org/10.1007/978-1-4020-3071-0

A Case-Based Perspective on Technology-Driven Transformation and ESG Integration

Agatha Lavinia, Ismail Khan, Muhammad Rahies Khan, Umathevi Munikrishnan, and Sibtain Sohair

Abstract This chapter explores Amazon's evolution from a traditional online bookstore to a global leader in e-commerce, logistics, and cloud computing, with a strong emphasis on digital transformation and Environmental, Social, and Governance (ESG) performance. It critically examines the limitations of Amazon's initial monolithic IT infrastructure, highlighting its inefficiencies, security vulnerabilities, and adverse environmental and social impacts. In response, Amazon undertook a significant transformation by integrating advanced technologies such as artificial intelligence (AI), machine learning, Internet of Things (IoT), blockchain, and Amazon Web Services (AWS). This chapter details how these innovations have enabled Amazon to enhance operational efficiency, automate supply chains, optimize warehouse management, and strengthen governance. Particular attention is given to how AWS facilitates real-time data processing, scalability, and sustainable practices. Furthermore, the use of robotics and smart systems has improved labor conditions and customer satisfaction, while blockchain enhances supply chain transparency and security. This chapter concludes with forward-looking recommendations for further

A. Lavinia
Tarumanagara University, Jakarta, Indonesia
e-mail: agatha.lavinia0698@gmail.com

I. Khan
Sunway Business School, Sunway University, Kuala Lumpur, Malaysia
e-mail: ismailwazir048@gmail.com

M. R. Khan (✉)
Department of Management, Bahria University, Karachi, Pakistan
e-mail: mrahies581@gmail.com

U. Munikrishnan
Faculty of Business and Management, UCSI University, Kuala Lumpur, Malaysia
e-mail: umathevi@ucsiuniversity.edu.my

S. Sohair
Computer Science Engineering and Maths, Middlesex University, London, UK
e-mail: sibtainsohair@hotmail.com

M. Venkatesh et al. (eds.), *Green Manufacturing Technologies in Industry 5.0*,
https://doi.org/10.1007/978-981-95-0363-6_10

investment in AI, renewable energy, employee well-being, and the development of private blockchain networks. Overall, Amazon's digital reinvention presents a model of technological innovation aligned with sustainability and corporate responsibility.

1 Introduction

In the past few years, Amazon has become a big name in technology, mainly in IT. It started as an online store, but now it does much more. Amazon's transformation highlights its adaptation to disruptive digital environments (Wessel and Christensen 2012), becoming a leader in the platform economy. The main reason for this change is better IT systems, smart business moves, and good use of resources. This change is not only making Amazon different, but it is also changing online shopping, cloud computing, and AI. This writing will discuss about the IT systems Amazon uses, the business plans they follow, and the resources that help them stay ahead in technology.

Amazon's shift from a bookstore to a global technology leader illustrates effective digital strategy formulation (Ross et al. 2017; Bharadwaj et al. 2013). Amazon started as an online bookstore in 1994. Over time, it grew into one of the largest e-commerce and technology companies in the world. The company expanded its services to include cloud computing, logistics, and artificial intelligence. As Amazon grew, it relied on an old monolithic IT system to support its operations. This system worked as a single structure where all functions were tightly connected. At first, it managed online orders, inventory, and customer data effectively. However, as the company expanded, the system faced problems that affected efficiency, security, and sustainability. Amazon's old system struggled with scalability. When more customers visited the platform, the system slowed down. During high-traffic periods, like holiday sales, website performance dropped. Pages loaded slowly, and some transactions failed. This frustrated customers and caused lost sales. The system could not handle the growing number of products, warehouses, and sellers. Updates took a long time to implement, making it hard to add new features. Security risks increased due to the monolithic system. All functions were linked, so a single vulnerability could expose the entire platform to attacks. Hackers could access customer data and payment details more easily. A breach in one area affected the whole system. Updating security measures was difficult because changes in one part required adjustments in others. This increased the risk of cyber threats and data leaks.

Not only that, Amazon's old system caused environmental, social, and governance (ESG) problems. High energy consumption was one of the biggest concerns. Traditional data centers required large amounts of electricity to run servers and cooling systems. Amazon's expanding operations led to higher carbon emissions. The inefficient system wasted energy because it could not adjust to demand. Unused server capacity still consumed power, contributing to environmental damage. Another thing, the warehouse operations were not efficient as their inventory management relied on manual processes and outdated technology. Workers had to sort and track products with limited automation. This increased errors and delays in the workplace and also employees had to work longer hours to meet demand, leading to poor working conditions. Labor-intensive processes caused higher injury rates and dissatisfaction among workers. The lack of proper tracking systems led to excess stock or shortages, creating waste. Moreover, their supply chain issues also affected ESG performance. The old system struggled to track shipments accurately. Packages were sometimes lost or delayed. Inefficient routing led to longer delivery times and increased fuel consumption. Delivery trucks often traveled unnecessary distances, producing more carbon emissions. Poor tracking also resulted in incorrect orders, increasing returns and waste.

Social responsibility was another concern. The outdated system made it hard to monitor supplier practices. Some suppliers operated under poor labor conditions. Amazon could not track these issues effectively, leading to ethical concerns. Customers and regulators expected better transparency, but the system lacked tools to provide detailed reports. Ensuring responsible sourcing and fair wages for workers was difficult. Governance issues arose due to limited data visibility. Decision-makers struggled to access real-time insights. Reports took longer to generate, making it hard to respond to market changes quickly. Fraud detection was weak, allowing unauthorized transactions to go unnoticed. The system lacked automation for compliance checks, increasing regulatory risks. Meeting sustainability goals was challenging without clear data on emissions, waste, and energy use. Product recommendations were less accurate. The old system used basic algorithms to suggest items to customers. These recommendations often failed to match customer preferences. As a result, users saw irrelevant products, reducing engagement and sales.

Without advanced machine learning, Amazon could not personalize shopping experiences effectively. This affected customer satisfaction and loyalty. The payment system was not optimized. Processing transactions took longer, especially during peak times. Some payments failed due to system overload. Refunds and chargebacks were harder to manage. Customers faced delays in resolving payment disputes. Fraud detection relied on manual checks, making it less effective. This created security risks and financial losses. The adoption of AI-driven compliance and detection can address regulatory challenges (Rajpoot and Raffat 2024). Customer support was inefficient. The system did not have AI-driven chatbots to handle basic inquiries. Customers had to wait longer to get help from human representatives. Tracking complaints and resolving issues took more time. Without automation, service quality varied, leading to inconsistent customer experiences. Marketing strategies lacked precision. The system could not analyze customer behavior in real time. Advertisements targeted broad audiences instead of individual preferences. This led to lower conversion rates. Without detailed insights, Amazon struggled to refine its marketing campaigns. Competitors with better data-driven strategies gained an advantage. Amazon's rapid growth made the old system unsustainable. The increasing number of users, products, and transactions exposed its limitations. The company faced challenges in maintaining efficiency, security, and ESG standards. To address these problems, a major IT transformation was necessary.

2 Amazon Transformation

2.1 Business Flow

Nowadays, Amazon has changed how they operate their business and the systems they use over the years to handle the issues of its previous practices. Amazon now implements cutting-edge technologies to enhance order processing, supply chain management, as well as delivery systems. It leverages a complex automated system based on AI, machine learning, and big data analytics in order to manage its inventory, anticipate customer demand, and optimize logistics. Amazon's extensive use of artificial intelligence reflects real-world AI applications in business (Davenport and Ronanki 2018). Its integration of robotics and automation aligns with current trends in digital transformation (Brynjolfsson and McAfee 2017; Bughin and Hazan 2017). Amazon's warehouses also use robots for sorting, packing, and shipping products, minimizing human error and increasing operational speed. The order flow from Amazon's website is now complex and efficient. When a customer submits an order on Amazon's site or app, the system checks product availability in real time. Amazon use their cloud system called AWS (Amazon Web Services) to handle the order, it then will search for stock availability, locations and chooses the nearest fulfillment center to deliver products quickly and with the lowest cost. Using advanced algorithms, AWS selects the ideal warehouse, considering stock, location, and shipping possibilities. Once the warehouse is selected, the order is routed to that fulfillment center. Amazon's smart warehouses run on automation and AI systems. Kiva, now known as Amazon Robotics, are robots that bring shelves full of products to cutting down on time spent walking across giant warehouses. Workers then will retrieve the items from the shelves and scan them to update inventory records, and then get them ready for packing. Automated conveyor belts carried the items to packing stations where they are packed into cartons and labeled for shipment.

The package is then directed to the sorting area, where AI tools analyze possible delivery routes, considering the package's destination, current traffic patterns, and the shipping speed selected such as Prime 1-day delivery. Packages are automatically sorted and then loaded onto trucks, vans, or planes. Amazon also has a delivery network, called Amazon Logistics, and employs Amazon Flex drivers for last-mile delivery. Flex drivers, who use their own vehicles to do this work, provide that flexibility during peak times. In some regions, Amazon also employs drones for small packages and third-party couriers for longer distance deliveries. Tracking updates are provided to customers during the process and all tracking data is managed by AWS and streamed to the Amazon app or site. With that, customers can track when their order is processed, packed, shipped, and delivered giving the transparency of the status of their order. Not only that, AWS also collects and handles customer comments and delivery confirmation. GPS is also attached in delivery vehicles to track where packages have been left exactly, to ensure accurate tracking and reduce the occurrence of lost deliveries.

Amazon's order flow is now powered with the most advanced technology in every steps of their business. First, amazon use AI to create forecasts demand, allowing warehouses to check the availability of each product to meet its demand. Second, the automation that amazon uses accelerates picking and packing products and minimizes human errors and costs. Amazon use cloud computing technologies to guarantees that data is processed swiftly and securely stored. Amazon warehouse also uses IoT (Internet of Things) sensors to accurate track inventory in real time and notifying managers when stock is low or when items are misplaced. Not only that, Amazon's support to implement ESG objectives in their business also can be seen on how they use automated systems to save the energy needed when running the warehouses.

Additionally, all operations at AWS data centers also run on renewable energy, allowing them to give green impact to the ecosystem. AI-based tools will then reduce waste by avoiding overstocking and ensuring deliveries are accurate. Blockchain technology then aids in tracking the journey of products through the supply chain, promoting transparency and reducing fraudulent practices. Amazon's new system increases their efficiency and promotes ESG factors through supply chain optimization, allowing them to create better condition for both employees and environment. From the ordering of items from websites to the final delivery, Amazon has it all backed with a combination of AI, automation, cloud computing, and logistics technology as they morph into a global leader across e-commerce and supply chain space as well. Amazon continuously enhances efficiency and optimizes logistics to reduce environmental impact, improve employee's condition with automating manual labor, and strengthen governance through properly managed operation.

With this improved service, we see the highs and lows of all-time customer satisfaction for amazon from 2004 through 2022. Amazon scored a high American Customer Satisfaction Index (ACSI) in 2010, indicating a high level of customer approval. But by 2020, that score had fallen to 65, the first time it had slipped below 80 since the company's founding. That number ticked up slightly in 2021 to 78 but remained below highs in the past. By 2022 it was 79%, a slight recovery, but still a long way from the earlier highs.

2.2 *Smart Warehousing*

Amazon made significant changes to its management, labor policies, technology system, and warehouse design, leading to improved operations and ESG performance. Amazon's warehousing improvements mirror broader shifts in e-commerce logistics, including automation and robotic management (Boysen et al. 2019). The use of AI for predictive maintenance and supply chain management aligns with best practices identified in logistics research (Min 2010; Dolgui et al. 2018). Previously, Amazon faced many challenges, such as management incompetence, information asymmetry, unsafe work environment, high employee turnover rate, etc. Managers were working with stale data, and departments were working in silos, causing delays and errors. Manual processes also put physical strain on workers, who were dissatisfied and preferred to leave the job. Due to that, now Amazon adopted advanced data management tools such as AWS cloud computing system, automation, and new governance policies to increase their productivity and efficiency.

Amazon uses AWS to scale cloud solutions for real-time access to data and allows seamless data sharing across departments. AWS helped amazon in sharing data between various departments, improving communication and decision-making. AWS also improves customer service as it manages faster data processing, operation of stable cloud storage, and order management. By enabling personalized recommendations, faster deliveries, and improved inventory management, this system boosted the customer engagement by 45% and the sales by around 35%. Additionally, with AWS, the products can be automatically restocked based on up-to-date sales data, preventing it from running out of stock. Not only that, amazon also introduced Amazon Chime as a communication platform for instant messaging, video calls, and file sharing, which reduced errors at work and increase efficiency. AWS also helped integrate AI systems that monitor customers' choices, recommend products based on the previous purchase and watch list, and help keep the supply chain running seamlessly. To enhance customer engagement, amazon also uses Amazon CloudFront as a powerful content delivery network, provides faster access to web content for customers across the globe and significantly enhancing user experience and engagement.

Automation, such as Kiva robots for warehouse operations and Pegasus robots for sorting, took physical strain off employees and improved accuracy. Kiva robots transport items within the warehouse, retrieving products from shelves and delivering them to packing stations, while Pegasus robots specialize in sorting and organizing smaller parcels. To ensure these robots could move swiftly, Amazon adjusted the layout of its warehouse, ensuring wide aisles for smooth robot traffic with adjustable shelf heights for robots and modular shelving systems which can be reconfigured according to demand. QR codes and sensors are spread throughout each level to guide navigation, and workstations are positioned so that travel distances are as short as possible, allowing robots to process items quickly and accurately, increasing overall warehouse efficiency. Multi-tiered shelving systems and automated lifts further improved storage space and minimized the time spent locating items. The company also launched smart conveyor belts and automated packaging systems to speed up processing and cut labor costs. The company integrated digital twin technology, creating virtual models of warehouses to simulate and improve operations before physical changes are made. Predictive maintenance systems that identify potential equipment failures which enable continuous operations also contributed additional layers of efficiency in this system.

Not only that, Amazon also placed warehouse location near major cities. By doing that, it could reduce distance for delivery and emissions, as a support to ESG project. Amazon use AI to create route optimization plan to reduce fuel consumption and use electric vehicles to reduce carbon footprint. Moreover, Amazon made use of speedy off-the-shelve delivery systems by automating sorting hubs with AI real-time shipment tracking in addition to advanced demand forecasting, further increasing the ability for quality implementation of delivery services with reduced delivery time. Machine learning algorithms were used to predict periods of high demand in advance so that Amazon could stock up on products that people were going to need. Not only that, Amazon Prime Air improved delivery speed and reduced environmental impact via drones for small package deliveries. To track down parcel location and stocking, AI-based inventory scanners is used to achieve precise counting of stocks and on-time re-filling of shelves to avoid the commodity getting lost or wastage. The investments of Amazon in smart logistics hubs that contained automated systems for loading and unloading goods made the operations further easier and the timelines to make the delivery better. The use of blockchain technology by Amazon improved supply chain transparency, allowing customers to know and track products delivery accurately. With frequent warehouse upgrades, technology advancements, and new renewable energy projects, these improvements led to lower overall operating costs and less impact to the environment.

To improve governance and policy, amazon also implemented employee training programs to ensure more experienced workforce. Wage and safety measures have also been focused on to prevent possible issues of labor exploitation. Amazon's old system had reported high injury rates due to long hours, repetitive tasks, and manual labor. Heavy lifting and fast-paced environments led to musculoskeletal injuries and accidents. In response, Amazon rolled out automated systems such as Kiva and Pegasus robots to decrease physical strain, creating more ergonomic designs to help prevent injuries, and integrated with real-time monitoring systems to identify hazards. To ensure compliance and reduce workplace accidents, most companies followed regular safety audits along with mandatory training programs, providing better and safer working environment for workers. Wearable technology for workers, including sensors to monitor their posture and fatigue, also helped to lower injury risks and ensure the well-being of workers. Amazon had enhanced employee satisfaction and retention rates with their investment in employee wellness programs, mental health support, and continuous training opportunities. This included expanded access to educational programs, career growth opportunities, and comprehensive healthcare benefits. Along with this, Amazon built a worker feedback system that enabled workers to communicate concerns and propose solutions, creating a collaborative space to work.

Amazon's ESG goals can be seen through an extensive use of renewable energy and effort to optimize logistics, policies, and commitment to increase corporate governance. With technological innovations and better management of the company could increase efficiency and create a safer environment for employees as well as creating a sustainable environment and creating an example for other e-commerce industry. Their investments in solar energy and wind power proved Amazon's commitment to reducing its impact on the environment and also improving the quality of service. By consistently adopting advanced technology and prioritizing ESG factors, Amazon has established itself as a leading company in the global marketplace, ensuring its competitiveness, innovation, and social responsibility to the industry. With AI to analyze demand, implementing intelligent warehouse design, and operating with renewable energy enabled Amazon to meet increasing customer expectations with zero environmental impact and improve worker health and safety. Amazon proved to successfully implement technology to address its challenges and embrace ESG factors. With this, Amazon not only focus on the short goal but also investing in long-term growth, sustainability, and operational excellence, launching it as an unstoppable force for digital disruption and a responsible business model on a global scale.

2.3 AWS Cloud Computing

Amazon uses cloud computing with Amazon Web Services (AWS) to improve its operations (Armbrust et al. 2010; Buyya et al. 2009). AWS is the main system for Amazon's cloud services. It offers storage, computing power, and data tools that help Amazon run its global business. AWS has many services that help Amazon process orders, store data, and manage supply chains. For example, Amazon Simple Storage Service (S3) stores large amounts of data, while Amazon Elastic Compute Cloud (EC2) provides virtual servers for computing needs. Amazon's delivery, warehousing, and cloud computing systems are connected. Smart warehouses use robots and AI to manage inventory, pick products, and pack orders and use robots like Kiva move shelves to workers for saving time and effort. Sensors will track goods in real time and send data to AWS then this data will be processed to manage stock availability, predict demand, and avoid delays.

AWS verifies inventory and selects the optimal warehouse for quick shipment when customers make a purchase, and robots will prepare the order while shipping labels are created automatically. Next, AI will schedule where the best delivery routes are to conserve gas and time. After that, parcel will be transported with trucks that also attached with IoT devices to track the routes they are using and parcel condition and AWS will collect that data to track deliveries. Leveraging AWS builds a single source across system where all data were integrated to enhance service excellence. With AWS, Amazon could easily modify its operations according to demand. Amazon uses additional computing power in busy times like Prime Day or holidays which helps Amazon to handle high traffic and keeps the services fast and reliable. With additional support Amazon could cover demand fluctuation and help save energy when the demand is low.

There are various features to support AWS cloud service such as Auto Scaling, SageMaker, CloudWatch, Glue, and Redshift to improve productivity and data accuracy. Auto Scaling allows Amazon to automatically adjust its computing resources to avoid over-use of resources and reduces costs. AWS CloudWatch watches over Amazon's operations and issues alerts when something goes wrong. Amazon uses AWS for big data that allows Amazon to optimize inventory availability, pricing, and customer recommendations. Tools like Amazon Redshift allow Amazon to rapidly analyze massive amounts of data. AWS Glue assists in linking data from multiple sources, for instance, Amazon analyzes purchasing patterns to recommend items consumers may purchase. Amazon SageMaker assists you in creating machine learning models to forecast demand, detect fraud, and enhance customer service.

A huge part of Amazon's warehouse and supply chain operation is powered by the Internet of Things (IoT). IoT devices, including smart sensors, RFID tags, and connected machinery help to track products in real time, monitor warehouse conditions including temperature, humidity, as well as manage inventory lifespan. These devices detect damage to products from early stage to minimize loss, prevent stock shortages, and provide transparency in operations by allowing real-time data to be shared across systems. IoT systems facilitated with predictive maintenance by monitoring equipment performance to reduce downtime, minimize unexpected breakdowns, and minimize energy consumption in warehouses through automated control systems and efficient energy management. Similarly, IoT devices help in steering around Pegasus and Kiva robots within the warehouse. These robots are monitored using real-time data transmission so their mobility can be managed, including steering them away from each other to speed up product pick-up and placement. These not only optimizes the supply chain by eliminating delays but also supports ESG through lower energy consumption and a more efficient allocation warehouse. The infrastructure of Amazon's IoT ecosystem also assists with optimizing the delivery route, therefore cutting down consumption of fuel and speeding up delivery, which enhances operational efficiency and contributes toward ESG sustainability goals as well.

Amazon uses AI across a wide range of applications, including personalized recommendations, voice-based assistants like Alexa, automated customer service, fraud detection, supply chain optimization, and customer sentiment analysis. AI analyzes wide datasets to forecast demand, checks dynamic pricing, and improves warehouse automation by supervising robots and inventory. AI now empowers Amazon to optimize customer satisfaction and operational performance by learning from recent data. It also enables Amazon to keep its competitive edge through quick identification of emerging market trends, enhancing automation for inventory management, and ensuring product availability. In turn, it facilitates Amazon's growth and ESG objectives through minimizing waste, lowering emissions, and creating a better service quality. Amazon's technological advancement are reflecting commitment on ESG factor with the usage of its cloud computing which decreases the use of physical servers so then creating a lower energy usage. Big data analytics aids in monitoring and lowering carbon emissions, managing waste, and maintaining regulatory compliance. Automation of processes in IoT and AI technologies leads to improved efficiency, less consumption of resources, and greater worker safety. Amazon is a pioneer of big data analysis, cloud computing, IoT, and even AI, and it continuously raises the bar for the e-commerce market in terms of technological innovation, operational efficiency, and environmental sustainability.

AWS helps Amazon's achieve its green goals as it uses more renewable energy like solar and wind. Cloud computing lessens the need for buildings and curtails energy consumption. AI saves resources by predicting how much power will be needed. Tools available from AWS enable Amazon to discover ways to reduce what is wasted, recycle materials, and reduce shipping pollution. Not only that, it also aids Amazon in improving working conditions. Robots do the heavy lifting and fast packing, which also means reduces hard work for employees. AI tools aid in forecasting staff needs and reducing stress, for instance, Amazon uses AI to plan work shifts to avoid overtime. Amazon Connect, a cloud service, assists in handling employee queries and resolving issues quickly. AWS makes Amazon's data more secure as well. AWS Key Management Service—secures data using encrypted keys. AWS Identity and Access Management manages access to systems. It monitors for security issues and safeguards data. AWS backup protects backups of data while AWS shield protects against cyberattacks. Amazon uses them to comply with data laws and protect customer data. Amazon also leverages artificial intelligence to identify fraudulent online payments and keep users safe from scams. The use of AWS, AI, IoT, and big data is what changed Amazon with respect to operations; operating faster, safer, and more efficiently. Amazon's delivery, warehousing, and cloud computing systems work hand in hand to enable efficient operations, quick deliveries, and enhanced service provision for customers.

2.4 Blockchain Integration

The application of blockchain in supply chain management resonates with contemporary research highlighting blockchain's transformational impacts on organizational processes (Gul et al. 2024; Khan et al. 2024a, 2024b; Masood et al. 2024). With AWS, Amazon has constructed its cloud system using numerous data facilities worldwide. AWS provides easy to use services such as storage, computing, and databases to assist businesses with their operations. Features such as EC2 for virtual computers and S3 for storage allow companies to operate more quickly. Amazon always try to find a new way to enhance its service such as integrating artificial intelligence with smart devices and big data analysis. Its IoT Core, for instance, enables devices to communicate with cloud systems, while Amazon Sage Maker helps developers build smart applications. AWS or Amazon web services is now using blockchain to increase efficiency, trust, and safety. The Amazon Managed Blockchain Service enables businesses to create and manage blockchain networks with little effort. This service can work on popular blockchain systems including Hyperledger and Ethereum. Nestlé, for instance, uses this service to trace coffee beans, from farm to store shelf, to ensure the product is genuine. Amazon is also utilizing blockchain in its supply chain where it tracks products to avoid fraud. Adding blockchain, however, is not simple to AWS, as blockchain store data in many different locations while AWS keeps data in one place for speed and accuracy. This difference poses challenges for integrating blockchain system with Amazon cloud system. As a solution to this problem, Amazon implements additional sidechains and uses them to perform some work separately.

Amazon takes security of data very seriously. It operates on secure smart contracts that constantly monitor systems and apply powerful security mechanisms such as AWS Key Management Service stores secret codes safely. Amazon is also transforming its cloud engagement with the interplanetary blockchain, which involves considerable expense and requires the top factor. It also has to abide by stringent data regulations around the globe like GDPR in Europe. Amazon states that their pre-made blockchain templates allow businesses to quickly deploy secure blockchain systems. However, one of the challenges is to process big data. Due to this, AWS capable of handling a lot of data while blockchain, on the other hand, poses challenges in the execution of many transactions all at once. Amazon requires systems that can deal with large workloads, but existing blockchain tools tend to slow down when too much information is present. With this challenge, Amazon must make changes in its cloud to support blockchain. That means purchasing new equipment, undertaking training, hiring experts and possibly pausing service during the switch. It has to abide by different laws governing data in different countries it operates, while also keeps data safe across border. They also need strong protection when connecting to the blockchain.

2.5 *Amazon Research Insight*

Amazon is one of the several system transformations, with lessons on digital transformation, innovation, and ESG (Environmental, Social, and Governance) commitment. Amazon's old system relied on manual work and simple software. Warehouses monitored products using spreadsheets and simple databases. Supply chains were sluggish and vulnerable of error. Workers filled orders manually, resulting in delays, lost products, and heavy costs. Customers had frequent experience with late deliveries and bad service. Those inefficient processes lead to the system using more energy, while poor inventory management creates more waste. Workers endured long hours and dismal conditions doing manual labor. Labor union had criticized Amazon for mistreating workers and maintaining unsafe conditions. Additionally, inadequate data management left Amazon unable to keep track of inventory correctly or meet customer demands. It led to waste, poor stock management, and bad decisions. Amazon began making changes to its operations to address these issues. To improve its systems, Amazon uses automation, AI, big data, and cloud computing to enhance productivity and efficiency. Warehouses management also now using robots such as Kiva to avoid manual work by moving shelves to the workers and assisting to pick up and pack items to minimize errors, save time and reduces costs. Automation reduced energy use by optimizing system runs and artificial intelligence helped with inventory management, delivery planning, and pricing. For example, AI tools analyzed sales data to predict what customers would buy, enabling Amazon to stock the right products at the right time.

AWS, or Amazon Web Services, Amazon's cloud computing platform became the core of Amazon's cloud system. AWS enabled Amazon to store substantial amounts of data, figure out complex procedures, and extend its systems during period when it was bustling, such as high season. By only using cloud resources,

AWS enabled Amazon to save energy. AWS Auto Scaling automatically adjusted computation power and AWS CloudWatch monitored systems to prevent issues. With this system, it enabled Amazon operations to be seamless and efficient. AWS enables amazon to have a real-time tracking of products by IoT sensors in warehouses to know where the items are located and their time of arrival. Sensors on delivery trucks tracked their locations and monitored their condition to avoid delays, minimize waste, and cut energy usage. AI tools devised the optimal delivery routes to conserve fuel and time. In warehouses, robots and sensors made sure that energy was not wasted and that operations ran smoothly. AWS also improved the safety of Amazon's operations. Data protected with Encrypt KMS (Key Management Service) and AWS Identity and Access Management (IAM) governed who could access systems. AI detected security threats and secured data from cyberattacks at Amazon. Through AWS features, Amazon compliance with data protection regulations, including GDPR (General Data Protection Regulation) and CCPA (California Consumer Privacy Act), makes sure that customer data was in safe hands and was not at any risk.

Amazon blockchain with cloud computing then added another layer of security and trust. Blockchain which is a digital ledger that securely records transactions is used to track products from suppliers to customers, reducing fraud and mistakes. Blockchain systems were slow because every transaction had to be verified by numerous points. Concepts such as hybrid models where combined version of private and public blockchains is being used and sidechains where relatively smaller connected blockchains helped Amazon optimize blockchain for better performance. As Amazon's blockchain business grew, AWS powered it with secure and scalable cloud services. Amazon Managed Blockchain is one of such AWS services that significantly simplified the creation and management of blockchain networks. Amazon has used Automation, AI, Cloud Computing, and Blockchain to improve its ESG performance where less human input meant less energy are being used and waste. Amazon's technology changes helped it grow, work better, and meet ESG goals by saving energy, reducing waste, and improving working conditions for employees.

3 Recommendation

Amazon has significantly advanced its technological infrastructure, yet there remains considerable room for further improvement. Continued investment in warehouse automation is essential to reduce manual labor stress and enhance operational efficiency. Deploying advanced AI-powered robots for sorting, packing, and inventory management can effectively minimize human error, accelerate processing times, and optimize warehouse workflows (Boysen et al. 2019; Min 2010). In addition, Amazon should prioritize enhancing employee working conditions beyond physical ergonomics. While automation can alleviate physical stress, the company should simultaneously invest in robust mental health support initiatives, ensure fair compensation, and institute regular breaks to mitigate worker fatigue and stress (Daugherty and Wilson 2018; Khan et al. 2024a, 2024b). Further adoption of AI-driven tools to monitor employee well-being without breaching privacy is recommended, for instance, employing intelligent systems that track workload levels and suggest timely breaks when stress indicators are detected (Geissbauer et al. 2015). Moreover, sustainability initiatives must be expanded. Amazon should enhance its renewable energy strategy beyond the AWS data centers by constructing additional solar and wind energy facilities dedicated to its warehouses and delivery networks, thereby significantly reducing its carbon footprint (Ali et al. 2023; Winston 2018). Employing electric delivery vans, particularly in densely populated urban areas, and installing extensive electric charging infrastructure at all warehouse locations would further advance sustainability goals. Additionally, warehouse rooftops can be equipped with solar panels to lower energy costs and environmental impact (Elkington 2018).

Blockchain technology presents another vital improvement opportunity. Although AWS Managed Blockchain offers substantial benefits, Amazon should establish an internal blockchain development team dedicated to crafting tailored blockchain solutions. A private blockchain network can ensure faster transactions, enhance data security, provide real-time inventory tracking, and maintain greater transparency across supply chains (Tapscott and Tapscott 2017; Treiblmaier 2018). Amazon's cloud computing capabilities would also benefit from expanding AWS infrastructure geographically. Establishing additional data centers in underserved regions can reduce latency, accelerate service delivery, and better accommodate the increasing global demand for cloud services (Armbrust et al. 2010; Buyya et al. 2009). Employee training programs focused on advanced AWS tools and certifications in cloud architecture and security would significantly bolster internal technical expertise and operational efficiency (Ross et al. 2017).

Leveraging AI to improve customer interactions is equally essential. AI chatbots equipped with advanced natural language processing and voice recognition capabilities can deliver more natural and effective communication, greatly enhancing customer service quality (Davenport and Ronanki 2018). Additionally, using AI to predict delivery windows by analyzing traffic patterns, weather conditions, and historical data can substantially improve delivery accuracy, customer satisfaction, and reduce the occurrence of missed deliveries (Brynjolfsson and McAfee 2017). Enhanced transparency regarding Amazon's environmental performance is critical for maintaining stakeholder trust. Publishing comprehensive and regular sustainability reports can foster greater confidence among consumers and investors. Furthermore, incentivizing suppliers who meet established sustainability benchmarks or utilize eco-friendly processes can further align Amazon's operations with ESG objectives (Winston 2018). Given the growing evidence that Environmental and Social risks adversely impact financial stability, as demonstrated through declining Altman's Z-Scores among KSE-30 firms (Chishti et al. 2024), it becomes imperative to examine how digital and technological innovations can facilitate ESG risk mitigation a perspective explored in the context of real-world corporate cases. With the increasing threat of cyberattacks, Amazon should expand its application of AI-based cybersecurity solutions, implementing sophisticated systems capable of detecting and preventing unauthorized data access proactively. Regularly scheduled cybersecurity training workshops for all employees can strengthen internal awareness and preparedness against evolving threats (Geissbauer et al. 2015). Exploring drone technology presents another promising avenue for Amazon's logistics enhancement. Investing in drones for rapid and eco-friendly delivery of smaller packages, particularly within urban areas, could considerably reduce vehicular congestion, lower fuel consumption, and expedite delivery times. Working collaboratively with local governments to establish designated drone delivery zones would be beneficial for efficient integration into existing urban frameworks (Boysen et al. 2019). Finally, adopting a circular economy model would significantly boost Amazon's sustainability profile. Utilizing reusable, recyclable, or biodegradable packaging, coupled with customer incentives for returning packaging materials, would dramatically reduce environmental waste and promote sustainable consumption (Elkington 2018). Moreover, intensified investments in AI research collaborations with leading academic institutions could further amplify Amazon's innovative capabilities, driving enhanced efficiencies in supply chain optimization, energy conservation, and inventory management (Davenport and Ronanki 2018). By embracing these strategic recommendations, Amazon can bolster its technological and operational excellence, enhance its ESG performance, and solidify its position as a trusted and responsible leader in the global market.

4 Conclusion

This chapter has examined Amazon's evolution from a simple online bookstore into a global technological powerhouse, emphasizing its comprehensive digital transformation and commitment to Environmental, Social, and Governance (ESG) performance. Initially burdened by a monolithic IT infrastructure, Amazon encountered substantial operational, security, and sustainability challenges. The transition toward advanced technologies such as artificial intelligence (AI), machine learning, Internet of Things (IoT), blockchain, and Amazon Web Services (AWS) effectively resolved these issues, significantly enhancing the company's operational efficiency, supply chain automation, warehouse management, and governance practices.

Amazon's investment in robotics and automation not only optimized logistical processes but also improved labor conditions by reducing manual strain and workplace hazards. The adoption of blockchain has reinforced transparency and security across its supply chain, establishing Amazon as a leader in integrating technology-driven solutions within operational frameworks. Additionally, AWS facilitated scalable real-time data management, thereby ensuring sustainable energy practices and reducing environmental impact. However, as highlighted, opportunities remain for further advancement. Amazon should continue to expand its warehouse automation strategies, employee well-being initiatives, and renewable energy investments to sustain and elevate its market leadership and ESG contributions. The development of private blockchain networks and enhancement of cybersecurity protocols represent additional areas for strategic improvement. By fostering transparency through sustainability reporting, incentivizing suppliers to adopt greener practices, and embracing emerging technologies such as drone deliveries and circular economy models, Amazon can further strengthen its position as an innovative, responsible, and sustainable global enterprise.

In essence, Amazon's case exemplifies how strategic technological adoption, integrated with robust ESG practices, can drive operational excellence and corporate responsibility. Moving forward, maintaining momentum in digital innovation and ESG commitment will be crucial for Amazon's sustained global leadership and the establishment of best practices that inspire industry-wide transformations.

References

Ali A, Khan MR, Ghouri AM (2023) Carbon neutrality concept and progress. In: Recent developments in green finance, green growth and carbon neutrality. Elsevier, pp 85–108

Armbrust M, Fox A, Griffith R, Joseph AD, Katz RH, Konwinski A et al (2010) A view of cloud computing. Commun ACM 53(4):50–58

Bharadwaj A, El Sawy OA, Pavlou PA, Venkatraman N (2013) Digital business strategy: toward a next generation of insights. MIS Q 37(2):471–482

Boysen N, De Koster R, Weidinger F (2019) Warehousing in the e-commerce era: a survey. Eur J Oper Res 277(2):396–411

Brynjolfsson E, McAfee A (2017) Machine, platform, crowd: harnessing our digital future, vol 4, No. 1. WW Norton & Company, New York, pp 1448–1476

Bughin J, Hazan E (2017) The new spring of artificial intelligence: a few early economies. VoxEU.org

Buyya R, Yeo CS, Venugopal S, Broberg J, Brandic I (2009) Cloud computing and emerging IT platforms: vision, hype, and reality for delivering computing as the 5th utility. Futur Gener Comput Syst 25(6):599–616

Chishti MF, Rao M, Raffat MW, Rafi S (2024) Estimating corporate risk and corporate value an application of Altman's Z-score on the Kse-30 index. Int J Contemp Issues Soc Sci 3(2): 2833–2841

Daugherty P, Wilson J (2018) Human + machine: reimagining work in the age of AI. Harvard Business Press

Davenport TH, Ronanki R (2018) Artificial intelligence for the real world. Harv Bus Rev 96(1): 108–116

Dolgui A, Ivanov D, Sokolov B (2018) Ripple effect in the supply chain: an analysis and recent literature. Int J Prod Res 56(1–2):414–430

Elkington J (2018) 25 years ago I coined the phrase "triple bottom line". Harv Bus Rev 25:2–5

Geissbauer R et al (2015) Industry 4.0: opportunities and challenges. Strategy& PwC, pp 1–52

Gul I, Raziq A, Khan MR (2024) Towards sustainable supply chains with Blockchain. In: Achieving secure and transparent supply chains with Blockchain technology. IGI Global Scientific Publishing, pp 18–37

Khan MI, Haq MAU, Khan MR, Ghouri AM, Farooqui R (2024a) Industry 4.0 or Pharma 4.0?: assessing suitability, benefits, challenges, and opportunities for healthcare supply chains. In: *Convergence of Industry 4.0 and supply chain sustainability*, pp 319–341

Khan MR, Masood F, Gul I, Khan MR (2024b) Blockchain technology in supply chain management: opportunities and challenges. In: *Convergence of Industry 4.0 and supply chain sustainability*, pp 275–295

Masood F, Khan NR, Khan MR (2024) Blockchain-based supply chain management: a challenge yet an opportunity to gauge. In: Achieving secure and transparent supply chains with Blockchain technology. IGI Global Scientific Publishing, pp 267–285

Min H (2010) Artificial intelligence in supply chain management: theory and applications. Int J Log Res Appl 13(1):13–39

Rajpoot MH, Raffat MW (2024) The AI-driven compliance and detection in anti-money laundering: addressing global regulatory challenges and emerging threats: AI-driven AML: compliance threat detection. J Comput Sci Appl (JCSA), ISSN: 3079-0867 (Online) 1(2)

Ross JW, Beath CM, Sebastian IM (2017) How to develop a great digital strategy. MIT Sloan Manag Rev 58(2):7–9

Tapscott D, Tapscott A (2017) How blockchain will change organizations. MIT Sloan Manag Rev 58(2):10–13

Treiblmaier H (2018) The impact of the blockchain on the supply chain. Supply Chain Manag Int J 23(6):545–559

Wessel M, Christensen CM (2012) Surviving disruption. Harv Bus Rev 90(12):56–64

Winston A (2018) The big pivot: radically practical strategies for a hotter, scarcer, and more open world. Harvard Business Review Press

Green Manufacturing: Challenges and Ethics

Zohaib Razzaq, Inayat Ul Haq Subhani, Muhammad Umar, and Muhammad Babar Usman

Abstract In today's era, green manufacturing has become increasingly critical due to environmental pressures, changing regulations, and new consumer pressures. This chapter highlights the diverse challenges and various ethical considerations faced by firms in adopting green manufacturing practices. This chapter also stipulate various technological and economic challenges that include high initial costs; retraining the workforce and integrating renewable energy systems; uncertain ROI and segmented market demand. This chapter illustrated regulatory complexity as the top most challenge as different standards, including the EU's tough carbon laws and China's changing life-cycle assessment framework, made compliance even more arduous, especially for multinational corporations. From an ethical standpoint, green manufacturing practices require an adherence to transparency, fair labor practices, and the management of products through their entire life cycle. Moreover, frameworks such as stakeholder theory and virtue ethics guide decision-making that strikes a balance between profitability and environmental well-being. Moreover, this chapter also highlighted that the rise of innovative solutions, AI-powered process efficiencies, sustainable materials, blockchain-facilitated traceability foreshadow a transformative future and operational tendencies toward collaboration across sectors, highlighting the necessity for agile and proactive approaches. Finally, this chapter contends that green manufacturing practices cannot be effectively implemented if organizations do not bridge with technological evolution, focus on ethical considerations, and inclusive policies as concentrating on these aspects will create a paradigm where economic performance and environmental protection are equal concerns. This chapter offers guidance for industry executives, policymakers, and scholars seeking to analyze the complexities of sustainable practices transformation.

Z. Razzaq · I. U. H. Subhani · M. Umar (✉) · M. B. Usman
Department of Management Sciences, Lahore Garrison University, Lahore, Pakistan
e-mail: dr.zohaibrazzaq@lgu.edu.pk; dr.inayatulhaq@lgu.edu.pk; muhammadumar@lgu.edu.pk; fa-24-ph.dms-006@lgu.edu.pk

M. Venkatesh et al. (eds.), *Green Manufacturing Technologies in Industry 5.0*,
https://doi.org/10.1007/978-981-95-0363-6_11

1 Introduction

In today's era, green manufacturing principles have become a crucial business strategy that ensures environmental responsibility, operational efficiency, and long-term profitability. Green manufacturing includes a wide variety of practices such as the utilization of eco-friendly raw materials, energy-efficient production, waste reduction techniques, and implementation of circular principles that includes reduce, reuse, and recycle (Umar et al. 2023). This technique assimilates green principles into production processes aimed at minimizing negative ecological impacts and confirming financial viability. Due to increasing ecological concerns, climate change, overuse of resources, and growing demand for sustainable goods, there is growing global pressure to reinforce green measures in production firms (Machado et al. 2020; Khan et al. 2024).

The evolution of green manufacturing is exhilarated by regulatory pressures, market demand, and corporate social responsibility (CSR) commitments. To ensure industrial practices are sustainable, governments and international organizations have established strict environmental regulations and policies forcing businesses to innovate and comply with materials and manufacturing processes aligned with green manufacturing standards (Dzage et al. 2024). Concurrently, consumers are increasingly in need for eco-friendly products, forcing companies to pursue green manufacturing to gain a competitive edge. In addition, ethical considerations have crucial role for businesses to minimize environmental harm and promote social and economic welfare (Khan et al. 2021). From an economic prospect, green manufacturing saves money by minimizing waste, maximizing resource usage, and plummeting energy usage. More importantly, organizations that adopt sustainable practices tend to enjoy greater brand reputation, enhanced customer loyalty and stake holders trust (Umar et al. 2023). Moreover, the transition to green technologies—for instance, the integration of renewable energy and smart manufacturing systems—is also in line with the wider global sustainability agenda, strengthening industries as essential contributors to long-term environmental performance (Machado et al. 2020).

This chapter discussed the challenges that various industries encounter while adopting green manufacturing practices and the prevailing ethical considerations as well as provides an inclusive overview of green manufacturing by addressing technological advancements, implications to economy, regulations, and case studies of successful implementation of green manufacturing practices. The argument will strike a balance between cutting-edge industrial practices and ethical considerations, giving concrete examples and a framework for actionable insights to firms looking to transform operations to be increasingly more sustainable.

This chapter aims to identify the complexities and challenges in adopting green manufacturing practices and emphasize the ethical responsibilities that should be coupled with the transition toward a sustainable future. It can be effectively utilized by industry professionals, policymakers, and scholars who want to study about the practical and theoretical sides of sustainable manufacturing. The importance of this study is to influence future practices and policies that promote sustainable development in the industrial sector.

This chapter is structured as follows; first, it provides the readers with the historical context and the technological developments that lead to green manufacturing. Then it investigates the technological, economic, and regulatory challenges. Afterwards, a detailed discussion of the ethical implications of green manufacturing is provided. Finally, this chapter concludes with some practical case studies that demonstrate the implementation of sustainable strategies in green manufacturing along with their future implications.

2 Literature Review

2.1 Historic Development of Green Manufacturing

In the late twentieth century, environmental movements initiated the concept of green manufacturing which was initially aimed at compliance with environmental regulations, the approach has developed into a broader commitment to sustainable development. This major change was significantly influenced by the issuance of works like "Our Common Future", which initially articulated the concept of sustainable development and emphasized the need to mainstream issues of the environment in economic policy (Brundtland 1987).

During the late 1980s and the early 1990s, businesses started adopting "cleaner production" practices that minimized waste and emissions in response to regulatory pressures and public interest in environmental issues. ISO 14001 became the international standard for environmental management systems in 1996, and marked an important formalization of green practices in manufacturing (Taylor and Morgan 2018).

2.2 Case of Early Adopters

Observing the experiences of some early adopters of green manufacturing can provide some of the challenges faced and successes that can be achieved from implementing these practices. As a concrete case, one major automotive manufacturer established a zero-waste policy for its factories (Johnson and Wang 2020). The company was able to reduce its environmental footprint and significantly saved costs by reengineering manufacturing parts to utilize recyclable materials and invest in advanced waste recovery systems.

Another case, a textile manufacturer that switched to using organic cotton and non-toxic dyes. Switching to eco-friendly materials not only reduced their products' environmental impact, but also had a positive effect on their brand image and customer loyalty, revealing that business success can be aligned with green practices (Garcia and Patel 2021).

2.3 Technological Developments Supporting Green Manufacturing

The implementation of green manufacturing practice has become adoptable and efficient due to technological innovations (Gul et al. 2024). Significantly, the environmental footprints of manufacturing processes have reduced considerably due to cleaner production technologies, waste minimization practices, and energy-efficient systems. The advent of precision manufacturing technologies has further increased material efficiency and minimized waste. Furthermore, innovation in renewable energy technologies like solar and wind has led to manufacturing plants' minimal dependence on fossil fuels.

As revealed by Chen and Zhao (2021), Masood et al. (2024) and Khan et al. (2024), the complementary use of digital technologies such as the Internet of Things (IoT), blockchain technology (BT), and artificial intelligence (AI) has enhanced energy optimization, resource flow, and environmental performance within the manufacturing process, usually named as Industry 4.0.

3 Issues in Green Manufacturing Implementation

3.1 Technological Challenges

Implementing green technologies into current manufacturing systems is a huge challenge (Anderson and Kumar 2020). Investments are usually substantial in terms of capital to enable equipment upgrades to energy-efficient machinery and to include advanced manufacturing technologies such as additive manufacturing (3D printing) (Chen et al. 2023). Moreover, existing workforce might not be trained to use new technologies with high-tech tools implemented.

In addition, the reliability and maintenance of modern environmental technologies also add more challenges to production processes. For instance, renewable energy sources like solar or wind power have variability in energy supply that limits adoption and requires backup systems or integrated with other traditional energy sources to maintain energy production levels overviewed.

3.2 Economic Challenges

Green manufacturing has several economic challenges. Green manufacturing requires enormous investments in technology, waste management systems, and sustainable sourced materials. For SMEs, which typically operate on tight margins, it can be a daunting task. One of the major barriers is high initial investment along with financial constraints making it difficult for SMEs to adopt technology. This study postulates that these investments have long payback duration, making them unattractive for companies caused by uncertainty regarding return on investment. Another study confirmed the previous notion and indicates that financial barriers to adoption related to restriction in access to finance and high implementation costs may result in long payback periods of multiple years especially for small firms (Ali et al. 2023).

Market dynamics also make an important difference in regions, a report of Market US (2023) highlights that demand is driven by consumer awareness and policy support in North America, Europe, and Asia Pacific; while the rest of the regions fall behind, generating market uncertainty for manufacturers. Although sectors such as cosmetics and pharmaceuticals experience ever-increasing demand: but predictability in demand is limited, due to market fluctuations.

There is a rising demand for green products, but it is not uniform across different sectors and regions. Furthermore, companies are at a disadvantage, due to competitive pressures from companies who do not believe in sustainable future and can provide lower priced products as they do not have to pay for the additional costs associated with green manufacturing. Moreover, government incentives and subsidies can mitigate some of these costs, yet their availability and sufficiency vary significantly across countries and different regions, adding an additional layer of complexity to the economic feasibility of green manufacturing (Greene and Patel 2022).

3.3 Market and Regulatory Compliance Challenges

Green manufacturing in the context of regulation and compliance is related to several challenges. Manufacturers need to act in accordance with a wide range of environmental regulations, which can fluctuate extensively between jurisdictions. Compliance, for instance, involves extensive documentation, periodic environmental audits, and continuous investment in pollution-controlling technologies (Lee et al. 2018). Primary regulatory challenge for multinational corporations (MNCs) includes variation in environmental regulations across different jurisdictions having different laws. Sustainability rules and standards for each country or region particularly vary significantly, resulting in a diverse compliance requirement. For instance, regulatory system in China, Germany, and the USA. Research identified a lack of clear green product evaluation standards in China, especially in terms of life cycle assessment (LCA) methods, including system limitations and life cycle stages (Chang et al. 2014). Contrarily industrial migration causes Germany and the USA find challenging to form a closed-loop national manufacturing industry chain, while in both countries life cycle evaluation within factory boundaries has not been concretely realized to a large extent. This variability requires MNCs to organize technologies and practices capable of meeting the stringent regulatory criterion in their global operations (Chen et al. 2023). MNCs frequently implement indirect strategies, such as shifting facilities to countries with minimal laws or accomplish strict regulations through green innovation. The same dual approach highlights the complexity and cost of responding to different regulations, relying on adaptable strategies and investments in order to maintain standards.

Consulting and Legal Compliance Costs for compliance with green manufacturing regulations can be extensive. Financial Times (2024) reports that the cost of European Union (EU) green regulations is increasing substantially, with few companies claimed to experience annual costs ranging to hundreds of thousands of dollars. Therefore, this financial burden is essential to avoid facing either fines or legal issues, especially in heavily regulated markets such as the EU, thus increasing operational complexity. Green manufacturing standards and regulations are constantly changing, necessitating manufacturers to remain apprised and flexible. China's performance evaluation system of local government officials views economic growth and environmental protection equally, requiring a continuous adjustment to new policies. In this variable landscape, proactive measures are critical for manufacturers, and manufacturers must aim for regulatory prospect and investment in regulatory intelligence and provision planning.

3.4 Limited Awareness and Understanding of Sustainability

One of the biggest challenges is that a lot of stakeholders believe that sustainable practices are not that important and they are resistant in adapting environment-friendly methods. These not only lead to wasting resources but also missing out on potential savings (Ferronato and Torretta 2019; Tortorella et al. 2023). This barrier can be addressed through education and awareness programs (Maryville 2019).

4 Ethical Deliberations in Green Manufacturing

4.1 Corporate Responsibility

In the context of green manufacturing, corporate responsibility requires more than following laws and regulations. It requires action toward environmental stewardship by reducing adverse effects on the environment and making sure that all business practices are innovative and sustainable (Khan et al. 2021). It includes everything from resource procurement and energy consumption, to managing waste and controlling emissions. Ethical responsibilities to stakeholders about environmental practices and the effects of production processes require transparency. Companies are more frequently held accountable by investors, consumers, and community groups who insist on the highest standards of environmental and social responsibility. CSR reports and third-party sustainability certifications have served as verification substitutes.

Being an ethical corporate citizen also involves a similar level of transparency with stakeholders on environmental policies and the real-life effects of production practices. Companies are no longer assessed only in terms of their financial performance; companies are now being assessed by their ability to create transparency and authenticity around their sustainability efforts. According to Hsu and Bui (2022), expectations from consumers, investors, and community service groups for high standards of environmental and social responsibility are increasing day by day. Consumers, for instance, are leaning toward brands that offer tangible proof of environmentally friendly operations like minimizing packaging waste and reducing carbon emissions, whereas investors focus primarily on companies with extensive environmental, social, and governance (ESG) measures to reduce long-term risks (Govindan 2024).

Third-party certifications including Leadership in Energy and Environmental Design (LEED) standard and the Global Reporting Initiative (GRI) aim to provide independent validation of a company's sustainability claims, thus increasing trust and competitiveness (Bag et al. 2022). These tools not only act as accountability mechanisms but also encourage innovation, as companies struggle to match or even exceed industry benchmarks. Some manufacturers, for instance, utilize advanced technologies such as machine learning to maximize the quality and marketability of remanufactured circular products by aligning sustainability with customers' expectations and the priorities of their investors (Govindan 2024).

4.2 Labor Practices

Green manufacturing is similarly relevant when it comes to ethical labor practices. It is important that green initiatives do not fragile workers' rights and safety, such as ensuring proper fair labor conditions through guaranteeing fair wages, safe working conditions, and employees' training in new technologies or processes (Sharma et al. 2023). Therefore, sustainable manufacturing can be an exercise for better working situations and prioritize employees' health and safety (Akundi et al. 2022). In addition, in green manufacturing, ethical labor practices not only focus on the environment, but also on the social impacts of manufacturing decisions. For instance, when sourcing its raw materials, a company must not only focus on the environmental impact, but also on human rights as well as the workplace conditions of the facilities involved in the supply chain, particularly located in countries with lenient labor laws and regulations.

4.3 Product Life Cycle

Ethically, companies must step back and examine the life cycle of their products from design to disposal. This means creating products that are durable, easy to repair, and recyclable, minimizing their environmental footprint and waste. Other ethical concerns involve the use of sustainable and eco-friendly materials to keep the products from damaging the environment at the time of disposal. Moreover, companies need to provide information of their products regarding environmental impacts over their entire life cycle. Such transparency can empower consumers to make responsible choices and lead to sustainable consumption practices.

4.4 Data Privacy and Confidentiality

The incorporation of technology including blockchain and AI requires strong precautions to secure customer data. Such breaches might damage trust, as experienced with leaks such as AMCA in 2018 (Agbo et al. 2019). Data should be encrypted and access has to be limited (Yaqoob et al. 2021).

4.5 Equitable Access to Sustainable Technologies

Green manufacturing innovations like 3D-printed personalized medications might be expensive for low-income areas to take advantage of. Ensuring that all ethnicities have equitable access is the ethical imperative for global health equity (Andreadis et al. 2022).

4.6 Environmental Justice

Sustainable practices must not shift environmental burdens to politically weak or economically vulnerable populations. For instance, production processes should minimize toxic waste through adoption of green chemistry having positive effects and resulting in environment protection (Haq et al. 2016; Puhlmann et al. 2024).

4.7 Transparency and Accountability

Blockchains part in battling the issue of counterfeit drugs would require that the supply chain should be transparent; however, stakeholders should not attain monopolist power over the data. Decentralized systems that empower everyone involved are needed for ethical implementation (Zakari et al. 2022; Khan et al. 2023).

5 Practical Case Studies

5.1 Textile Industry: A Case Study of Patagonia's Supply Chain Transparency

Patagonia, an outdoor clothing manufacturer is known for its sustainability. The firm has launched efforts like utilizing recycled materials for its clothing appearances and ensuring fair labor practices throughout its supply chain. Patagonia's Worn Wear program encourages individuals to repair and recycle gear, rather than buy new, thus reducing waste and advocating for a circular economy in the textile industry.

Patagonia focuses on creating durable, versatile, and repairable designs. Raw materials are procured for recycling, renewable, organic, and regenerative purposes. Ethical sourcing ensures compliance with human rights and animal welfare. Patagonia emphasizes energy efficiency, worker well-being, and supply chain transparency. While the brand has implemented more eco-friendly practices by reducing hazardous materials used in its products, industry-standard practices around Durable Water Repellent (DWR) coatings remains a challenge. Meanwhile, the company follows strict sustainability norms amid at improvised global environmental legislation. Patagonia is transparent, and its stakeholders get information on social and environmental impacts regularly. Patagonia's use of green manufacturing practices stands as a strong illustration of how companies can be profitable while making decisions that consent fewer harmful environmental and social footprint. Essentially, the company's focus on environmentally conscious sourcing, its production methods, labor practices, and circular economy principles represent a leading illustration of sustainable supply chain management.

5.2 Electronics Industry: The Environmental Responsibility of Apple

Apple has set standard in electronics industry to commit to sustainability through the significant reduction of its carbon footprint. The company also has programs to ensure all its operations utilize 100% renewable energy, and aims for net-zero climate impact by 2030 for all products and supply chain life cycle. Therefore

Apple also prioritizes material recovery and the recycling of its devices, using robots such as Daisy to take apart their iPhones and recover valuable material (ESG Today 2024).

6 Green Manufacturing: Key Emerging Trends & Innovations

6.1 Innovations in Sustainable Materials and Circularity

The manufacturing sector is witnessing a radical shift in its paradigm of sustainable material usage, fueled by the growing urgency of circularity. This move is toward material choice, product design, and end-of-life management rather than a recycling process. A key part of this journey is the emerging field of bio-based materials. Consequently, bio-based materials sourced from renewable feed stocks (plants and microorganisms) offer a viable alternative to traditional materials derived from fossil fuels. For illustration, advances in biopolymer technology are producing biodegradable plastics with properties similar to traditional plastics that can help tackle the ongoing challenge of plastic waste. At the same time, advanced recycling technologies are making it possible to recover valuable materials from multifarious waste streams. Chemical recycling, for instance, is breaking down polymers into the monomers from which they were made, enabling to create virgin-quality plastics from reprocessed sources (Ma et al. 2024).

Digital technologies are becoming essential enablers of circularity (Rafi and Sulman 2025). In fact, blockchain technology is now developing material traceability from the logistics platform-supply chain, which is one of the auxiliary means to truly comprehend transparency and responsibility after recycling. This transparency builds trust between stakeholders and rewards the investment in embracing the circular practices. In addition, life cycle assessment (LCA) tools are becoming more advanced, considering more extensive datasets and improved methods. To make informed choices about material use and design, manufacturers have to balance this and obtain a better holistic picture of the product life cycle emissions and other environmental impacts. Industrial symbiosis, in which waste from one industrial process becomes a useful input to another, is building real momentum. It is more resource- and waste-efficient and better for the planet. On an industry-wide level, companies are also looking more toward design for disassembly, to ensure the recovery of component parts for reuse at the end of the product life (Korhonen et al. 2022).

This drive is not just about climate change concerns, even economic considerations are also very much at play. Resource scarcity, volatile commodity prices, and stringent regulatory pressures are driving manufacturers toward more sustainable material practices. Adopting circularity allows companies to lessen their dependence on virgin resources, decrease supply chain risks, and improve competitiveness (Nikseresht et al. 2024).

6.2 Digitalization and Smart Technologies for Green Manufacturing

Digitalization is being integrated with smart technologies such as the Internet of Things, to fundamentally change the green manufacturing landscape with unprecedented levels of efficiency and sustainability. The introduction of Industry 4.0 concepts, notably artificial intelligence (AI), the Internet of Things (IoT), and digital twins are initiating a paradigm shift in the production processes (Umar et al. 2023;). Machine learning, in turn, helps streamline production activities through optimal scheduling, waste reduction, and energy conservation—transforming factories into Ancillary factories. AI tools smart enough to look at terabytes of data can spot patterns, as well as anomalies (Raffat & Ahmad, 2025)—allowing for predictive maintenance and mitigating the risk of equipment's breaking down failure in the first place. The use of IoT sensors is more important to monitor factories in real time, enabling manufacturers to be more precise about streamlining their production processes. Their real-time data creation through data extraction assists in resolving coordination issues by providing a comprehensive view of processes, which is vital for integrating diverse logistics and transport systems. This immediacy allows the manufacturer to make necessary adjustments to enhance resource usage and decrease the environmental impact on the system (Bai et al. 2022).

Another revolution to innovate product and process design is digital twin technology. Manufacturers leverage digital twins to simulate and fine-tune production processes before any physical implementation through virtual replicas of real-world assets. This also reduces the need for physical prototypes, which lowers material waste and energy consumption. Moreover, digital twins allow predictive maintenance, minimizing interruption and prolonging the life span of the (industrial) outfit. Smart grid technologies and energy management systems enable the adoption of renewable energy sources. It can power the whole manufacturing facility including linking distributed renewable energy generation (solar, wind, etc.) to its energy supplies. Data analytics is used through energy management systems providing information about energy consumption for efficient and effective utilization (D'Eusanio and Petti 2024).

Automation and robotics use enhances manufacturing process agility mitigating the risk of resource wastage. Cloud computing and big data also enable supply chain optimizations that avoid the unnecessary transportation of materials. This digital transformation in manufacturing will not only lead to better environmental performance but also greater productivity, enhanced cost efficiency, and increased innovation. Adoption of smart technologies enables manufacturers to realize a sustainable and competitive future (Ye et al. 2023).

6.3 Manufacturing Decarbonization and Energy Transition

In green manufacturing evolution, decarbonization and the energy transition play an integral part toward a sustainable future, which entails a holistic approach that encompasses minimizing greenhouse gas emissions associated with the entire manufacturing value chain. This involves a pivotal switching to renewable energy sources. Manufacturers are investing in on-site solar and wind power generation, as well as procuring renewable energy from the off-site efficiency sources. The electrification of industrial processes is also growing rapidly, as companies switch out fossil fuel-powered equipment to electric versions. In addition, the option of other fuels such as green hydrogen is gaining even more popularity. Green hydrogen, produced from renewable energy sources, offers a carbon-neutral alternative to traditional fossil fuels.

Energy efficiency also represents a key aspect of decarbonization. Manufacturers are adopting advanced energy management systems and data to optimize energy usage. Sensors and monitors that track energy flows in real time pinpoint opportunities for improvement and automate energy-saving steps. Many manufacturers are focusing on their emissions, compelling all the stakeholders to mitigate emissions across their value chain. This includes policies that encourage sustainable sourcing, optimize transportation routes, and engage suppliers in their own decarbonization strategies. Carbon capture, utilization, and storage (CCUS) technologies are potentially useful to limit emissions in some of these hard-to-abate industrial sectors. CCUS technologies, where CO2 emissions from industrial processes are captured and utilized for production via either underground storage or producing other valuable products.

Sustainability, with an emphasis on ESG (Environmental, Social, and Governance) goals (Ullah et al. 2024), is another motivator for decarbonization initiatives. It is increasingly demanded from companies by investors, customers, and other stakeholders to prove their commitment to sustainability. Vow to reduce emissions, manufacturers report their environmental performance and integrate sustainability into their core business strategies. There are also strong pushes toward a circular economy, which reduces the need for the extraction of raw materials and thus reducing the energy which is needed for creation and manufacture of those products ((IEA 2023).

7 Future Perspectives of Green Manufacturing

7.1 Bio-Integrated and Regenerative Manufacturing

Bio-integrated/regenerative manufacturing is the future of green manufacturing. This means going beyond simply halting harm to actually repairing and restoring natural systems. Bio-manufacturing—exploiting living organisms or biological processes—will also become more mainstream. It comprises making materials and products via microbial fermentation or enzymatic synthesis and cellular agriculture. These technologies hold the promise to develop materials with novel properties/functionalities and to reduce dependence on traditional resource-intensive means of manufacturing. Regenerative manufacturing will create closed-loop systems that rely on natural models of operation, where particles return to the same ecosystem for lifespan extending purposes. All of them include the manufacture of self-healing materials, bio-based composites, or the so-called living architecture, capable of adapting and evolving over time (Bai et al. 2022).

In addition, synthetic biology will become more prominent to create microorganisms that can transform renewable feedstock's directly into valuable chemicals and materials. This will enable the development of sustainable substitutes to fossil fuel-based products. Manufacturing processes will also increasingly incorporate ecological principles, with businesses structuring their operations to imitate natural cycles and enhance biodiversity. Further refinement of "cradle-to-cradle" design will become a "beyond 2023" initiative. Additionally, this bio-integrated and regenerative manufacturing will be enabled by the use of digital technologies. Leveraging AI and machine learning will enhance biological processes, creating new bio-materials and predictive models of interaction between ecosystems. Bio-based supply chains will take advantage of blockchain technology to improve traceability and transparency confirming the sustainable and ethical sourcing of the material. Eventually, factories will be embedded into their environments, functioning with natural systems toward a sustainable, circular future (Zhou et al. 2024; Korhonen et al. 2022).

7.2 *Distributed Manufacturing Network (Hyper Local)*

Like bio-integrated/regenerative manufacturing, hyper-localized distributed manufacturing networks have also been seen as the future of green manufacturing. That is, moving away from large, centralized production facilities to smaller and more nimble manufacturing "hubs" that are located nearer to consumers. This not only reduces transportation distances and carbon emissions, but also contributes to supply chain resiliency. The ability to print on-demand is also critical to making distributed manufacturing a reality, which will be facilitated by additive manufacturing (aka 3D printing). Such components will shape the future of adaptable and flexible production, such as micro-factories and modular manufacturing systems. These manufacturing hubs will be connected through digital platforms creating collaborative networks capable of sharing resources and optimizing production. If there is one certain thing, it is that localized material sourcing will become far more crucial, with firms utilizing what is available in their region, rather than relying on a global supply chain. Urban manufacturing will also undergo a revival, encompassing factories being incorporated into urban locales in order to minimize transportation expenses and promote local economies (Luo et al. 2025).

The cottage industry of micro manufacturing will be facilitated by "maker spaces," and hyper/localized productions will take off, allowing individuals and small businesses to create their own products. Manufacturing will have potential to be more widely accessible through digital fabrication tools and open-source designs. Digital twins and simulation tools will help in optimizing distributed manufacturing networks to enable efficient allocation of resources and production planning. Blockchain technology will promote supply chain transparency, enabling greater transparency and traceability between suppliers and consumers, which means companies will code on their supply chain and ensure that the materials are ethically and sustainably sourced. This will reduce the environmental impact by creating a more resilient and sustainable production network with reduced inter-industrial transportation and promote local economic development (Wang et al. 2025; Ok and William 2025).

7.3 AI and Autonomous Systems in Integrated Production

Keywords such as autonomous systems green manufacturing and artificial intelligence (AI) will play a transformative role in manufacturing production processes which will be optimized using AI-powered systems creating less waste and using less energy. By getting familiarity with big data: Machine learning algorithms that analyze large datasets to find patterns and anomalies will support predictive maintenance and prevent devices from breaking down. Automation will generate autonomous robots and drones that will automate tasks creating efficiency and safety. Creating digital twins will make it viable to virtually simulate production processes to optimize designs and operations before transitions to physical. The use of AI will also impact the development of smart materials that can change in response to their environment leading to improved product lifespan and less waste (Vijay Kumar and Shahin 2025).

With supply chain management, AI will allow monitoring of materials as well as products as they move worldwide, making the entire process transparent and sustainable. AI-driven systems will bring efficiency to logistics and transportation, cutting carbon emissions and minimizing waste. Moreover, intelligent energy management systems utilizing AI will be created to align energy consumption and assimilate renewable sources. Autonomous systems will also be employed for environmental monitoring and remediation, such as robots and drones collecting data and performing tasks in hazardous or inaccessible environments. Biotech is now the gold standard in the manufacturing industry. AI will be leveraged to supplement life cycle assessments; creating more accurate, more available, and more relevant data about our environmental impact. Blending AI with intelligence will build an integrated closed-loop system of factories in the new manufacturing paradigm (Vishwakarma et al. 2025).

8 Conclusion

This chapter has navigated the historical evolution of green manufacturing with all the current challenges and ethical considerations that frame this crucial field. Explaining the simple compliance with environmental regulations and complex practices that are sustainable shows a radical change in the way industries do business, instead of being purely about economics, they are increasingly sympathetic to environmental and social priorities. The cases of businesses like Patagonia and Apple emphasized that adopting green practices do not only address damages posed but they also improve market standing and brand recognition. The aforementioned cases emphasized the practical benefits and challenges in practical sustainability practice and demonstrate how models become exhibited actual situations. Additionally, these cases illustrate the approach that green manufacturing initiatives are being extensively involved around the world and how limited efforts contribute to global sustainability initiatives such as the United Nations Sustainable Development Goals (SDGs).

For future interpretations, looking forward how the combination of emerging tools including AI, IoT, big data analytics, and new material sciences with green manufacturing will change product design, production, and recycling process (Rafi and Sulman 2025). Critical focus on innovations in sustainability can drive sustainable practices to be more efficient, cost-effective, and scalable across industries. Nevertheless, regulatory frameworks change and increase in consumer awareness, manufacturers need to be flexible, constantly modifying their strategies to fulfill new regulations and market demands. As regulations tighten and supply chain transparency becomes critical prospect, companies will face pressure to comply and innovate collectively. Conversion of manufacturing processes to low carbon footprints is associated with several challenges ranging from economic, technological, and organizational fragment. A successful adoption of sustainable manufacturing requires a comprehensive strategy that includes ethical considerations, encourages human-centered design, empowers association, and precautions that economic development is consistent with environmental protection and social welfare.

References

Agbo C, Mahmoud Q, Eklund J (2019) Blockchain technology in healthcare: a systematic review. In: Health Care, vol 7, No. 2. MDPI, p 56

Akundi A, Euresti D, Luna S, Ankobiah W, Lopes A, Edinbarough I (2022) State of Industry 5.0 analysis and identification of current research trends. Applied system. Innovations 5(1):Article 27. https://doi.org/10.3390/asi5010027

Ali A, Khan MR, Ghouri AM (2023) Carbon neutrality concept and progress. In: Recent developments in green finance, green growth and carbon neutrality. Elsevier, pp 85–108

Anderson P, Kumar S (2020) Challenges in green manufacturing: cost, workforce, and technological adaptation. J Sustain Manuf 12(3):210–225

Andreadis I, Gioumouxouzis CI, Eleftheriadis GK, Fatouros DG (2022) The advent of a new era in digital healthcare: a role for 3D printing technologies in drug manufacturing? Pharmaceutics 14(3):609

Bag S, Dhamija P, Bryde DJ, Singh RK (2022) Effect of eco-innovation on green supply chain management, circular economy capability, and performance of small and medium enterprises. J Bus Res 141:6072

Bai C, Dallasega P, Orzes G, Sarkis J (2022) Industry 4.0 technologies assessment: a sustainability perspective. Int J Prod Econ 247:108456

Brundtland GH (1987) Our common future. Oxford University Press

Chang D, Lee CKM, Chen CH (2014) Review of life cycle assessment towards sustainable product development. J Clean Prod 83:48–60

Chen X, Li Z, Gallagher KP, Mauzerall DL (2023) Status and challenges of green manufacturing: comparative analysis of China and other countries. Resour Conserv Recycl 197:107051. https://doi.org/10.1016/j.resconrec.2023.107051

Chen L, Zhao X (2021) Technological innovations in green manufacturing: from waste reduction to industry 4.0. Ind Ecol Rev 34(2):198–215

Companies count the cost of compliance with green regulation (2024, January 15) Financial Times. https://www.ft.com/content/276028dc-a492-44ec-9eee-58d9fb51c159

D'Eusanio M, Petti L (2024) Blockchain technology and social life cycle assessment: synergies and implications. Int J Life Cycle Assess:1–12

Dzage EJ, Hussain MR, Dapaah PO, Mustapha Y (2024) Corporate social responsibility, sustainable environmental practices and green innovation; perspectives from the Ghanaian manufacturing industry. Int J Corp Soc Responsib 9:4. https://doi.org/10.1186/s40991-024-00090-2

ESG Today (2024, February 21) Apple reduces carbon footprint of new iPhone by 30%. ESG Today. https://www.esgtoday.com/apple-reduces-carbon-footprint-of-new-iphone-by-30/

Ferronato N, Torretta V (2019) Waste mismanagement in developing countries: a review of global issues. Int J Environ Res Public Health 16(6):1060

Garcia M, Patel R (2021) Sustainable textile manufacturing: innovations and consumer perceptions. J Sustain Bus 29(4):310–328

Govindan K (2024) Unlocking the potential of quality as a core marketing strategy in remanufactured circular products: a machine learning enabled multi-theoretical perspective. Int J Prod Econ 269:109123. https://doi.org/10.1016/j.ijpe.2023.109123

Greene WH, Patel JK (2022) Economic impediments to green manufacturing: a sectoral analysis. Econ Model 93:346–360

Gul I, Raziq A, Khan MR (2024) Towards sustainable supply chains with Blockchain. In: Achieving secure and transparent supply chains with Blockchain technology. IGI Global Scientific Publishing, pp 18–37

Haq MA, Khan NR, Parkash R, Jabeen A (2016) Impact of JIT, waste minimization, and flow management on operational performance of manufacturing companies. Qual-Access Success 17(153):48

Hsu C-C, Bui T-L (2022) Corporate social responsibility and consumer behavior: a meta-analysis of sustainable practices in manufacturing. Sustainability 14(19):12345. https://doi.org/10.3390/su141912345

International Energy Agency (IEA) (2023) Net zero by 2050: a roadmap for the global energy sector. IEA

International Organization for Standardization (1996) ISO 14001: environmental management systems specification with guidance for use. ISO

Johnson K, Wang H (2020) Green manufacturing strategies: lessons from early adopters. Routledge

Khan MR, Masood F, Gul I, Khan MR (2024) Blockchain technology in supply chain management: opportunities and challenges. In: Convergence of Industry 4.0 and supply chain sustainability, pp 275–295

Khan SAR, Yu Z, Umar M (2021) How environmental awareness and corporate social responsibility practices benefit the enterprise? An empirical study in the context of emerging economy. Manag Environ Qual Int J 32(5):863–885

Khan MR et al (2023) Blockchain supply chain management and supply chain sustainability. In: *Blockchain driven supply chain management. Springer Nature Singapore, Singapore, pp 155–180

Korhonen J, Honkasalo A, Seppälä J (2022) Circular economy: the concept and its limitations. Ecol Econ 198:107452

Lee J, Kao HA, Yang S (2018) Industry 4.0: smart manufacturing for the future. Engineering 4(1): 11–20. https://doi.org/10.1016/j.eng.2018.02.002

Luo Q, Deng L, Zhang Z, Wang H (2025) The impact of digital transformation on green innovation: novel evidence from firm resilience perspective. Financ Res Lett 74:106767

Ma Y, Feng GF, Chang CP (2024) From traditional innovation to green innovation: how an occurrence of natural disasters influences sustainable development? Sustain Dev 32(3): 2779–2796

Machado CG, Winroth M, da Silva EHDR (2020) Lean-green manufacturing practices and their link with sustainability: a systematic literature review. J Clean Prod 253:119848. https://doi.org/10.1016/j.jclepro.2019.119848

Market US (2023) Green technology and sustainability market size & trends, growth analysis & forecast. https://market.us/report/green-technology-and-sustainability-market/

Maryville University (2019) The importance of environmental awareness when running a business

Masood F, Khan NR, Khan MR (2024) Blockchain-based supply chain management: a challenge yet an opportunity to gauge. In: Achieving secure and transparent supply chains with Blockchain technology. IGI Global Scientific Publishing, pp 267–285

Nikseresht A, Golmohammadi D, Zandieh M (2024) Sustainable green logistics and remanufacturing: a bibliometric analysis and future research directions. Int J Logist Manag 35(3):755–803

Ok E, William J (2025) Harnessing the power of digitalization to drive circular economy in supply chain management

Puhlmann N et al (2024) Designing greener active pharmaceutical ingredients. Eur J Pharm Sci 192: 106614

Raffat MW, Ahmad A (2025) Enhancing anti-money laundering systems with machine learning: A comparative analysis of supervised models. Journal of Computational Informatics & Business, 2(2):1–7

Rafi S, Sulman M (2025) Post-pandemic insights: evaluating the impact of big data analytics, circular economy practices, and digital marketing on firm performance. J Comput Inf Bus 1(1): 8–16

Sharma D et al (2023) A comprehensive study on Industry 4.0 in the pharmaceutical industry for sustainable development. Environ Sci Pollut Res Int 30(39):90088–90098

Taylor E, Morgan J (2018) From regulations to sustainability: the evolution of green manufacturing. J Environ Manag Sustain 22(3):455–473

Tortorella GL, Prashar A, Antony J, Fogliatto FS, Gonzalez V, Godinho Filho M (2023) Industry 4.0 adoption for healthcare supply chain performance during COVID-19 pandemic. Oper Manag Res 17(2):389–405

Ullah A, Waqar M, Nazir SS, Adnan A, Khan MA, Raffat MW, Rafi S (2024) The impact of information communication technology and financial innovation on the financial performance of Chinese commercial banks. Remit Rev 9:364–383

Umar M, Yu Z, Muhammad Z, Khan SAR (2023) Unleashing the role of Blockchain technology and government support in green supply. LogForum 19(2):195–209

Vijay Kumar V, Shahin K (2025) Artificial intelligence and machine learning for sustainable manufacturing: current trends and future prospects. Intell Sustain Manuf 2(1):10002

Vishwakarma LP, Singh RK, Mishra R, Kumari A (2025) Application of artificial intelligence for resilient and sustainable healthcare system: systematic literature review and future research directions. Int J Prod Res 63(2):822–844

Wang J, Dong TT, Peng DH (2025) Green innovation in human-centric smart manufacturing. Manag Decis

Yaqoob I et al (2021) Blockchain for healthcare data management: opportunities, challenges, and future recommendations. Neural Comput Appl 34(14):11475–11490

Ye P, Ray D, Song C (2023) Circular economy for CO2 utilization and hydrogen production in China. In: Cyclic economy, p 37

Zakari N et al (2022) Blockchain technology in the pharmaceutical industry: a systematic review. PeerJ Comput Sci 8:e840

Zhou J, Li X, Zhang Z, Hou T, Xu J, Wang Y et al (2024) Bio-based and bio-degradable nanofiber materials: a sustainable platform for energy, environmental, and biomedical applications. Chem Eng J 491:152105.

Future Trends and Research Directions in Green Manufacturing

Prio Utomo, Tat-Huei Cham, Ahsan Anwar, Shukhrat Mamatkulov, and Risidaxshinni Kumarusamy

Abstract Green manufacturing is gaining traction across industries as a response to intensifying global environmental concerns and rising demand for sustainable production methods. The growth of green manufacturing research can be attributed to the misalignment between traditional manufacturing industries' characteristics and global environmental conservation, as these sectors face mounting pressure to reduce their ecological footprint while maintaining (or increasing) their economic

P. Utomo
Department of Technology Management, Universitas Multimedia Nusantara, Tangerang, Indonesia
e-mail: prio.utomo@umn.ac.id

T.-H. Cham (✉)
Swinburne University of Technology Sarawak Campus, Kuching, Malaysia

Faculty of Business, Sohar University, Sohar, Oman

Tashkent State University of Economics, Tashkent, Uzbekistan

International College, Krirk University, Bangkok, Thailand

Universitas Multimedia Nusantara, Tangerang, Indonesia
e-mail: tcham@swinburne.edu.my; jaysoncham@gmail.com

A. Anwar
Centre for Business Informatics and Industrial Management (CBIIM), UCSI Graduate Business School, UCSI University, Kuala Lumpur, Malaysia

Advanced Research Centre, European University of Lefke, Lefke, Turkey

Korea University, Seoul, South Korea

Department of Business Administration, ILMA University, Karachi, Pakistan
e-mail: ahsananwar@ucsiuniversity.edu.my; ahsananwar786@gmail.com

S. Mamatkulov
Tashkent State University of Economics, Tashkent, Uzbekistan
e-mail: shuhrat.mamatqulov@tsue.uz

R. Kumarusamy
Putra Business School, Selangor, Malaysia
e-mail: risi@putrabs.edu.my

M. Venkatesh et al. (eds.), *Green Manufacturing Technologies in Industry 5.0*,
https://doi.org/10.1007/978-981-95-0363-6_12

impact. This article aims to develop current knowledge on the challenges, opportunities, and future directions in the green manufacturing literature. A total of 78 articles from the fields of business, management, and economics were retrieved from the Scopus Database and evaluated using thematic analysis. This study uses eight themes for thematic analysis such as Green Supply Chain Management; Green Innovation and Product Development; Lean, Agile, and Green Integration; Circular Economy and End-of-Life Management; Digital Transformation and Technology in Green Manufacturing; Sustainability Performance and Environmental Management Systems; Green Finance and Policy Interventions; and Challenges in Adopting Green Practices in SMEs. The researchers, policymakers, and stakeholders of the manufacturing industry can use the findings of this study as a valuable reference for attaining the targets of sustainable development Goal (SDG) 13 (climate change), SDG 08 (decent economic growth), SDG 09 (green innovations) and environmental, social, and economic benefits.

1 Introduction

The manufacturing industrial sector has typically raised environmental pollution due to high production levels and overexploitation of energy resources (Jie 2017). To cope with this environmental pollution, the manufacturing sector is under high pressure to control their contribution to control the ecological footprint (Leong et al. 2019; Rehman et al. 2021). All these required a transitional change mainly depending on the innovations and the adoption of technologies, which not only will preserve the profitability of the industries but will protect the environment also (Cham et al. 2022; Huynh et al. 2020; Wang 2020; Zhang and Cham 2024). This makes the adoption of technologies complex to maintain a balance between economic growth and environmental sustainability (Choudhary et al. 2019; Wang et al. 2021; Zhang et al. 2023).

Due to the dire need to deal with this trade-off situation and follow the transitional path of sustainability, green manufacturing plays a pivotal role (Kannan et al. 2022; Micheli et al. 2020; Naz et al. 2024). Green manufacturing, which is known as sustainable production, works in an innovative way for the production units causing high pollution emissions. The process consists of the integration of modern technologies ensuring sustainability in all the manners—economic, social, and environmental. The protection of the environment and the preservation of resources make this concept novel. The whole process makes the industrial units focus on 3Rs—reduce, reuse, and recycle up to the 5Rs consisting of repair, reuse, refurbish, rebuild, and recycle (Haleem et al. 2023). The new production strategy works to resolve all the issues faced by conventional manufacturing techniques (Al-Hakimi et al. 2022; Bui et al. 2024; Prasad et al. 2022). The rising importance of green manufacturing has made it insensible to maintain a balance between economic expansion and environmental sustainability as per the mission of government and business organisations of this era (Afum et al. 2020; Lacap et al. 2021; Primandaru et al. 2023).

Though the current bulk of research has highlighted the importance of green technology, the evolving trends of this innovative technology are still to be discussed in detail to secure sustainability in the environment. For instance, technology relating to Industry 4.0, like the Internet of Things (IoT), usage of artificial technology (AI), and analytics of big data, is useful for the optimum utilisation of the resources and control of waste production. The integration of these tools in green production practices has not yet been discussed. There is also a lack of standard approach addressing the needs of industries specifically in geographical context. Additionally, most existing research has examined incremental improvements rather than systemic transformations consisting of interdependent factors like policy measures, supply chain networks, and social demands.

To guide scholars in understanding and addressing these gaps, this article comprehensively analyses trends in the green manufacturing literature, focusing on challenges, opportunities, and future directions. By synthesising recent developments in digital technologies, sustainable materials, and policy mechanisms, we seek to offer insights into the interplay among technological innovation, environmental performance, and economic feasibility. Overall, this article emphasises how collaborations among industry, academia, and governing parties can drive the manufacturing sector's holistic transformation towards sustainable practices. We intend for these findings and recommendations to provide a reference for researchers, practitioners, and policymakers committed to pioneering the future landscape of environmentally responsible manufacturing.

2 Research Methods

To effectively categorise the green manufacturing literature, we conducted a systematic literature review. The four-step procedure we adopted is described below:

2.1 Literature Search

First, we executed a literature search in the Scopus Database using the search query: "green manufacturing". The predetermined inclusion criteria were journal articles with "green manufacturing" in the title, abstract, and/or keywords, published between 2019 and 2024, and in the subject areas of business management, accounting, economics, econometrics, and/or finance. A total of 78 eligible articles were obtained from the search.

2.2 Data Extraction and Organisation

The bibliographic details and abstracts of each of the 78 articles were extracted for comparison and coding. This structured approach facilitated consistency and traceability across our review of numerous articles.

2.3 Thematic Analysis and Coding

After data extraction, we performed coding and thematic analysis to group similar concepts into overarching themes. Referring to Clarke and Braun (2017) and Byrne (2022), the analysis procedure involved the following steps: (i) familiarising with each article's main points, focus area, and takeaways; (ii) open coding; (iii) developing initial themes to understand recurring and interrelated patterns; and (iv) refining and restructuring themes and sub-themes.

2.4 Data Synthesis and Reporting

Following the guidelines of Sigala et al. (2024), we reported the final results in a specific order for each theme: key research challenges, research opportunities, future research directions, and potential research questions. This structure provides scholars a clear basis for further studies.

3 Results and Discussion

Thematic analysis of the 78 articles identified eight themes: Green Supply Chain Management; Green Innovation and Product Development; Lean, Agile, and Green Integration; Circular Economy and End-of-Life Management; Digital Transformation and Technology in Green Manufacturing; Sustainability Performance and Environmental Management Systems; Green Finance and Policy Interventions; and Challenges in Adopting Green Practices in SMEs. The following section deals with the opportunities, relating challenges, and prospects of these in detail.

3.1 Green Supply Chain Management

In the supply chain area, the green supply chain management (GSCM) is a new approach enhancing sustainability with efficiency. This modern manufacturing strategy covers both the economic goals and the environmental objectives. The supply chain mechanism involving buying, producing, distributing, and logistics is all integrated with environmental concerns. In short, the GSCM mechanism works for minimising the environmental damages relating to the supply chain without reducing financial benefits.

3.1.1 Challenges

One of the main challenges relating to GSCM is the complicated coordination among the large number of stakeholders dealing in manufacturing, supplying, and logistics (Liu et al. 2019; Naz et al. 2024; Wang et al. 2021). The issue arises with the lack of collaboration among the decision-makers while implementing eco-friendly techniques. This makes hindrance in sharing data and maintaining transparency to adopt GSCM for sustainability (Micheli et al. 2020; Ramli et al. 2022).

3.1.2 Opportunities

The GSCM mechanism develops a comprehensive framework of supply chain management dealing with buying, designing, and distributing things (Ngouapegne et al. 2024). The tracking system also works efficiently in the adoption of technology as it ensures the applicability of the conditions and collaboration level (Prasad et al. 2022). Moreover, supply-side participation can be strengthened by promoting environmentally conscious purchasing habits as it will make suppliers conscious of the environment (Micheli et al. 2020; Naz et al. 2024).

3.1.3 Future Research Directions

Further areas are yet to be explored in the field of GSCM, like utilising digital tools and tracking the product life cycle to raise transparency and make the system accountable. Moreover, the governmental regulatory bodies, by implementing environmental policies and certifications make the application of GSCM efficient. The following research questions work in this regard:

RQ01: How can digital innovations (e.g. blockchain) be effectively integrated into traditional supply chains to improve data sharing and transparency without overburdening smaller firms financially?

RQ02: What strategies best align multiple supply chain stakeholders (manufacturers, logistics providers, policymakers) to balance cost efficiency with environmental objectives?

RQ03: How do different cultural and regulatory contexts influence the effectiveness of GSCM frameworks, and what adaptations are needed to ensure broader adoption?

RQ04: Which incentive mechanisms (subsidies, tax breaks, etc.) most effectively encourage the adoption of green procurement and eco-design among suppliers in emerging economies?

RQ05: How can GSCM performance be measured consistently across diverse sectors to allow cross-industry comparisons and benchmarking for sustainability improvement?

3.2 Green Innovation and Product Development

Green innovation involves adopting novel eco-friendly technologies, designs, and processes in the product lifecycle to promote sustainable industrial growth and reduce firms' carbon footprint.

3.2.1 Challenges

The key challenges in green innovation lie in its high R&D cost, uncertain returns, knowledge gaps, limited adoption, and complex stakeholder needs. Businesses, particularly small- and medium-sized enterprises (SMEs), face challenges in balancing the upfront costs of innovation with uncertain market demand (Andersén 2021; D'Angelo et al. 2023). SMEs are deficient in implementing and having knowledge of eco-friendly technology (Primandaru et al. 2023). Despite this, the increasing trend in consumer demand for green products is also putting pressure on manufacturers to adopt cost-effective technologies promoting environmental sustainability (Andersén 2021).

3.2.2 Opportunities

Green technology holds several opportunities to integrate lifecycle mechanisms in the manufacturing and design of the products (D'Angelo et al. 2023; Rehman et al. 2021). The adoption of green technology would also decrease the R&D expenditures as stakeholders consciously share the information and resources (Wang et al. 2021). Through the provision of incentives and trade regulations, the government regulatory bodies can also play an effective role in this regard (Lu et al. 2022; Tan et al. 2019; Wang et al. 2021).

3.2.3 Future Research Directions

Prospects of green manufacturing are several in nature. Future studies can delve into the features of green production designs as per ISO 14001 to motivate green innovations like the use of AI and IoT along with other latest analytics. This will enhance product design optimisation and sustainability monitoring. Moreover, the effect of consumer behaviour in this regard relates to the willingness to pay for green products and the market of green products. These possibilities can be summarised in the following research questions:

RQ06: Which collaboration models between suppliers, manufacturers, and consumers most effectively reduce the high R&D costs associated with green product development?
RQ07: How can policymakers or industry associations design clear product certification standards to guide eco-design and enhance consumer trust in green innovations?
RQ08: What roles do consumer preferences and willingness to pay play in shaping the market success of novel green products, and how can firms accurately assess this demand?
RQ09: In what ways can integrating lifecycle assessment tools at the earliest stages of product design yield both environmental benefits and competitive advantage for manufacturers?
RQ10: How do organisational culture and leadership styles interact with technology adoption to accelerate or hinder green innovation capabilities among SMEs and large firms?

3.3 Lean, Agile, and Green Integration

This theme refers to the combination of lean, agile, and green (LAG) practices in the manufacturing sector, with the aim to enhance cost efficiency, waste reduction, and environmental sustainability while remaining responsive to market fluctuations.

3.3.1 Challenges

The main challenges in LAG achievement are aligning conflicting goals, measuring performance, and allocating limited resources. Balancing the opposing goals of the LAG components is problematic, as "lean" strives for efficiency, "green" champions sustainability, and "agile" prioritises flexibility (Choudhary et al. 2019; Schmitt et al. 2021). Quantifying the success of LAG in process improvement and goal achievement is another complex task (Choudhary et al. 2019). Moreover, the availability of resources to channel towards advanced LAG systems is low, particularly among SMEs which are often characterised by financial, technical, and managerial limitations (Prasad et al. 2022; Primandaru et al. 2023).

3.3.2 Opportunities

The LAG concept offers valuable opportunities to promote lean-green manufacturing goals. Potential areas for development are: integrated LAG operational systems that align cost, time, and environmental metrics (Prasad et al. 2022; Shokri et al. 2022), employee empowerment and green quality community-building to leverage low-cost, high-impact environmental initiatives (Goyal et al. 2022), and joint collaborations and knowledge-sharing platforms among functional teams and external partners (Choudhary et al. 2019).

3.3.3 Future Research Directions

The development of more sophisticated measurement systems, technology-driven synergies, and industry-specific adaptations in LAG principles are among the areas that require further study in green manufacturing settings. Based on this direction, research questions that indicate high potential are as follows:

RQ11: What implementation frameworks can simultaneously balance lean's cost-reduction goals, agility's responsiveness needs, and green manufacturing's ecological objectives?

RQ12: How do organisations measure the joint performance of lean–green methods without creating conflicting metrics or duplicative efforts?

RQ13: Which employee engagement mechanisms, including "green quality circles", are most effective in transitioning from lean-focused projects to holistic lean-green transformations?

RQ14: What are the major knowledge and cultural barriers that impede managers and shop-floor employees from harmonising LAG approaches?

RQ15 How can digital tools (AI, real-time analytics) help organisations dynamically balance lean efficiency with agile responsiveness and ecological targets in unpredictable markets?

3.4 Circular Economy and End-of-Life Management

The circular economy paradigm aims to extend product lifespans, reuse materials, and recycle end-of-life components. It advocates strategies such as remanufacturing, recycling, and reuse to foster sustainable manufacturing systems.

3.4.1 Challenges

Supply chain coordination complexities, technological infrastructure gaps, and financial and marketing constraints are key challenges in this domain. The collection, inspection, and remanufacturing of used products require close collaboration among diverse stakeholders (Hosseini-Motlagh et al. 2020; Rosado and Kalmykova 2019), which can create complications in their coordination. Additionally, technology is vital in ensuring the efficacy of the circular economy and end-of-life management. However, the waste management system is not developed yet, and the recycling technology is also inefficient, particularly in emerging economies of the world (Ramli et al. 2022; Schmitt et al. 2021). The uncertainty relating to the acceptance of regenerated goods and the high initial investment expenditures are to be proved major hindrances in the way of circular production process (Andersén 2021; Bui et al. 2024).

3.4.2 Opportunities

Streamlining the end-of-life process will help in maintaining the balance between economic and environmental goals (Al-Hakimi et al. 2022; Rosado and Kalmykova 2019). In this regard, reserving logistics, reducing waste, and circulation of raw material prove helpful (Kamble et al. 2021; Ricardianto et al. 2022). The role of government could not be ignored in implementing policies and giving incentives like carbon taxes and subsidies. These initiatives make the business firms work consciously and efficiently to protect the environment as well.

3.4.3 Future Research Directions

Future research avenues in this domain include circular economy models beyond manufacturing, reverse logistics for better used product management in global supply chains, and technologies to monitor product lifecycles for real-time decision-making on reuse or recycling points. Thus, several research questions can be developed from these directions:

RQ16: Which policy instruments (subsidies, tax incentives) most effectively motivate manufacturers to incorporate recyclable materials and design for remanufacturing on a scale?

RQ17: How can supply chains be structured to manage product returns efficiently, ensuring maximum value recovery through reuse, remanufacturing, or recycling?

RQ18: What technological innovations (e.g. IoT, sensor-based sorting) significantly reduce the cost of disassembly and improve the feasibility of circular production systems?

RQ19: How do consumer perceptions of remanufactured or refurbished products influence the market acceptance and profitability of circular business models?

RQ20: Which collaborative approaches between policymakers, manufacturers, and recyclers can standardise product returns, create secondary markets for recovered materials, and expand circular economy networks?

3.5 Digital Transformation and Technology in Green Manufacturing

Although technology is a recurring concept in other domains, the unique traits, features, and attributes of digital technology require further specific studies pertaining to its implications for green manufacturing. By enhancing transparency, data accuracy, and traceability across supply chains, digital technologies (e.g. AI, IoT, blockchain, and simulations) are increasingly crucial for monitoring and optimising sustainability.

3.5.1 Challenges

Apart from investment costs, the management, security, and privacy of data are major challenges in this domain. Digital technologies generally handle large, diverse datasets, which makes data protection crucial, especially for SMEs with limited expertise (Jiao et al. 2020; Prasad et al. 2022). Implementing IoT or blockchain solutions also demands significant investment (Daoud et al. 2022; Huynh et al. 2020), while relying on digital networks in a supply chain poses serious security and privacy concerns (Meža et al. 2021).

3.5.2 Opportunities

Opportunities for digital transformation lie in tapping into its various potential benefits. For example, blockchain improves transparency in material sourcing, waste management, and end-to-end tracking, which can foster trust and efficient resource allocation (Prasad et al. 2022). Sensor usage, machine learning, and real-time data further facilitate dynamic resource distribution and minimal emission (Choudhary et al. 2019; Meža et al. 2021). Apart from that, "digital twins" (i.e. virtual representations of physical operations) serve as real-time simulations that optimise processes, reduce scrap, and detect inefficiencies early (Meža et al. 2021; Prasad et al. 2022).

3.5.3 Future Research Directions

Future research on digital technology and digital transformation can look into cross-sector collaboration, scalable digital solutions, and ethical and regulatory frameworks for data governance, privacy, and security. As technology adoption expands, green manufacturing scholars may seek to answer the following questions:

RQ21: What cost-effective digital solutions can smaller firms implement to track and optimise energy usage and emissions data while mitigating cybersecurity risks?

RQ22: How does blockchain-based traceability influence consumer trust and overall performance in supply chains aiming to be more transparent and sustainable?
RQ23: What roles do digital twins and AI-driven simulations play in the real-time monitoring of production systems for waste and emission reduction?
RQ24: Which regulatory frameworks or data privacy standards ensure responsible and equitable use of advanced digital technologies in green manufacturing settings?
RQ25: How can big data analytics or cloud-based platforms guide dynamic, multi-tier supply chains to rapidly adapt to environmental disruptions and policy changes?

3.6 Sustainability Performance and Environmental Management Systems

Sustainability performance and environmental management systems (EMS) constitute performance metrics that guide organisations in measuring and continuously improving their economic, social, and environmental performance.

3.6.1 Challenges

Inconsistencies in evaluation metrics, adoption levels, and compliance rates are key barriers to EMS implementation. The lack of standardised metrics for measuring sustainability performance makes industry benchmarking difficult (Leong et al. 2019; Micheli et al. 2020). EMS implementation also demands intensive training, documentation, and managerial commitment, which may overwhelm smaller organisations and heighten their resistance towards such systems (Kraus et al. 2020; Leong et al. 2019).

3.6.2 Opportunities

EMS has potential in terms of comprehensive frameworks to measure lifecycle impacts using environmental, social, and economic performance indicators (Micheli et al. 2020). Technologies to simplify EMS implementation and collect real-time data also represent an opportunity to facilitate measurement standardisation.

3.6.3 Future Research Directions

The literature on EMS in green manufacturing opens several avenues for future studies. For instance, there is a need to develop global standards that align international certifications with best-practice exchanges and enhanced competitiveness. Other topics worth examining are performance-based incentives to drive SMEs' EMS adoption and collaboration among governments, NGOs, and business leaders. Moving forward, researchers can attempt to explore the research questions below:

RQ26: How can EMS frameworks be tailored to meet the specific financial, technical, and human resource capacities of SMEs?
RQ27: What standardised indicators or key performance metrics enable cross-industry comparisons of environmental outcomes without oversimplifying unique operational contexts?
RQ28: To what extent do digitised EMS solutions improve real-time reporting, regulatory compliance, and continuous improvement processes in large multinational enterprises?
RQ29: Which stakeholder engagement approaches best encourage top management to invest in and rigorously maintain EMS certifications in regions with weak enforcement?
RQ30: How do integrated social, economic, and environmental performance measurements under EMS frameworks influence a company's strategic decisions and innovation efforts?

3.7 *Green Finance and Policy Interventions*

Green finance mechanisms, such as bonds, loans, and carbon credits, are critical for funding green manufacturing projects. Effective green financing policies can improve environmental quality (Sharif et al. 2024) by adopting the low-carbon technologies and green supply chain practices.

3.7.1 Challenges

With regard to green finance challenges, clean technologies tend to be perceived as high-risk and involve long payback periods, leaving SMEs struggling to secure funds (Park et al. 2021; Primandaru et al. 2023). Transparency gaps and regulatory complexities also limit firms' understanding of green subsidies and related credits. Additionally, performing carbon accounting across a multi-tier supply chain is problematic, leading to unreliable data for policy compliance.

3.7.2 Opportunities

In light of the challenges mentioned above, green finance opportunities are evident in encouraging public-private partnerships to lower financing barriers for green technologies. There is also vast potential in innovating green finance instruments, as well as in leveraging corporate social responsibility and consumer awareness to attract green investment and foster policy compliance.

3.7.3 Future Research Directions

Future research on green finance should pay attention to standardised green finance's impact on reducing transaction costs and increasing investment confidence; adaptive policies that tighten or relax requirements based on sector-specific capability and innovation maturity; and empirical methods to measure policy interventions' short- and long-term effects on environmental performance and financial stability. The following research questions emerge from this direction:

RQ31: What new financing instruments (e.g. blockchain-based green bonds) can effectively lower the risk premium for investing in clean technologies, especially in emerging economies?

RQ32: How can policymakers design adaptable environmental regulations that encourage green investments without stifling R&D for environmentally friendly products?

RQ33: Which combination of subsidies, tax incentives, and carbon credit schemes maximise ROI for firms adopting advanced green manufacturing processes?

RQ34: How should public-private partnerships be structured so that SMEs gain equitable access to green credit and do not remain at a competitive disadvantage?

RQ35: Which economic and environmental indicators best measure the long-term impact of green financing on a country's industrial sustainability performance?

3.8 Challenges in Adopting Green Practices in SMEs

Many articles point out that SMEs face unique hurdles in implementing green manufacturing practices, achieving lean-green transformations, adopting advanced technologies, and complying with environmental standards.

3.8.1 Challenges

SMEs often have limited resources, insufficient knowledge, and a lack of specialised staff, hampering their ability to adopt green practices (Primandaru et al. 2023; Zain et al. 2024). Unfortunately, government initiatives and external support pertaining to green manufacturing tend to focus on large companies.

3.8.2 Opportunities

For SMEs to overcome this challenge, opportunities lie in the development of low-investment green solutions like energy audits or simplified lean-green tools that can be quickly monetised with minimal capital. SMEs can also collaborate to share knowledge and facilities, as well as co-finance green projects to alleviate the financial burden on individual firms. These opportunities can be accelerated via localised policies that incentivise SMEs and develop their capabilities.

3.8.3 Future Research Directions

Future research should conduct in-depth case analyses of successful green implementation in SMEs to develop best practices regarding cost-effectiveness and environmental compliance. Studies can also investigate the integration of policy support, training, and financing for sustainability transitions in local SMEs, as well as the development of simple, cloud-based platforms to track emissions, manage recycling, and facilitate communication among SME stakeholders. The research questions below highlight areas for further exploration in the SME domain of green manufacturing:

RQ36: Which low-cost, high-impact green initiatives (e.g. energy audits, eco-efficiency training) have proven most effective for resource-limited SMEs in emerging markets?

RQ37: What roles do local governments and industry associations play in supporting or hindering the transition to green manufacturing among SMEs through policies and capacity-building programmes?

RQ38: How do cultural attitudes towards sustainability and environmental responsibility shape the pace at which SMEs adopt green processes in different regions?

RQ39: What practical tools or frameworks can standardise sustainability reporting for SMEs, ensuring transparency without imposing excessive administrative burdens?

RQ40: How can partnership models—such as cluster-based approaches or shared green infrastructures—mitigate capital and knowledge barriers that typically limit SME progress in environmental initiatives?

4 Conclusion

Manufacturing activities have been widely scrutinised for damaging the environment through resource extraction, energy consumption, carbon emissions, and waste generation. Consequently, the urgency to align economic growth with ecological stewardship has driven myriad sustainable manufacturing initiatives in today's global market. This study broadens the isolated and narrow view of sustainability by comprehensively reviewing the challenges, opportunities, and future directions in green manufacturing research. Based on a thematic analysis of 78 Scopus-indexed articles, we identify eight green manufacturing themes and 40 research questions to guide future studies.

First, the "GSCM" theme sets the stage for the integration of eco-friendly principles across supply chain procurement, logistics, and reverse operations, ultimately calling for collective action among diverse supply chain partners. Second, "Green Innovation and Product Development" highlights the centrality of eco-design, lifecycle assessments, and stakeholder collaboration in reducing resource consumption and addressing market demand for greener offerings. Next, "LAG Integration" emphasises reconciling efficiency, flexibility, and environmental stewardship, enabling firms to balance cost reduction with ambitious sustainability targets. This theme is complemented by "Circular Economy and End-of-Life Management", wherein remanufacturing, recycling, and extended product lifespans not only lower environmental impact but also reveal new economic opportunities through closed-loop strategies.

Digital transformation emerges as a powerful enabler throughout our discussion. The theme of "Digital Transformation and Technology in Green Manufacturing" highlights how smart sensors, blockchain software, and advanced analytics can boost transparency and resource optimisation. However, the adoption of these technologies is hindered by high costs and technical challenges, especially among SMEs. For performance measurement and continuous improvement, "Sustainability Performance and EMS" refers to structured approaches to integrate daily operational functions with long-term environmental goals. Nonetheless, the next theme, "Green Finance and Policy Interventions" signals that meaningful progress depends on reducing investment barriers, incentivising the latest technologies, and rewarding responsible corporate behaviour through regulations and novel funding schemes. Finally, the "Challenges in Adopting Green Practices in SMEs" theme is a timely reminder that SMEs' relative disadvantages in resources, expertise, and stable returns can prevent them from engaging in sustainability initiatives.

To conclude, accelerating the green transformation of the manufacturing industry essentially requires supply chain partners to adopt low-cost knowledge-sharing solutions across their multi-sector networks. Accordingly, the themes and research questions identified herein provide insightful references and contextual bases for manufacturing industry stakeholders, policymakers, and academicians aiming to assimilate environmental, social, and economic gains via green manufacturing.

References

Afum E, Agyabeng-Mensah Y, Sun Z, Frimpong B, Kusi LY, Acquah ISK (2020) Exploring the link between green manufacturing, operational competitiveness, firm reputation and sustainable performance dimensions: a mediated approach. J Manuf Technol Manag 31(7):1417–1438. https://doi.org/10.1108/JMTM-02-2020-0036

Al-Hakimi MA, Al-Swidi AK, Gelaidan HM, Mohammed A (2022) The influence of green manufacturing practices on the corporate sustainable performance of SMEs under the effect of green organizational culture: a moderated mediation analysis. J Clean Prod 376:134346. https://doi.org/10.1016/j.jclepro.2022.134346

Andersén J (2021) A relational natural-resource-based view on product innovation: the influence of green product innovation and green suppliers on differentiation advantage in small manufacturing firms. Technovation 104:102254. https://doi.org/10.1016/j.technovation.2021.102254

Bui T-D, Nguyen TTV, Wu K-J, Lim MK, Tseng M-L (2024) Green manufacturing performance improvement under uncertainties: an interrelationship hierarchical model. Int J Prod Econ 268: 109117. https://doi.org/10.1016/j.ijpe.2023.109117

Byrne D (2022) A worked example of Braun and Clarke's approach to reflexive thematic analysis. Qual Quant 56(3):1391–1412. https://doi.org/10.1007/s11135-021-01182-y

Cham TH, Cheah JH, Memon MA, Fam KS, László J (2022) Digitalization and its impact on contemporary marketing strategies and practices. J Mark Anal 10(2):103–105

Choudhary S, Nayak R, Dora M, Mishra N, Ghadge A (2019) An integrated lean and green approach for improving sustainability performance: a case study of a packaging manufacturing SME in the U.K. Prod Plan Control 30(5–6):353–368. https://doi.org/10.1080/09537287.2018.1501811

Clarke V, Braun V (2017) Thematic analysis. J Posit Psychol 12(3):297–298. https://doi.org/10.1080/17439760.2016.1262613

D'Angelo V, Cappa F, Peruffo E (2023) Green manufacturing for sustainable development: the positive effects of green activities, green investments, and non-green products on economic performance. Bus Strateg Environ 32(4):1900–1913. https://doi.org/10.1002/bse.3226

Daoud AO, Othman AAE, Ebohon OJ, Bayyati A (2022) Overcoming the limitations of the green pyramid rating system in the Egyptian construction industry: a critical analysis. Archit Eng Des Manag 18(2):114–127. https://doi.org/10.1080/17452007.2020.1802218

Goyal A, Agrawal R, Kumar Sharma A (2022) Green quality circle: achieving sustainable manufacturing with low investment. Resour Conserv Recycl Adv 15:200103. https://doi.org/10.1016/j.rcradv.2022.200103

Haleem A, Javaid M, Singh RP, Suman R, Qadri MA (2023) A pervasive study on green manufacturing towards attaining sustainability. Green Technol Sustain 1(2):100018. https://doi.org/10.1016/j.grets.2023.100018

Hosseini-Motlagh S-M, Nami N, Farshadfar Z (2020) Collection disruption management and channel coordination in a socially concerned closed-loop supply chain: a game theory approach. J Clean Prod 276:124173. https://doi.org/10.1016/j.jclepro.2020.124173

Huynh TLD, Hille E, Nasir MA (2020) Diversification in the age of the 4th industrial revolution: the role of artificial intelligence, green bonds and cryptocurrencies. Technol Forecast Soc Chang 159:120188. https://doi.org/10.1016/j.techfore.2020.120188

Jiao J-L, Zhang X-L, Tang Y-S (2020) What factors determine the survival of green innovative enterprises in China?– A method based on fsQCA. Technol Soc 62:101314. https://doi.org/10.1016/j.techsoc.2020.101314

Jie X (2017) Research on green manufacturing innovation based on resource environment protection. In: IOP Conference Series: Earth and Environmental Science, vol 94. IOP Publishing, p 012112. https://doi.org/10.1088/1755-1315/94/1/012112

Kamble SS, Belhadi A, Gunasekaran A, Ganapathy L, Verma S (2021) A large multi-group decision-making technique for prioritizing the big data-driven circular economy practices in the automobile component manufacturing industry. Technol Forecast Soc Chang 165:120567. https://doi.org/10.1016/j.techfore.2020.120567

Kannan D, Shankar KM, Gholipour P (2022) Paving the way for a green transition through mitigation of green manufacturing challenges: a systematic literature review. J Clean Prod 368:132578. https://doi.org/10.1016/j.jclepro.2022.132578

Kraus S, Rehman SU, García FJS (2020) Corporate social responsibility and environmental performance: the mediating role of environmental strategy and green innovation. Technol Forecast Soc Chang 160:120262. https://doi.org/10.1016/j.techfore.2020.120262

Lacap JPG, Cham TH, Lim XJ (2021) The influence of corporate social responsibility on brand loyalty and the mediating effects of brand satisfaction and perceived quality. Int J Econ Manag 15(1):69–87

Leong WD, Teng SY, How BS, Ngan SL, Lam HL, Tan CP, Ponnambalam SG (2019) Adaptive analytical approach to lean and green operations. J Clean Prod 235:190–209. https://doi.org/10.1016/j.jclepro.2019.06.143

Liu Y, Zhang Y, Batista L, Rong K (2019) Green operations: what's the role of supply chain flexibility? Int J Prod Econ 214:30–43. https://doi.org/10.1016/j.ijpe.2019.03.026

Lu F, Zhou G, Liu Y, Zhang C (2022) Ensemble transfer learning for cutting energy consumption prediction of aviation parts towards green manufacturing. J Clean Prod 331:129920. https://doi.org/10.1016/j.jclepro.2021.129920

Meža S, Mauko Pranjić A, Vezočnik R, Osmokrović I, Lenart S (2021) Digital twins and road construction using secondary raw materials. J Adv Transp 2021:8833058. https://doi.org/10.1155/2021/8833058

Micheli GJL, Cagno E, Mustillo G, Trianni A (2020) Green supply chain management drivers, practices and performance: a comprehensive study on the moderators. J Clean Prod 259:121024. https://doi.org/10.1016/j.jclepro.2020.121024

Naz F, Samadhiya A, Kumar A, Garza-Reyes JA, Kazancoglu Y, Kumar V, Upadhyay A (2024) The effect of green supply chain management practices on carbon-neutral supply chain performance: the mediating role of logistics eco-centricity. J Manuf Technol Manag 35(7):1375–1396. https://doi.org/10.1108/JMTM-09-2023-0401

Ngouapegne CNM, Chinomona E, Okoumba WVL (2024) Green supply chain management: an empirical study in enhancing green economic performance. J Transp Supply Chain Manag 18:1–10. https://doi.org/10.4102/jtscm.v18i0.1087

Park JG, Lee D-H, Kim H-S, Yeo W-J, Jeon M, Bae JY, Kim DU, Lee K-S, Kim G-H, Chang KS, Kim IJ (2021) Novel approach to improve the optical performance by machining process without surface finishing. Int J Precis Eng Manuf Green Technol 8(5):1381–1392. https://doi.org/10.1007/s40684-021-00329-4

Prasad S, Rao AN, Lanka K (2022) Analysing the barriers for implementation of lean-led sustainable manufacturing and potential of Blockchain technology to overcome these barriers: a conceptual framework. Int J Math Eng Manag Sci 7(6):791–819. https://doi.org/10.33889/IJMEMS.2022.7.6.051

Primandaru N, Kusuma MW, Nasution OB (2023) Antecedents of green manufacturing implementation by local MSMEs in Indonesia. Environ Econ 14(2):103–113. https://doi.org/10.21511/ee.14(2).2023.08

Ramli EMM, Zainudin DB, Islam R (2022) Explore the research trends of green supply chain in the manufacturing industry: a bibliometric analysis. Oper Supply Chain Manag 15(3):345–358. https://doi.org/10.31387/oscm0500351

Rehman SU, Kraus S, Shah SA, Khanin D, Mahto RV (2021) Analyzing the relationship between green innovation and environmental performance in large manufacturing firms. Technol Forecast Soc Chang 163:120481. https://doi.org/10.1016/j.techfore.2020.120481

Ricardianto P, Kholdun AI, Fachrey KR, Agusinta L, Setiawan EB, Abidin Z, Purba OR, Perwitasari EP, Endri E (2022) Building green supply chain management in pharmaceutical companies in Indonesia. Uncertain Supply Chain Manag 10(2):453–462. https://doi.org/10.5267/j.uscm.2021.12.006

Rosado L, Kalmykova Y (2019) Combining industrial symbiosis with sustainable supply chain management for the development of urban communities. IEEE Eng Manag Rev 47(2):103–114. https://doi.org/10.1109/EMR.2019.2911060

Schmitt T, Wolf C, Lennerfors TT, Okwir S (2021) Beyond "Leanear" production: a multi-level approach for achieving circularity in a lean manufacturing context. J Clean Prod 318:128531. https://doi.org/10.1016/j.jclepro.2021.128531

Sharif A, Sofuoglu E, Kocak S, Anwar A (2024) Can green finance and energy provide a glimmer of hope towards sustainable environment in the midst of chaos? An evidence from Malaysia. Renew Energy 223:119982. https://doi.org/10.1016/j.renene.2024.119982

Shokri A, Antony J, Garza-Reyes JA (2022) A new way of environmentally sustainable manufacturing with assessing transformation through the green deployment of lean six sigma projects. J Clean Prod 351:131510. https://doi.org/10.1016/j.jclepro.2022.131510

Sigala M, Ooi KB, Tan GWH, Aw ECX, Cham TH, Dwivedi YK, Kunz WH, Letheren K, Mishra A, Russell-Bennett R, Wirtz J (2024) ChatGPT and service: opportunities, challenges, and research directions. J Serv Theory Pract 34(5):726–737. https://doi.org/10.1108/JSTP-11-2023-0292

Tan JX, Cham TH, Zawawi D, Aziz YA (2019) Antecedents of organizational citizenship behavior and the mediating effect of organization commitment in the hotel industry. Asian J Bus Res 9(2): 121–139

Wang C (2020) The control method of green production efficiency based on multi-layer and complex network under the background of export trade transformation. Int J Prod Dev 24(3): 163–181

Wang X, Cho S-H, Scheller-Wolf A (2021) Green technology development and adoption: competition, regulation, and uncertainty — a global game approach. Manag Sci 67(1):201–219. https://doi.org/10.1287/mnsc.2019.3538

Zain RM, Ramli A, Zain MZM, Yekini LS, Musa A, Rahim MNA, Dirie AN, Aziz NIC (2024) An investigation of the barriers and drivers for implementing green supply chain in Malaysian food and beverage SMEs: a qualitative perspective. WSEAS Trans Bus Econ 21:2169–2189. https://doi.org/10.37394/23207.2024.21.179

Zhang Y, Cham TH (2024) From slogans to actions: How does social media marketing and collectivism drive green consumption behaviour? Mark Intell Plan 43(4):725–754. https://doi.org/10.1108/MIP-02-2024-0093

Zhang Y, Pek CK, Cham TH (2023) The effect of social media exposure, environmental concern, and consumer habits in green consumption intention. J Strateg Manag 16(4):747–766. https://doi.org/10.1108/JSMA-01-2023-0006

MIX
Papier aus verantwortungsvollen Quellen
Paper from responsible sources
FSC® C105338

If you have any concerns about our products,
you can contact us on
ProductSafety@springernature.com

In case Publisher is established outside the EU,
the EU authorized representative is:
Springer Nature Customer Service Center GmbH
Europaplatz 3, 69115 Heidelberg, Germany

Printed by Libri Plureos GmbH
in Hamburg, Germany